Characterization of Semiconductor Materials

TEXAS INSTRUMENTS ELECTRONICS SERIES

Crawford ▪ MOSFET IN CIRCUIT DESIGN

Delhom ▪ DESIGN AND APPLICATION OF TRANSISTOR SWITCHING CIRCUITS

The Engineering Staff of
Texas Instruments Incorporated ▪ CIRCUIT DESIGN FOR AUDIO, AM/FM, AND TV

The Engineering Staff of
Texas Instruments Incorporated ▪ SOLID-STATE COMMUNICATIONS

The Engineering Staff of
Texas Instruments Incorporated ▪ TRANSISTOR CIRCUIT DESIGN

Hibberd ▪ INTEGRATED CIRCUITS

Hibberd ▪ SOLID-STATE ELECTRONICS

Kane and Larrabee ▪ CHARACTERIZATION OF SEMICONDUCTOR MATERIALS

Runyan ▪ SILICON SEMICONDUCTOR TECHNOLOGY

Sevin ▪ FIELD-EFFECT TRANSISTORS

Characterization of Semiconductor Materials

Philip F. Kane

Director, Central Analysis & Characterization Laboratory
Texas Instruments Incorporated

Graydon B. Larrabee

Manager, Research Branch
Central Analysis & Characterization Laboratory
Texas Instruments Incorporated

McGRAW-HILL BOOK COMPANY

New York St. Louis San Francisco London Sydney
Toronto Mexico Panama

CHARACTERIZATION OF SEMICONDUCTOR MATERIALS

 Library of Congress Catalog Card Number 79-81607

33273

1234567890 HDBP 7543210

To our wives, Sybil and Joyce, for their patience and understanding.

Foreword

It seems most appropriate that the authors have undertaken to summarize the methods of characterization that in many cases have now become an indispensable part of any well-based semiconductor effort — be it manufacturing or research in nature.

Noteworthy benefits have flowed from twenty years of intensive and large-scale investigation stimulated by the invention and development of the transistor. New and improved techniques of measurement have been devised for characterizing solids, and extensive measurements have been made. The focusing of most of the studies on nearly perfect single crystals of germanium, silicon, and germanium-like materials produced under a high degree of control has undoubtedly contributed strongly not only to the design and fabrication of exceptional devices but also to the usefulness of the data and to the striking advances achieved in the understanding of the effects of minute impurities and imperfections on the physical properties of solids.

Early recognition of characterization of materials in the semiconductor field as an activity vital to technological progress in semiconductor devices has undoubtedly had an impact on its acceptance in investigations in other areas in which materials are involved.

It is apparent that characterization will continue to be an essential interest to many individuals involved in the vast semiconductor industry. Philip F. Kane and Graydon B. Larrabee are writing particularly for these. They are well qualified to speak authoritatively in this field, both by their own personal attainments and by virtue of the great variety and number of characterization problems that flow to them from the large semiconductor technical effort of Texas Instruments. Philip F. Kane is Director of the Central Analysis and Characterization Laboratory in TI's Central Research Laboratories. Graydon Larrabee is associated with him in this responsibility.

The present volume is being published as part of the Texas Instruments Electronics Series to make available to the technical community the scientific techniques that are essential to improvements in future materials and devices and to continued growth of the semiconductor industry.

Gordon K. Teal
Vice President and Chief Scientist
for Corporate Development

Preface

Sometime during the latter half of 1964, Philip F. Kane was asked to contribute a chapter on semiconductors to "Standard Methods of Chemical Analysis." In compiling information for this, it became increasingly apparent that, first, there was no existing volume that fully covered this subject and, second, that the term analytical chemistry excluded many very important methods of evaluation from consideration. We felt that there was a distinct need for a treatise that would collect into one volume all the techniques, both compositional and structural, applied in assessing material quality in the many stages of producing modern semiconductor devices.

Commercially, only germanium, silicon, and the III-V compounds are of any great significance and accordingly only these basic materials are dealt with in this book. The various ancillary materials such as leads, solders, encapsulants and so on can be dealt with by conventional analytical methods and are not included; however, derivatives which are an essentially intrinsic part of the device, e.g. oxide films and epitaxial films, are covered since the methods are similar to those for bulk material. Surfaces are also considered an integral part of the device and are dealt with at some length. Raw materials for the production of the basic materials have been included since these give valuable information on the possible impurities in the semiconductors.

In March, 1967, almost a year after we had started on this work, the Materials Advisory Board issued a report (MAB-229-M) entitled "Characterization of Materials" which we felt crystallized most ably what we had been groping towards in our approach to semiconductors. As a result, we adopted the term characterization to describe the content of our book and we are grateful to the MAB committee for defining a term which we feel eminently acceptable to the scientists of different disciplines currently working in the field.

It is our hope that the information we have compiled will be of value to analytical chemists, quality control engineers, materials scientists, and production engineers. Those directly engaged in semiconductor manufacture may find it a useful sourcebook for more detailed reading and those occasionally involved with devices may find sufficient information to tackle a difficult evaluation with some degree of confidence. We hope, too, that it may prove useful as a textbook in the new departments of materials science now at several of our leading universities.

We would like to express our thanks to Drs. G. R. Cronin, R. A. Reynolds, and

G. K. Teal for their critical reading of portions of the manuscript, to David L. Carroll and Walter L. Behringer for many of the photographs and drawings, to Mrs. Leah L. Childress for typing the manuscript, and to Mrs. Helen L. Clark for much of the secretarial assistance. Finally, we must acknowledge the assistance of many of our colleagues at Texas Instruments Incorporated with whom we discussed several of the topics.

Philip F. Kane
Graydon B. Larrabee

Contents

1

Introduction

1-1. EARLY DEVELOPMENT

A semiconductor is conventionally defined as a material whose conductivity falls intermediate between that of metals, 10^4 to 10^6 ohm^{-1} cm^{-1}, and that of insulators, 10^{-22} to 10^{-10} ohm^{-1} cm^{-1}. This range of 10^{-9} to 10^3 ohm^{-1} cm^{-1} is sometimes subdivided at its higher end into a class termed *semimetals*, but this is a somewhat tenuous distinction. The fundamental differences between these three classes will be better described later in terms of the energy-band theory; however, one distinguishing feature of a semiconductor is its increasing conductivity with temperature. Metals, on the other hand, decrease in conductivity with temperature.

This phenomenon of inverse temperature coefficient and the associated effect of photoconduction have been known since the last century in compounds such as silver sulfide[1,†] and selenium.[2] In the early days of radio, the detector in all receivers was the crystal rectifier, usually a galena crystal, as shown in Fig. 1-1. The rectifying contact was made by a so-called "cat's whisker," a fine wire held in light contact with the crystal. It required frequent adjustment to find sensitive spots on the crystal, and it was displaced about 1925 by the introduction of cheap and reliable thermionic tubes.

While the tube almost entirely replaced the crystal in communications, in two areas semiconductors retained their importance. In 1926, the copper-oxide thin-film rectifier was developed by Grondahl,[3] and almost simultaneously the selenium rectifier by Presser.[4] The selenium rectifier has maintained its position for high currents to the present day.

Returning to the crystal rectifier, we find that one field in which it was not replaced by tubes was in microwave receivers. The vacuum tubes designed for this purpose were noisy, and this noise increased with frequency. The crystal rectifier was superior in this respect, and development continued in the point-contact diode. Southworth and King[5] described an early application of the silicon rectifier. The first commercial units were made in England by the British-Thomson-Houston Co., Ltd., using commercial silicon of about 98 percent purity some time early in 1940.

†Superscript numbers indicate References listed at the end of the chapter.

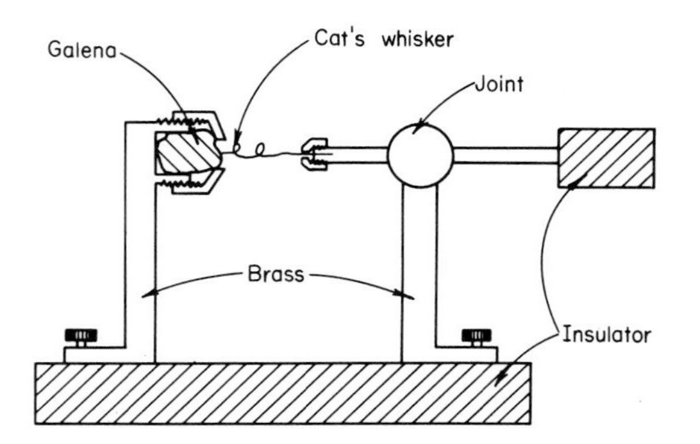

Fig. 1-1. Early cat's whisker rectifier.

Since radar systems relied heavily on these mixer diodes, considerable effort was expended in attempts to improve on these rather erratic BTH units. An important step forward was made by the General Electric Co., Ltd., of England, who introduced, in 1941, the so-called "red dot" crystal, prepared from purified silicon containing a fractional percentage of aluminum or beryllium. This device was able to withstand comparatively high electrical loads without deteriorating and was termed a *high-burnout crystal*. The success of this device stimulated other research along the same lines. In 1942 Du Pont, working in collaboration with the University of Pennsylvania semiconductor group led by Seitz,[6] succeeded in preparing silicon of better than 99.9 percent purity, and rectifiers prepared from this material were also high-burnout devices.

In 1943, Theuerer[7] found that the addition of boron to silicon at about 10 ppm resulted in a mixer of improved sensitivity (a mixer is a diode that converts high-frequency signals to lower frequencies, for example, microwaves to radiofrequencies, which can be more readily handled). Boron doping became widely used in silicon crystals.

The importance of these silicon detectors in wartime radar systems prompted a search for alternatives, and germanium was studied extensively by a group at Purdue University under the leadership of Prof. K. Lark-Horovitz. The discovery there by Benzer,[8] in 1944, that a high-inverse-voltage rectifier could be formed from this material especially stimulated additional work on germanium. They found that a large number of dopants could be added to the germanium to produce the device but that tin was the best. They subsequently found similar effects in silicon.

The position in 1948 was that semiconductors had a significant but not very large share of the electronics market. Selenium rectifiers were widely used, but silicon and germanium were used only in the specialized microwave field. A typical silicon mixer of the period is shown in Fig. 1-2. As can be seen, it really differs very little from the cat's whisker of the 1920s. The crystal is polycrystalline, and the device is just a point-contact diode with the whisker vibration damped with wax. Although North[9] had achieved better stability with germanium by using a welded contact, most commercial devices were still of the point-contact variety. The effects of surface preparation and of doping were little understood, and the manufacture of these diodes was as much an art, based on empirical methods, as a science. The state of the art in 1948 was reviewed in detail in a book by Torrey and Whitmer,[10] and this should be consulted for further details.

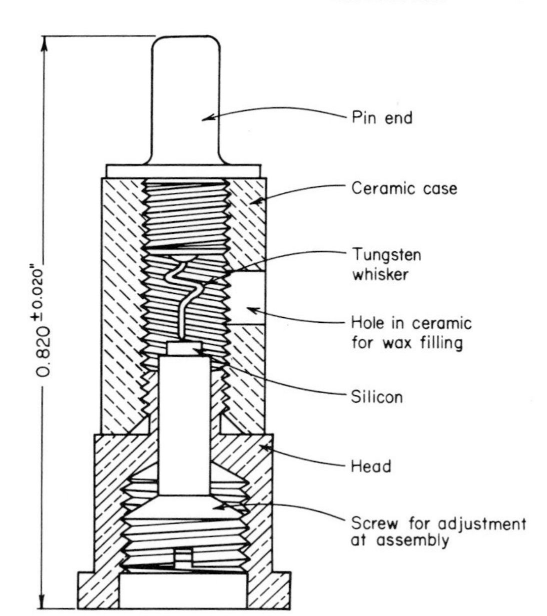

Fig. 1-2. A silicon diode of the 1948 period. (*After Torrey and Whitmer.*[10] *Courtesy of General Telephone & Electronics.*)

1-2. THE GERMANIUM TRANSISTOR

Semiconductor devices and the phenomenon of semiconduction were subjects of investigation for many years at Bell Telephone Laboratories. A broad and fundamental study of semiconduction was undertaken in 1945 by a group under Shockley. This work culminated in the announcement on June 30, 1948, of the invention of the transistor. It was first described by Bardeen and Brattain[11] in a letter to the *Physical Review* dated June 25 and published in the July 15 issue. Their device was a triode version of the point-contact diode, and their schematic is shown in Fig. 1-3. The contacts were placed very close together, about 0.005 to 0.025 cm apart. The germanium was the same as that used for high-inverse-voltage rectifiers, and its preparation was described by Scaff and Theuerer[12] in 1945 and is included in Torrey and Whitmer.[10] It was a gradient-cooled, polycrystalline n-type material of about 10 ohm-cm resistivity.

It is difficult to overestimate the importance of this discovery since it heralded a new era in electronics. Now, instead of being a curiosity with a very specialized application, the semiconductor device invaded the field held until this time by the vacuum tube: the field of amplification. Shockley[13] has discussed the events leading

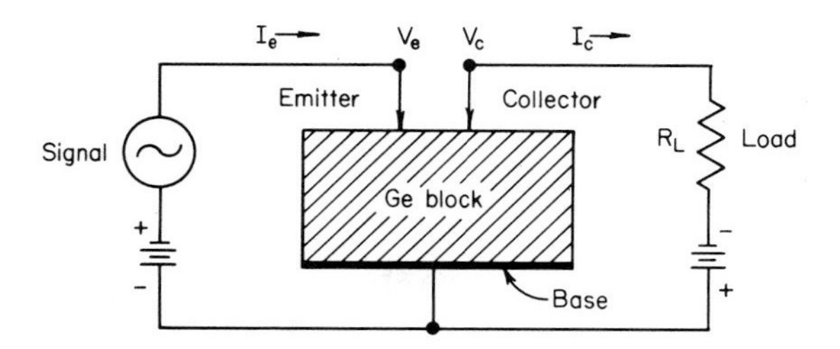

Fig. 1-3. The first transistor. (*After Bardeen and Brattain.*[11])

up to this achievement and pointed out that this was not a chance discovery but the direct result of Bardeen's theory of surface states published[14] over a year earlier. Bardeen and Brattain[15] later gave a fuller description of the development of the transistor and a theoretical treatment of its action.

In July, 1949, a series of papers was published in the *Bell System Technical Journal*[16-21] which dealt with several aspects of the newly discovered device; in particular, Shockley's paper[21] discussed the p-n junction and gave a theoretical treatment for the p-n-p transistor. This, unlike the first point-contact transistor, depends for its action on bulk properties of the germanium rather than surface effects. However, its practical development was due in no small part to the preparation by Teal and Little[22] of high-purity single-crystal material. In Shockley's opinion,[13] "There was probably no more important scientific development in the semiconductor field in the early days following the announcement of the transistor than the development of high-quality single crystals of germanium." In late 1950, Teal et al.[23] prepared p-n junctions by introducing dopants during the growth of the crystal, and the agreement of the junction properties with Shockley's theory[21] was demonstrated.[24] In 1951, Shockley, Sparks, and Teal[25] described the n-p-n transistor and its action; in particular, they stressed the difference between this eminently controllable device, inasmuch as the carrier concentrations could be controlled, and the point-contact transistor, which relied for its action on the contacts between the leads and the crystal, an unpredictable phenomenon at best. At the same time, Wallace and Pietenpol[26] published data on the performance of a number of experimental transistors and described their use in a variety of circuits.

At this point, the modern transistor had been achieved, and it stands as a remarkable achievement by Bell Telephone Laboratories, a triumph of interdisciplinary research. Shockley, Bardeen, and Brattain received the Nobel prize in 1956 for their key contributions to the success of this program, but it was the culmination of years of effort that started before the war and was the product not only of the physicists but also of the chemists and metallurgists. Without the materials of Teal and others there could have been no progress made; they founded a new branch of science, materials science, that now occupies centers at many leading universities.

1-3. COMMERCIAL GERMANIUM DEVICES

Although the transistor had been demonstrated and the grown-junction transistor developed, the preparation of the material was difficult; impurity levels were hard to control. It had been noted by Pearson et al.[27] that impurities concentrated in the liquid phase of germanium, and this fact was applied by Pfann[28] to its purification by a new technique: zone refining. This relatively simple technique simplified the preparation of material of very high purity and facilitated better control of dopant levels. The method was widely adopted by manufacturers, and by 1954 several commercial devices were available from RCA, GE, Raytheon, and others.

1-4. DEVICES FROM OTHER MATERIALS

The first paper by Bardeen and Brattain[11] had mentioned that, although only germanium was being described for the transistor, the same effect had been noted with silicon. The subsequent emphasis had been on germanium devices since the material is easier to grow in good single crystals. However, as early as 1952, Teal and Buehler[29] had reported the preparation of single crystals of silicon with grown p-n junctions. The earliest report of a grown-junction silicon transistor appears to be in March, 1954, when Raytheon[30] gave a brief account of such a device, but the first production units were reported on by Teal[31] on May 10, 1954. He gave detailed device characteristics, due to Adcock et al.,[32] and announced that two types of silicon transistor were available commercially from Texas Instruments.

In 1952, it was pointed out by Welker[33] that the III-V compounds (combinations of elements of Groups IIIA and VA of the periodic table) were also semiconductors. This stimulated considerable effort by many workers on a number of these compounds, and Welker and Weiss[34] in 1956 listed no less than 116 references. However, development has been relatively slow, and it is only recently that the III-V intermetallics have begun to encroach on the domination of silicon and germanium. New developments in silicon are now fewer, and germanium devices are experiencing a downward trend. The III-V intermetallics, led by gallium arsenide, while still far behind silicon, are being used for varactors, transistors, microwave diodes, light-emitting diodes, injection lasers, bulk microwave power devices, and bulk-effect integrated circuits; although of these, only varactors, Schottky-barrier microwave diodes, light-emitting diodes, and injection lasers are being made in production quantities. While the future is bright, the technological problems are formidable, and materials characterization must contribute in large part to their solution.

1-5. INTEGRATED CIRCUITS

The integrated circuit is really the outcome of the fusion of two approaches to microminiaturization: the printed circuit and the solid-state device. For some time prior to 1958, the Diamond Ordnance Fuze Laboratories, along with several other groups, had been working on the problem of reducing the size of the conventional printed circuits. In the fall of 1958, they sponsored a symposium on this subject, and the proceedings were subsequently published.[35] Several of the papers included descriptions of the photolithographic technique for preparing passive components (i.e., resistors, capacitors, etc.). In this process, a ceramic substrate is coated with a plastic resist which is sensitive to ultraviolet light. On exposure, the resist polymerizes. The unexposed portion can be dissolved in an organic solvent, and the exposed pattern remains as a coating. Metal may then be vacuum-evaporated to form a layer of any required geometry. This process can be repeated to build up, for example, a capacitor.

In the same symposium, a paper by Lathrop et al.[36] described application of this same technique to a transistor. The connections to the base and emitter were made by stripes through a resist mask; the collector contact was made through the underside to a base plate by a soldered joint. This must be one of the first introductions of an active device into an integrated circuit; however, it was not a true integrated circuit but what is termed a *hybrid*.

The first truly integrated circuit is due to Kilby, who, in the summer of 1958, fabricated a phase-shift oscillator from a single silicon bar. This device is shown in

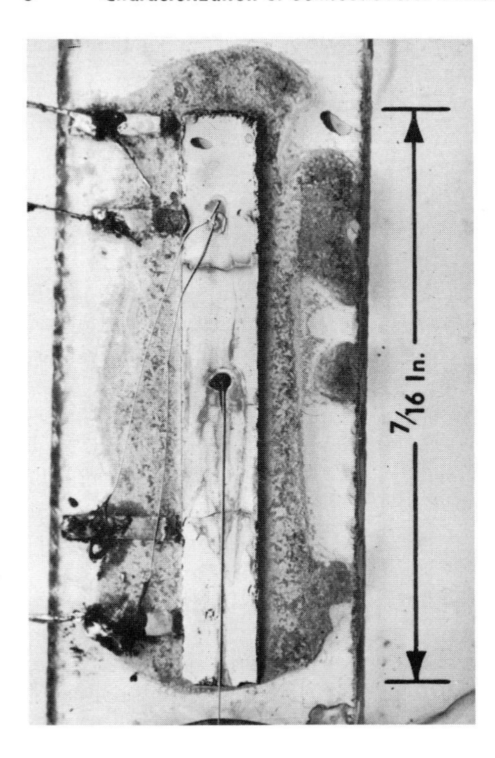

Fig. 1-4. Kilby's original integrated circuit. *(Photo courtesy of Texas Instruments Incorporated.)*

Fig. 1-4. In May, 1959, Kilby wrote a short description[37] of the techniques used. This device requires no interconnections from one component to another; the electrical path is through the silicon. It is a true integrated circuit, or, as it is often termed, a *monolithic integrated circuit.*

In 1960, Texas Instruments Incorporated announced the earliest product line of logic circuits, the SOLID CIRCUIT® Series 51. The technology involved in their production was described by Lathrop et al.[38] in May, 1960. It involved the photolithographic technique described above to form protective oxide masks on the silicon; diffusion through the oxide windows to form resistors, diodes, or transistors; and deposition of metal through resist windows to form contacts and capacitors. This is, essentially, the same process in current use. The patterns have become more complex and the steps more numerous, but the basic approach is the same. Figure 1-5 is an example of the current generation of circuits; the pack is about the same length as Kilby's device.

1-6. THE ROLE OF MATERIALS CHARACTERIZATION

Samuel Johnson once said, apropos of a woman preaching, that it was "like a dog's walking on its hind legs. It is not done well but you are surprised to find it done at all." In the early days of the transistor, something of the same atmosphere prevailed. It was such a remarkable achievement that it seemed almost ungrateful to demand rigorous specifications. The device had been fabricated from material

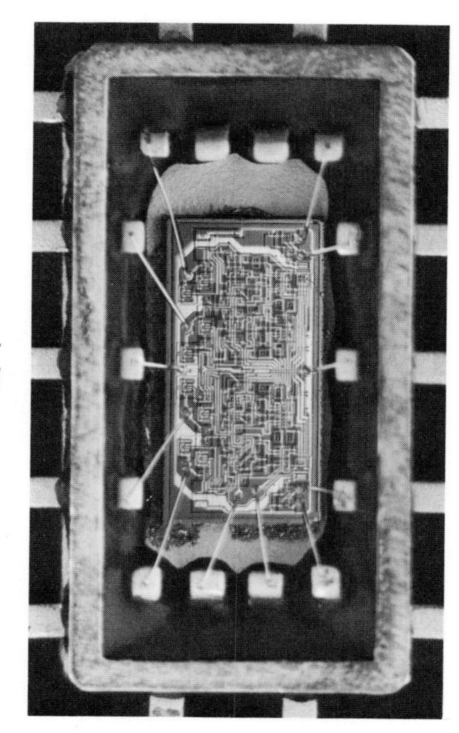

Fig. 1-5. A modern integrated circuit; a TTL BCD decade counter, SN5490. *(Photo courtesy of Texas Instruments Incorporated.)*

which had been characterized by only one property, its resistivity, but this was adequate for the first generation. However, in the late 1950s, increasing sophistication in transistor circuitry led to demands by the engineer for closer standards and better electrical characteristics. It was in this period that the concept of materials characterization was born, although not described as such. The analytical chemist, metallurgist, solid-state physicist, and crystallographer, working closely together, were called on to help relate the physical and chemical properties to the device characteristics.

Prior to 1948, the semiconductor diode was made from polycrystalline material, and this fact alone was enough to obscure many of the desirable properties of the material. Grain boundary effects were probably overriding, and the effect of chemical impurities could not be detected. Most materials that were analyzed at all were described as "spectroscopically pure." In a method described by Rick and McKinley[39] in 1944 and quoted by Torrey and Whitmer,[10] a method for high-purity germanium gave a product containing 0.2 percent zinc and about 0.05 percent other impurities, determined spectrographically. With these sorts of impurity levels, no great demands were placed on the analyst. However, with the preparation of single crystals by Teal and Little[22] in 1950 and the introduction of zone refining by Pfann[28] in 1952, the situation changed dramatically, and the analytical chemist was faced with a challenge that he has still not fully met.

Probably one of the earliest attempts to determine a dopant at a realistic level was by Smales and Brown,[40] who in 1950 described a method for arsenic in germanium dioxide; they specifically mentioned that it was carried out as a result of the use of germanium in transistors. In the same year, the first work on diffusion in germanium using radioisotopes was published;[41] this is still the preferred method for investigating diffusion when a suitable isotope is available. Honig[42] applied mass spectrometry to the analysis of germanium in 1953, but it was Hannay[43] who in 1954 designed an instrument specifically to meet the needs of semiconductor materials research, combining high sensitivity with broad coverage. In the same year, Hannay and Ahearn[44] published some results for this instrument, the solids mass spectrograph, and showed that analyses below 1 ppm were possible. In subsequent applications, it was found to give sensitivities down to 1 ppb for most impurities in all the semiconductor materials of interest. The commercial version of this instrument by Associated Electrical Industries, Ltd., has become almost standard in any laboratory dealing in high-purity materials, and it is difficult to overestimate the importance this machine has had in the successful commercial production of bulk semiconductor materials.

Since the introduction of integrated circuits in 1960, there has been an increasing interest in topographical analyses, concerned not so much with the bulk properties as with the surfaces and the thin films deposited on them. This has necessitated a reappraisal of the role of the analytical chemist in this field and his relationship to other scientists who are involved in characterizing semiconductor materials.

The crystal perfection of the substrate and film are of considerable interest, and x-ray topography and electron microscopy must be utilized in determining this property. The film thickness must be measured, perhaps by ellipsometry or by some other optical method. The distribution of elements in the surface requires an electron-probe microanalysis to achieve the scale of the photolithography. In short, a wide spectrum of tools, many of them requiring complex instrumentation and skilled personnel, must be employed in collaboration with the materials researcher and device technologist to advance the art in this highly sophisticated technology. The concept of materials characterization has been clarified recently by a committee of the Materials Advisory Board of the National Research Council and reviewed most succinctly and clearly in their report.[45] Their definition is as follows: "Characterization describes those features of the composition and structure (including defects) of a material that are significant for a particular preparation, study of properties, or use, and suffice for the reproduction of the material."

It is the object of this book to describe the current state of the art within this framework of characterization as it applies to semiconductor materials. It will have become apparent from this introduction that, while a large number of compounds and elements are semiconducting, only a relatively small group are of interest to the semiconductor industry. This group consists of germanium, silicon, and the III-V compounds, more specifically, gallium or indium with arsenic or antimony; and only these materials will be considered in the following chapters.

REFERENCES

1. Faraday, M.: *Diaries, Note* 317 (1833).
2. Smith, W.: *Nature,* **7**:303 (1873).
3. Grondahl, L. O.: *Science,* **64**:306 (1926).
4. Presser, E.: *Funkbastler,* **1925**:558.
5. Southworth, G. C., and A. P. King: *Proc. IRE,* **27**:95 (1939).

6. Seitz, F.: *Univ. Penn. NDRC* 14–112, 1942.
7. Theuerer, H. C.: *BTL Rep.* MM–43–120–74, 1943.
8. Benzer, S.: *Purdue Univ. NDRC* 14–342, 1944.
9. North, H. Q.: *General Electric Co. NDRC* 14–427, 1945.
10. Torrey, H. C., and C. A. Whitmer: "Crystal Rectifiers," McGraw-Hill Book Company, New York, 1948.
11. Bardeen, J., and W. H. Brattain: *Phys. Rev.*, **74**:230 (1948).
12. Scaff, J. H., and H. C. Theuerer: *BTL NDRC* 14–555, 1945.
13. Shockley, W.: in C. F. J. Overhage (ed.), "The Age of Electronics," p. 135, McGraw-Hill Book Company, New York, 1962.
14. Bardeen, J.: *Phys. Rev.*, **71**:717 (1947).
15. Bardeen, J., and W. H. Brattain: *Phys. Rev.*, **75**:1209 (1949).
16. Anon.: *Bell System Tech. J.*, **28**:335 (1949).
17. Shockley, W., G. L. Pearson, and J. R. Haynes: *Bell System Tech. J.*, **28**:344 (1949).
18. Ryder, R. M., and R. J. Kircher: *Bell System Tech. J.*: **28**:367 (1949).
19. Herring, C.: *Bell System Tech. J.*, **28**:401 (1949).
20. Bardeen, J.: *Bell System Tech. J.*, **28**:428 (1949).
21. Shockley, W.: *Bell System Tech. J.*, **28**:435 (1949).
22. Teal, G. K., and J. B. Little: *Phys. Rev.*, **78**:647 (1950).
23. Teal, G. K., M. Sparks, and E. Buehler: *Phys. Rev.*, **81**:637 (1951).
24. Goucher, F. S., G. L. Pearson, M. Sparks, G. K. Teal, and W. Shockley: *Phys. Rev.*, **81**:637 (1951).
25. Shockley, W., M. Sparks, and G. K. Teal: *Phys. Rev.*, **83**:151 (1951).
26. Wallace, R. I., and W. J. Pietenpol: *Bell System Tech. J.*, **30**:530 (1951).
27. Pearson, G. L., J. D. Struthers, and H. C. Theuerer: *Phys. Rev.*, **77**:809 (1950).
28. Pfann, W. G.: *Trans. AIME,* **194**:747 (1952).
29. Teal, G. K., and E. Buehler: *Phys. Rev.*, **87**:190 (1952).
30. Anon.: *Electronics,* **27**:8 (March, 1954).
31. Teal, G. K.: IRE National Conference on Airborne Electronics, Dayton, Ohio, May 10, 1954.
32. Adcock, W. A., M. E. Jones, J. W. Thornhill, and E. D. Jackson: *Proc. IRE,* **42**:1192 (1954).
33. Welker, H.: *Z. Naturforsch.*, **7a**:744 (1952).
34. Welker, H., and H. Weiss: in F. Seitz and D. Turnbull (eds.), "Solid State Physics," vol. 3, p. 1, Academic Press Inc., New York, 1956.
35. Horsey, E. F. (ed.): "Proceedings of the Symposium on Microminiaturization of Electronic Assemblies," Hayden Publishing Company, New York, 1959.
36. Lathrop, J. W., J. R. Nall, and R. J. Anstead: in E. F. Horsey (ed.), "Proceedings of the Symposium on Microminiaturization of Electronic Assemblies," p. 86, Hayden Publishing Company, New York, 1959.
37. Kilby, J. S.: *Electronics,* **32**:110 (Aug. 7, 1959).
38. Lathrop, J. W., R. E. Lee, and C. H. Phipps: *Electronics,* **33**:69 (May 13, 1960).
39. Rick, C. E., and T. D. McKinley: *OSRD Progr. Rep.* NWP-P-44-3K, Contract OEMSr-1139, 1944.
40. Smales, A. A., and L. O. Brown: *Chem. Ind.* (London), **1950**:441.
41. Fuller, C. S., and J. D. Struthers: *Phys. Rev.*, **87**:526 (1952).
42. Honig, R. E.: *Anal. Chem.*, **25**:1530 (1953).
43. Hannay, N. B.: *Rev. Sci. Instr.*, **25**:644 (1954).
44. Hannay, N. B., and A. J. Ahearn: *Anal. Chem.*, **26**:1056 (1954).
45. Characterization of Materials, *Nat. Acad. Sci. — Nat. Acad. Eng. NRC Rep.* MAB-229-M, Washington, D.C., 1967.

2

Semiconductor Principles

2-1. INTRODUCTION

To appreciate the problems involved in semiconductor materials research, some knowledge of the principles underlying semiconduction is essential. For the reader coming fresh to the field, this chapter is intended as an introduction which hopefully will allow him to follow the reasoning behind the research in the following chapters. It is not intended to be a comprehensive review of solid-state physics; for such a treatment the reader is referred to books by Dunlap,[1,†] Kittel,[2] or Shockley's classic.[3] An excellent, readable, and essentially nonmathematical treatment is given by Warschauer.[4]

2-2. CONDUCTION IN SOLIDS

Electricity can be conducted through solids by one of two mechanisms: ionic or electronic. In a crystal such as sodium chloride, the lattice sites are occupied by ions, alternately positive or negative, and their positions are governed by electrostatic forces acting mutually. Repulsion by like ions is exactly balanced by the attraction of unlike ions, and the lattice points are equilibrium positions. Conduction is an electrolytic process, brought about by the migration of ions through the solid.

Electronic conduction is the commoner method of transfer of electricity and is encountered in solids in which the atoms are held together not by ionic attraction and repulsion but by coordinate bonding. It is this class of solid which is of interest in semiconductors, and the properties of these materials are best described by the energy-band theory.

2-3. ENERGY BANDS

Consider the electronic configuration of the silicon atom. It consists of an inner shell (K level) of two electrons, a second shell (L level) of eight electrons, and an outer or valence shell (M level) of four electrons. In the usual convention, it is

†Superscript numbers indicate References listed at the end of the chapter.

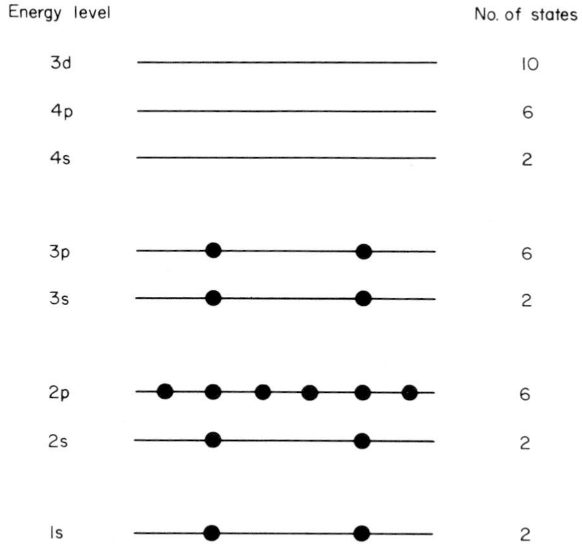

Energy level	No. of states
3d	10
4p	6
4s	2
3p	6
3s	2
2p	6
2s	2
1s	2

Fig. 2-1. Energy-level diagram for silicon.

described as $1s^2 2s^2 2p^6 3s^2 3p^2$, and its energy-level diagram is shown in Fig. 2-1. This represents the situation when the silicon atom is at an infinite distance from other atoms. Suppose we now start bringing this closer to other silicon atoms. At some point, the outer shells begin to overlap and the energy levels shift slightly. Since a number of atoms are all mutually interacting, the effect is to split the energy level into a number of closely spaced energy levels termed a *band*. The situation is then similar to that shown in Fig. 2-2. At infinity, the levels are all identical. At about 4 Å, the $3p$ and $3s$ levels begin to interact, and a number of sublevels are generated to form bands. At about 2.5 Å, the two levels overlap and then separate again into two bands. This is the point at which the bonds are formed and the electrons drop into the lower band. At distances close to zero, the inner shells interact, but for our

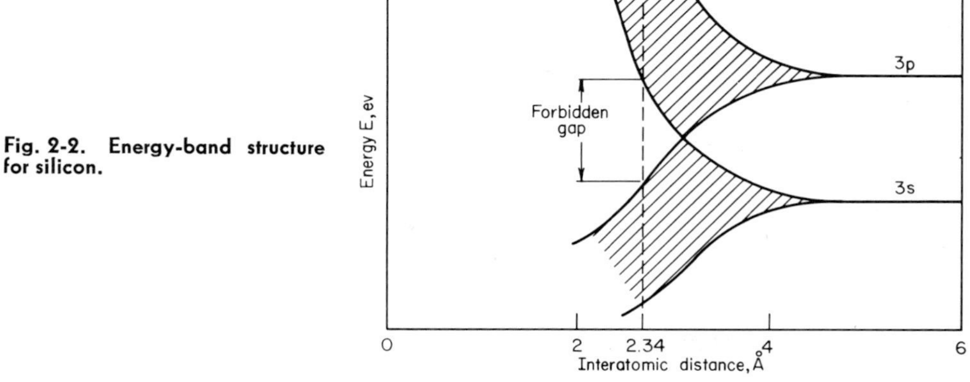

Fig. 2-2. Energy-band structure for silicon.

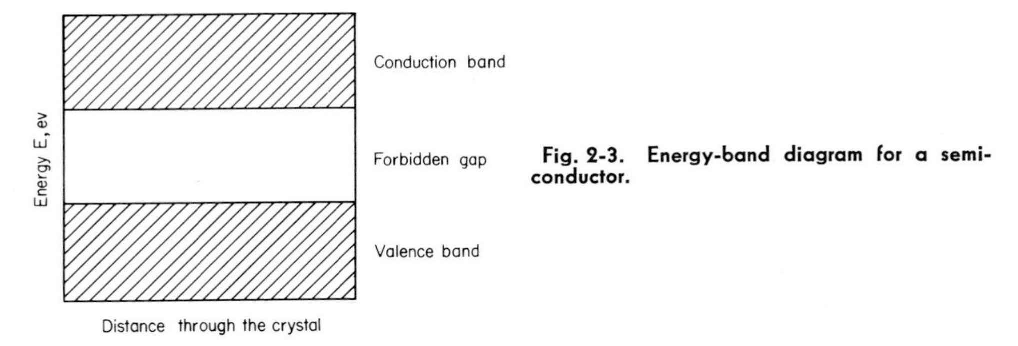

Fig. 2-3. Energy-band diagram for a semiconductor.

purposes this can be ignored and they are not shown in this diagram. For a silicon crystal under normal conditions of temperature and pressure, the interatomic distance is 2.34 Å, and at this distance the band structure will be represented by the vertical line drawn through this point or, by using distance through the crystal as the abscissa, where the interatomic distance is constant, by the conventional energy diagram of Fig. 2-3. The two bands contain four quantum levels each. The lower band, which is filled, contains the four valence electrons and is therefore termed the *valence band*. The upper band, which is empty, is termed the *conduction band* for reasons which will soon become obvious. They are separated by the forbidden gap, that is, a band in which no quantum levels can exist.

2-4. CHARGE CARRIERS

The energy-band diagram of silicon represents the condition of the solid in its ground state, that is, at absolute zero. The conduction band is completely empty. If we consider a two-dimensional representation, the silicon lattice will look like Fig. 2-4. All the electrons are in the valence state; or, put in another way, each

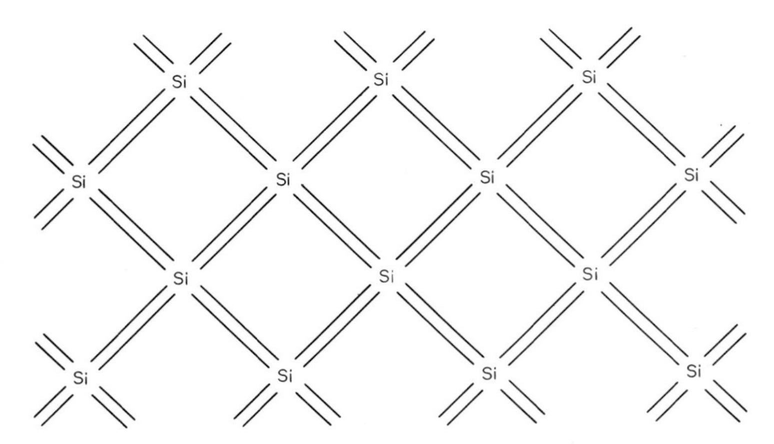

Fig. 2-4. Silicon lattice in the ground state.

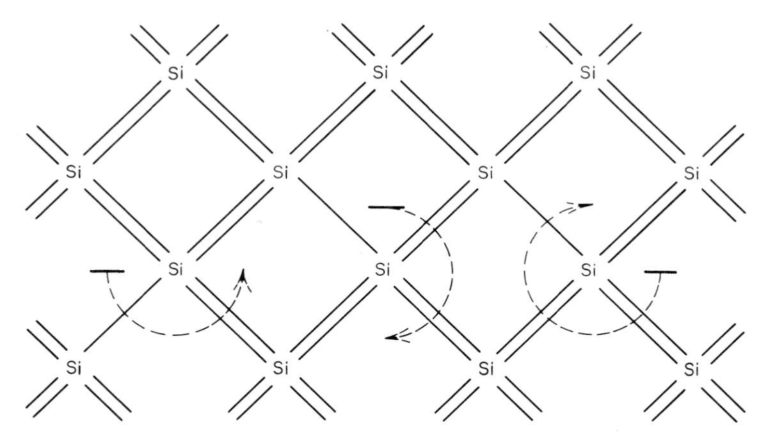

Fig. 2-5. Silicon lattice in the excited state.

silicon atom is bonded covalently to four other silicon atoms. There are no free or conduction electrons since all the available electrons are bound firmly to atom sites. Suppose, however, we supply energy to the lattice, either as heat or as light. Then some electrons will receive enough energy to allow them to jump the forbidden gap and enter the conduction band. For silicon, the forbidden gap is 1.1 ev. The position then is as shown in Fig. 2-5. These conduction electrons can move very readily from one atom to another and in effect wander through the lattice at random. If an electric field is applied to the lattice, the movement will cease to be random. Electrons will move against the field; that is, they become directional and a current flows.

It follows that if an electron jumps from the valence to the conduction band, it must leave a vacancy in the valence band. In other words, some bonds must be short an electron. In Fig. 2-5 these bonds are shown as single bonds. Under the influence of an applied field, the electrons move from negative to positive. It follows that since the single bonds represent a deficiency of electrons, they will appear to move with the field; that is, they act as positive charges. These positive charges are termed *holes*.

Conduction of the type described, in which the current is carried by electrons and holes derived only from the silicon atoms and not from any foreign atoms, is termed *intrinsic conduction*. The electrons and holes are collectively termed *charge carriers*. It is apparent that the more energy is supplied, the more charge carriers are generated. This explains one of the characteristic properties of semiconductors: their decreased resistivity with temperature, the so-called "inverse temperature coefficient."

2-5. CONDUCTION IN INSULATORS

An insulator has an energy-band diagram essentially identical to that for a semiconductor. Its valence band is filled and its conduction band empty. It differs in that the forbidden gap is so wide that thermal energy cannot excite electrons across it. No charge carriers result.

2-6. CONDUCTION IN METALS

For metals, the valence band is not completely filled. This is due to an overlap of two or more energy bands. For example, if in Fig. 2-2 the 3*s* and 3*p* bands, instead of separating immediately after overlapping, had continued as a combined band, we would have had a valence band with eight quantum states, only four of which were filled. In such a case, electrons can move very easily since there are quantum levels very close to the lowest ones. Very little energy is necessary for them to move from one atom to the next, and essentially all the valence electrons are available for conduction. However, conduction in this case is through the valence band. The conduction band, which in our hypothetical case would be the 4*s*, is always empty.

For metals, the number of electrons carrying the current is so high that the resistivity is essentially independent of temperature. The temperature coefficient is governed in this case by the mobility.

2-7. MOBILITY

If an electron in a vacuum is subjected to an electric field, it is accelerated linearly, and its velocity at any time is governed by Newton's laws of motion. In a solid, however, it suffers many collisions, and its resulting motion is random in all directions except that it will tend to move against the field. Although its movement is erratic, the force acting on it will eventually move it toward the positive terminal. This drift mobility is defined in terms of the velocity per unit field, i.e.,

$$\text{Drift mobility } \mu_D = \frac{v_D}{E}$$

where v_D = average velocity of electrons, cm/sec
E = applied field, volts/cm

Thus, the units for μ_D are square centimeters per volt per sec. Since current can also be carried by holes, the mobility can equally well be applied to all charge carriers.

As the temperature increases, the electrons travel faster. This would tend to suggest that the mobility increases, but in fact the reverse is true since the electron suffers many more collisions in unit time. Moreover, the atoms with which it is colliding are vibrating more, also tending to increase the number of collisions. The net result is that mobility decreases with temperature, and it is this fact that leads

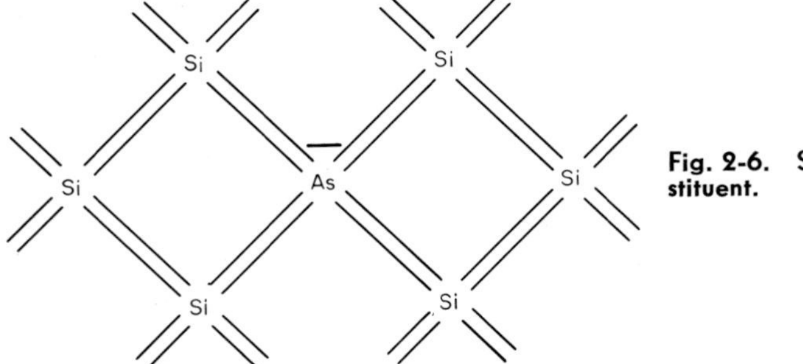

Fig. 2-6. Silicon lattice with donor substituent.

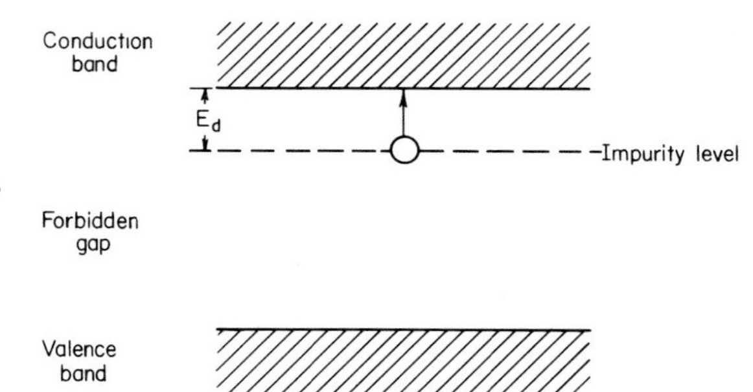

Fig. 2-7. Energy-band diagram for donor impurity.

to an increase in resistance with temperature for metals. In semiconductors, this effect is generally not enough to overcome the increasing conductivity brought about by the transfer of electrons to the conduction band.

2-8. EXTRINSIC CONDUCTION

In Sec. 2-4, the charge carriers were described as either electrons or holes, and in the case of intrinsic conduction were derived from the semiconductor material only. It follows that for a perfect crystal, the number of holes and electrons are equal. Suppose, however, we substitute for a silicon atom in the lattice an arsenic atom. The position will be as shown in Fig. 2-6; four of the arsenic bonds will be used to satisfy the surrounding silicon atoms, but the fifth bond will be in essence a free electron. Its energy level will be close to that of a conduction electron, and it will readily function as such. In terms of the energy diagram, the electron energy is in the forbidden gap, as shown in Fig. 2-7. The impurity level is a distance E_d, the activation energy, below the conduction band; and since this is significantly less than the energy necessary to cross the gap, such an electron readily contributes to the conduction band. Since it donates an electron to this band, such an atom is termed a *donor*.

Suppose, instead of arsenic, we substitute indium in the silicon lattice. Then the situation is as shown in Fig. 2-8, where one bond is unsatisfied. This represents a site which can readily capture an electron, an energy level considerably lower than the conduction band. In terms of the energy diagram, the situation is as shown in

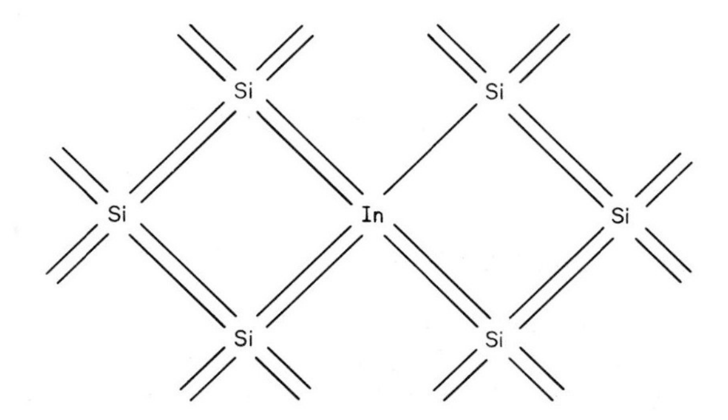

Fig. 2-8. Silicon lattice with acceptor substituent.

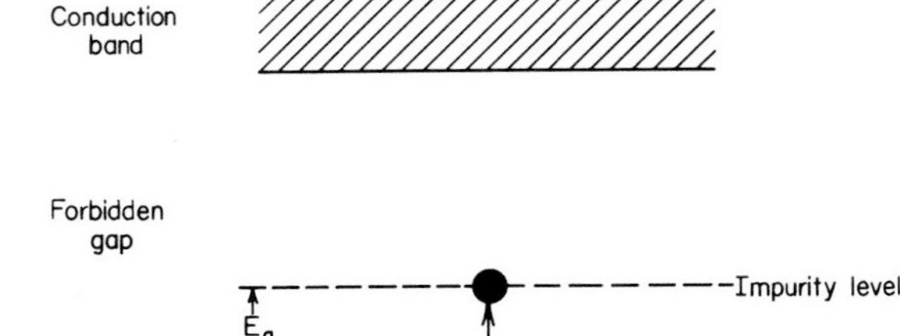

Fig. 2-9. Energy-band diagram for acceptor impurity.

Fig. 2-9. The activation energy E_a necessary to raise an electron from the valence band to the impurity level is very much less than that necessary to cross the forbidden gap. Consequently, holes are generated very much more readily. Since this impurity level readily accepts electrons, atoms with this property are termed *acceptors*.

Donors, which generate electrons, are termed *n-type impurities*. Acceptors, which generate holes (or positive charges), are termed *p-type impurities*. Conduction which is chiefly due to either acceptors or donors is termed *extrinsic conduction*, and it is this phenomenon which allows the properties of semiconductor materials to be tailored to device parameters.

In tailoring these extrinsic materials, dopants are added so that substitution of one atom in the lattice generates one electron or one hole, and a correspondence is assumed between dopant concentration and carrier concentration. In semiconductor materials, therefore, it is conventional to refer all impurity concentrations to an atoms per cubic centimeter basis rather than to the weight-weight basis familiar to chemists. The relationship between these two is given by the expression

$$\text{Parts per billion} = \text{atoms/cm}^3 \times \frac{M \times 10^9}{A \times d}$$

where M = atomic weight of impurity
A = Avogadro's number
d = density of bulk material

As an example, for boron, a p-type dopant, in silicon at a level of 10^{14} atoms/cm^3,

$$M = 10.8$$
$$A = 6.0 \times 10^{23}$$
$$d = 2.4$$

i.e.,

$$\text{boron content} = 10^{14} \times \frac{10.8 \times 10^9}{6.0 \times 10^{23} \times 2.4} = 0.75 \text{ ppb}$$

2-9. COMPENSATION

If both donors and acceptors are present in the material, the energy diagram will be a composite of Figs. 2-7 and 2-9, as shown in Fig. 2-10. In this case, the donor electron will drop to the acceptor level and will not be available as a current carrier;

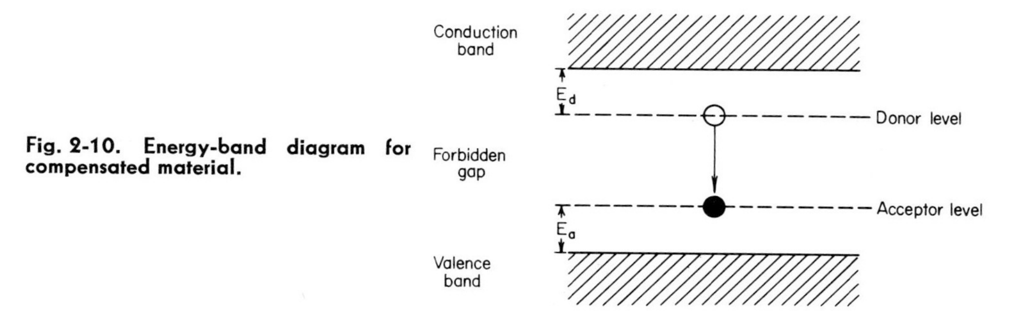

Fig. 2-10. Energy-band diagram for compensated material.

the donor is said to be compensated by the acceptor. If the number of donors and acceptors were equal, intrinsic material would result, but in practice an equal number is almost impossible to achieve. One species will predominate, and either n- or p-type material will result. If a donor is present to a greater degree, electrons will be dominant and the material will be n type. The majority carrier will be the electrons. Conversely, if an acceptor predominates, the holes will be the majority carrier and the material is p type.

2-10. DENSITY OF CARRIERS

For an intrinsic material at any particular temperature, it is apparent that, within the limits of statistical probability, there will always be the same number of electrons in the conduction band and, by corollary, an equal number of holes in the valence band. In point of fact, it can be shown by Fermi-Dirac statistics that the product of these two values is always a constant at any one temperature, i.e.,

$$np = \text{const}$$

where n and p represent the number of electrons and holes, respectively. The condition is entirely analogous to that of pure water, where

$$[H^+][OH^-] = \text{const}$$

dependent only on temperature. It is an equilibrium constant, and the law of mass action applies. For water, this constant is 10^{14}, and if we increase $[H^+]$, we must decrease $[OH^-]$. Similarly, for silicon at 300°K, the constant is $2.6 \times 10^{20}\ \text{cm}^{-6}$, and if we increase n by the addition of donors, we automatically decrease p. This constant gives us a means for determining the minority-carrier concentration if the majority-carrier concentration is known.

2-11. CARRIER CONCENTRATION

It is obvious that the conductivity of a semiconductor depends on two properties, the number of charge carriers and the mobilities, again entirely analogous to the transport of electricity through an aqueous solution. It is very simply expressed as

$$\sigma = ne\mu_n + pe\mu_p$$

where n, p = number of electrons and holes, respectively
e = charge on electron
μ_n, μ_p = drift mobility of electrons and holes, respectively

In practice, the material will be either n or p type, that is, either n or p will be large and the other correspondingly small. In the case of n-type material, the expression will reduce to

$$\sigma = ne\mu_n$$

Since σ (or its reciprocal, resistivity) can be measured, we can determine the number of majority carriers if the drift mobility can be determined.

2-12. HALL MOBILITY

It is possible to determine drift mobility by generating electrons (and holes) by exposure to light. The electrons are drifted down a length of the material under the influence of a field of known strength, and their arrival at some point detected by a collector. If the light is attenuated by a shutter, the time between the light exposure and the arrival of the electrons at the collector can be used to determine the drift mobility.

In practice, it is usually more convenient to determine a mobility termed the *Hall mobility*. For silicon and germanium, these two mobilities are approximately equal. If a charge moves in a magnetic field, then it experiences a force acting at right angles to both its direction and the direction of the magnetic field. This is the well-known left-hand rule, or motor rule, for the force exerted by a magnetic field on a conductor carrying a current. If the thumb and first and second fingers of the left hand are made mutually perpendicular, and if the forefinger indicates the direction of the magnetic field and the second finger the direction of the current, then the thumb indicates the direction of the resultant force. Since the conventional current flow is from high to low potential, this is also the direction of movement of positive charges.

If we now consider the carriers in a semiconductor, they will move under the influence of an electric field: electrons against the field, holes with the field. If we

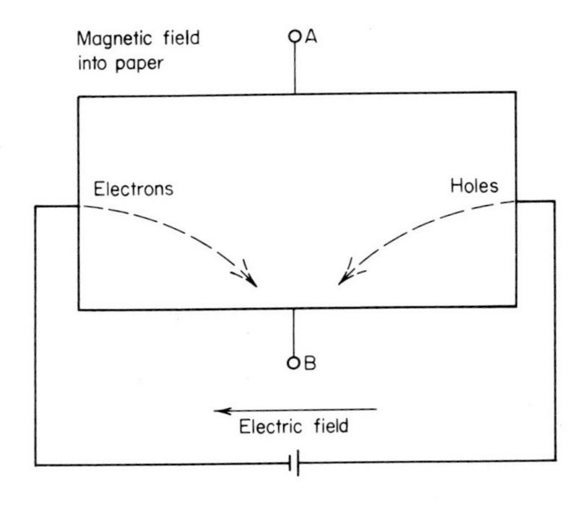

Fig. 2-11. The Hall effect.

apply a magnetic field at right angles to the electric field, an additional vector will be introduced. Instead of traveling straight toward the ends of the bars, the electrons or holes will also move in a direction mutually perpendicular to the two fields. They will take paths shown by Fig. 2-11. Both holes and electrons will move transversely in the same direction since, although the action of the magnetic field is reversed, the motions laterally are in opposite directions. A space charge builds up at the lower surface, and at equilibrium an electric field is set up across the crystal such that its magnitude and direction balance the space charge and the charge carriers flow straight through the crystal. If we connect a potentiometer across this field, i.e., at A and B, the polarity will indicate whether holes or electrons are the majority carrier, and the magnitude of the potential difference will indicate the Hall field induced.

As might be expected, the force exerted on the charge is proportional to the magnetic field and to the electric field and the velocity of the charge. In fact, these last two are related — the stronger the field, the higher the velocity for any particular crystal — and this is the relationship used to define mobility in Sec. 2-7, i.e.,

$$\mu_D = \frac{v_D}{E}$$

If we consider the condition necessary for charge carriers to move undeflected through the crystal, then the Hall field E_H must exactly compensate the force exerted by the magnetic field; i.e., for any charge carrier

$$eE_H = ev_DH$$

where e = charge on electron
v_D = velocity of charge carrier
H = magnetic field

or

$$E_H = vH$$

The Hall voltage will depend on the Hall field and the distance across it, i.e.,

$$V_H = E_Hd$$

By combining these various equations, it can be seen that

Hall mobility $$\mu_H = \frac{V_H}{EHd}$$

or, since

$$E = \frac{V_a}{l}$$

where V_a = applied voltage
l = length of bar

$$\mu_H = \frac{V_Hl}{V_aHd}$$

Since all the parameters on the right can be measured, the Hall mobility can be calculated.

In practice, it is more usual to measure the current through the bar rather than the applied voltage. The equation relating Hall voltage with the velocity of the electron was based on the forces on any one charge carrier. However, E_H must also depend on the number of charge carriers moving through the bar, i.e.,

$$E_H = R_H iH$$

where i = current density
R_H = a proportionality constant called the Hall coefficient

Now
$$i = \frac{I}{wd}$$

and
$$I = \frac{V_a}{R} = \frac{V_a \sigma wd}{l}$$

Combining,

$$R_H = \frac{E_H}{iH} = \frac{E_H l}{V_a \sigma H} = \frac{V_H l}{V_a \sigma H d}$$

But
$$\mu_H = \frac{V_H l}{V_a H d}$$

so that
$$R_H = \frac{\mu_H}{\sigma}$$

Returning to

$$R_H = \frac{E_H}{iH}$$

we can write this alternatively as

$$R_H = \frac{V_H w}{IH}$$

All the parameters on the right can be measured, and this is the usual method for determining the Hall coefficient. The value above would be in electrostatic units; in practical units

$$R_H = \frac{V_H w \times 10^8}{IH} \qquad \text{cm}^3/\text{coul}$$

where V_H is in volts, w is in centimeters and is the width in the direction of the magnetic field, I is in amperes, and H is in gauss. The value is conventionally negative for n-type and positive for p-type material.

In Sec. 2-10, we saw that

$$\sigma = ne\mu_D$$

where n = number of carriers
μ_D = drift mobility

If

$$\mu_D = \mu_H$$

then

$$R_H = \frac{\mu_D}{\sigma} = \frac{\mu_D}{ne\mu_D} = \frac{1}{ne}$$

Since e is the charge on the electron, 1.6×10^{-19} coul, n, the charge-carrier density per cubic centimeter, can be calculated.

Determination of the Hall coefficient and resistivity will give the majority-carrier density and the Hall mobility.

2-13. MEASUREMENT OF RESISTIVITY AND HALL COEFFICIENT

The commonest method for measuring resistivity is by the four-point probe, and this is described in more detail in Sec. 4-16. Four equidistant contacts are pressed against one surface. Across the outer two a voltage is impressed sufficient to maintain a flow of current I. The potential difference V between the two inner probes is measured with a potentiometer. Then it can be shown[5] for the resistivity ρ that

$$\rho = \frac{2\pi V}{I} a$$

where a is the spacing between the probes.

The Hall coefficient may be measured on a specially prepared bar by making contact to the two sides and two ends of the bar. A magnetic field, say 2,000 gauss, is passed through the bar perpendicular to the length and the plane of the Hall contacts. By measuring the current through the bar and the voltage induced across the Hall contacts, the Hall coefficient can be calculated by means of the equation given in Sec. 2-12.

In practice, several difficulties arise. One is due to the misalignment of the Hall contacts; even without a magnetic field a voltage difference will be detected. Moreover, passage of the current causes heating, and thermoelectric potentials are generated. These factors can be overcome to some extent by reversing the current flow and magnetic field direction in turn. The four measurements are combined to give a mean value. Alternatively, the magnetic field and current can be altered at different frequencies. In either case the Hall voltage can be deduced free from interfering potentials. The procedure is given in detail in Sec. 4-17.

2-14. LIFETIME

One other important property of a charge carrier is its lifetime. As a charge wanders through the lattice, there is a finite probability that it will meet a charge of the opposite sign and recombine. For majority carriers, this probability is very small; but for minority carriers it will be very high, that is, their lifetime will be short. When excess carriers are generated by, say, light, then the disappearance of the excess minority carriers follows the usual decay pattern. The number decaying during a period dt is proportional to the number n present at that time. This is an exponential decay:

$$\frac{dn}{dt} = \frac{1}{\tau} n$$

where $1/\tau$ is a constant. Alternatively, this may be written as

$$n = n_0 e^{-t/\tau} \qquad \text{cm}^{-3}$$

If $\tau = t$, then n is $1/e$ that of n_0; or τ, the lifetime, is the time for the number of carriers to decay to $1/e$ of its original value.

Lifetime can be measured by the same method described for drift mobility in Sec. 2-12. In this case, the shape of the carrier pulse as it arrives at the collector can be used to calculate the lifetime.

This property is important in many device applications, particularly in transistors, and is drastically modified by both physical and chemical defects in the crystal. It is discussed in detail in Sec. 4-18.

2-15. THE p-n JUNCTION

If a region of n-type material is made to adjoin a region of p-type material in the same crystal, then a p-n junction is formed at the boundary. This is an extremely important element in semiconductor technology.

Initially, the two regions may be represented by the diagram of Fig. 2-12. The circled charges represent the ionized donors, in the case of the n-type, or acceptors, in the case of the p-type material; and the uncircled charges represent the majority carriers. Since each material is electrically neutral, the charge carriers will equal the ionized dopants in each region. As the carriers move around the crystal, however, some holes will enter into the n-type region and some electrons into the p-type

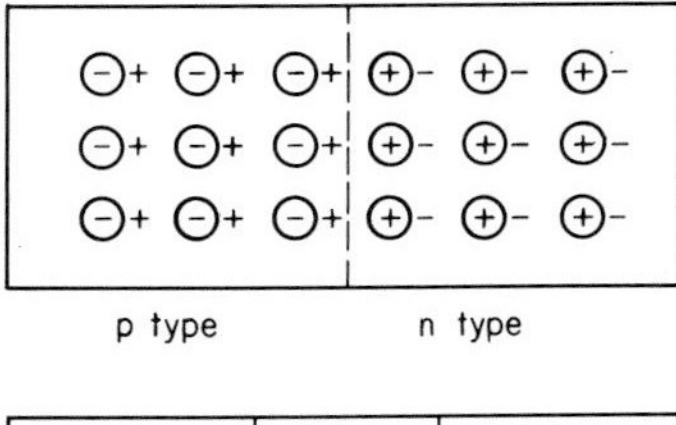

Fig. 2-12. The p-n junction—initial condition. (*After Warschauer.*[4])

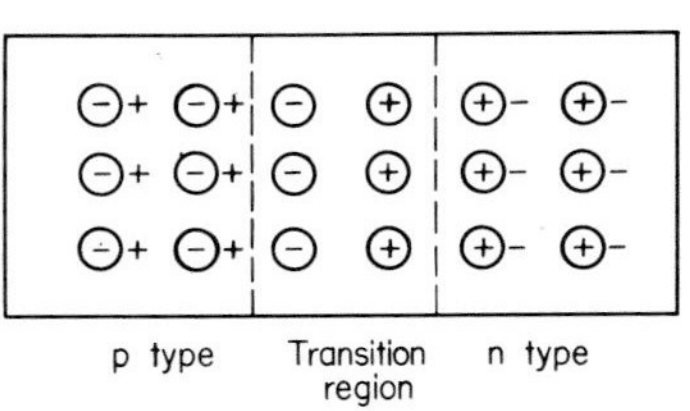

Fig. 2-13. The p-n junction—equilibrium condition. (*Adapted from Warschauer.*[4])

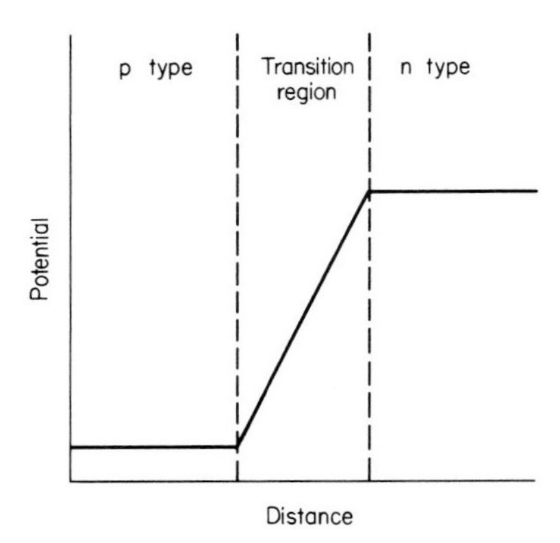

Fig. 2-14. Potential distribution across the p-n junction. (*Adapted from Warschauer.*[4])

region. In either case, they will then be minority carriers and their lifetimes will be very short. Recombination is very readily achieved; in fact, a region at the boundary, called the transition region, rapidly becomes depleted of carriers, as shown in Fig. 2-13.

Although the transition region is devoid of carriers, it does contain ionized dopants which are now no longer neutralized. Consequently, a charge double layer is set up which gives rise to a potential difference, as shown in Fig. 2-14. This effectively prevents any further movement of electrons into the p region or of holes into the n region. An equilibrium is set up dependent on the number of charge carriers in each region, their mobilities, and their lifetimes. In order to inject minority carriers into one or other region, an outside potential must be applied to overcome the effect of the potential difference across the transition region. If the potential difference is removed by making the n region negative and the p region positive with an applied voltage, the carriers will cross the transition region easily

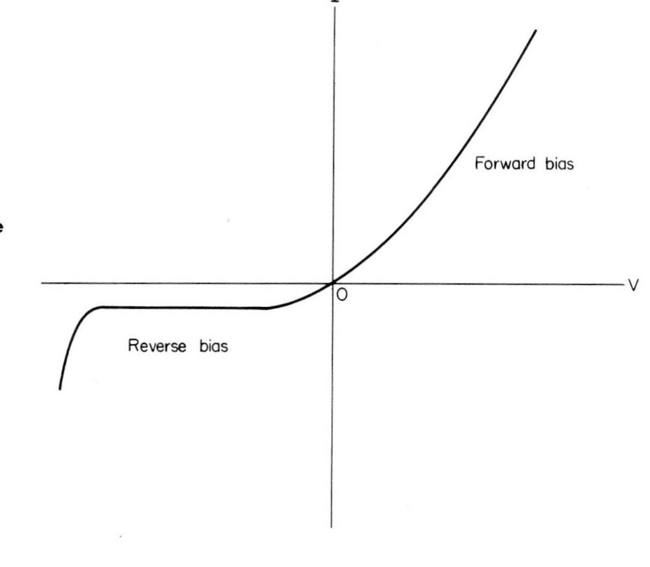

Fig. 2-15. Current-voltage curve or a p-n junction.

and a current will flow. This is termed *biasing in the forward direction*. Conversely, if the n region is made positive and the p region negative, the potential difference is increased still further, making it even more difficult for the carriers to cross the transition region. This mode is termed *reverse bias*. The overall current-voltage curve is shown in Fig. 2-15. The small current in the reverse-bias mode is due to the thermally generated minority carriers which are always present in each region and which, of course, find the potential difference favorable to their transfer across the transition region. At higher voltages in reverse bias, breakdown sets in.

The p-n junction can also be described in terms of the energy-band structure. Before doing this, we must introduce a new concept, the Fermi level. This is defined as that energy level which, if it existed, would have a 50 percent probability of being filled. In semiconductors, it is not usually a real energy level; it is purely a statistical concept. If we consider a pure intrinsic material, then at absolute zero the valence band would be 100 percent filled, the conduction band 0 percent filled. If the forbidden gap were nonexistent, that is, if all energy levels between the valence and conduction band existed, then the energy level which would have a 50 percent chance of being filled would be halfway between the two bands. This is the Fermi level, and the fact that it cannot exist since it is in the forbidden gap does not invalidate its use; it is a probability concept. Donor and acceptor levels do exist, although they are in the forbidden gap, because they are not in the same system as the intrinsic material.

If we add donors to a material, this increases the number of electrons in the conduction band. The probability that higher energy levels will be filled is increased, and consequently the Fermi level is raised. Conversely, if acceptors are introduced, a level just above the valence band is filled and the probability increased for levels being filled at lower energies. In this case the Fermi level is lowered. Generally, p-type material has a low Fermi level, n-type material a high Fermi level.

In considering the p-n junction, the Fermi level must be the same on both sides of the junction. If it were not so, electrons in different areas of the same crystal would have different energies. This would obviously be an unstable situation which would rapidly equalize, and the result would be that which we predicated: the Fermi level would be the same throughout the crystal. However, if the Fermi level is the same across the crystal, it follows that the band levels in the n- and p-type regions are not.

Consider the energy bands for n-type and p-type materials shown in Fig. 2-16.

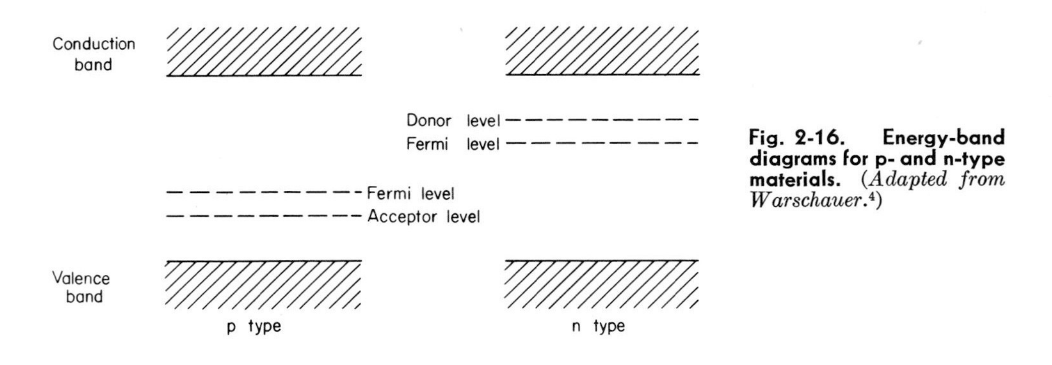

Fig. 2-16. Energy-band diagrams for p- and n-type materials. (*Adapted from Warschauer.*[4])

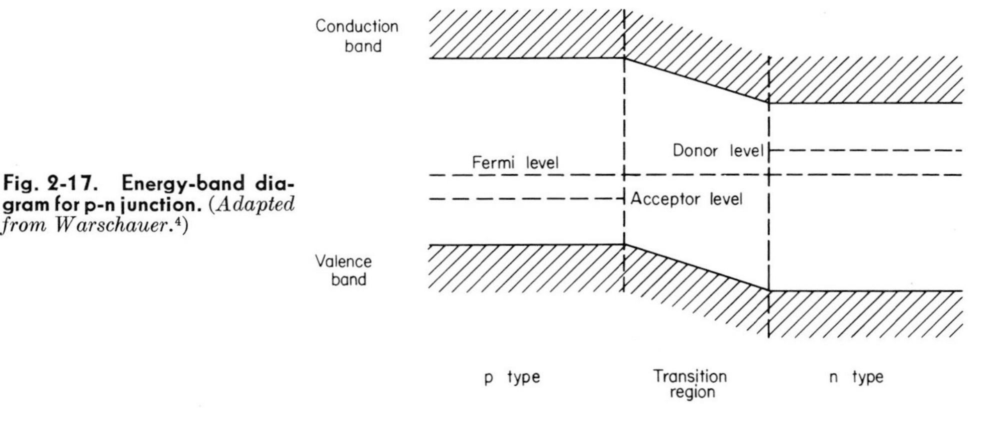

Fig. 2-17. Energy-band diagram for p-n junction. (*Adapted from Warschauer.*[4])

As we have already seen, the Fermi level in the n-type material is higher than that in the p-type. Now suppose we make a p-n junction between these two materials. The Fermi level must be the same on both sides, and it follows, as shown in Fig. 2-17, that the energy bands in the p region must be higher than those in the n region. A smooth gradation is assumed at the boundary to form the transition region. Since electrons tend to lower energy levels (and conversely holes to higher levels), there will be no movement of majority carriers across the transition region.

If we apply a voltage across the junction, in the forward direction, we are supplying energy which will enable many electrons, or holes, to overcome the energy barrier represented by the transition region. In terms of the energy band, we raise the whole energy diagram of the n-type region relative to that of the p-type, as shown in Fig. 2-18, where V represents the applied voltage. There is now no energy barrier to the diffusion of majority carriers across the transition region, and current flows. Conversely, if we make the n-type region positive, the energy diagram of the n-type region is lowered relative to the p-type, and the energy barrier is increased.

2-16. JUNCTION TRANSISTOR

The current-voltage curve shown in Fig. 2-15 is characteristic of a diode, and it is easy to see how rectification can be obtained by such a device. With an alternating

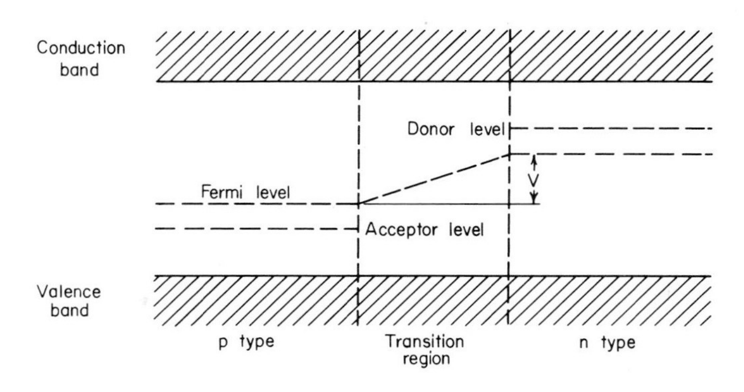

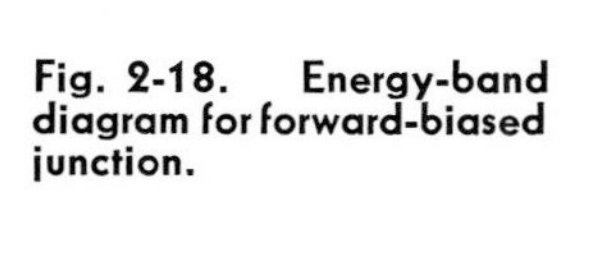

Fig. 2-18. Energy-band diagram for forward-biased junction.

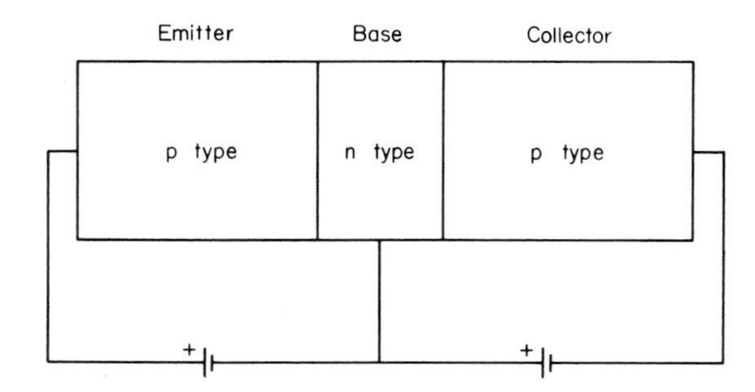

Fig. 2-19. Amplification by the p-n-p transistor. (*After Warschauer.*[4])

current imposed across the device, the applied voltage varies sinusoidally from reverse to forward bias. In the reverse mode, no current passes; in the forward mode, there is no barrier to current flow. A half-wave rectification results. The diode is said to have low impedance in forward bias, high impedance in reverse bias.

Let us make a structure with two p-n junctions, as shown in Fig. 2-19, and connect it in an external circuit. The region on the left is forward-biased, and holes are injected into the n-type center region. Since this acts as a source of holes, this region is termed the *emitter*. The center region is common to both parts of the external network and is termed the *base*. If the lifetime of the holes is long enough and the base narrow enough, almost all the holes injected from the emitter to the base will travel across the base as minority carriers and, since the other junction is reverse-biased, drop easily into the right-hand region, which for this reason is termed the *collector*.

What this means is that the same current passed by the emitter will also be passed by the collector. However, the emitter-base junction is of very low impedance, so that the voltage necessary to pass this current is comparatively low. On the other hand, since the collector-base junction is reverse-biased, this is of very high impedance and for the same current will require a very much larger voltage. This is the basis of the use of the p-n-p transistor as an amplifier.

The n-p-n transistor works on exactly the same principles. The emitter and collector are n type with the base p type, and the polarity of the external circuit is reversed.

REFERENCES

1. Dunlap, W. C.: "An Introduction to Semiconductors," John Wiley & Sons, Inc., New York, 1957.
2. Kittel, C.: "Elementary Solid State Physics," John Wiley & Sons, Inc., New York, 1962.
3. Shockley, W.: "Electrons and Holes in Semiconductors," D. Van Nostrand Company, Inc., Princeton, N.J., 1950.
4. Warschauer, D. M.: "Semiconductors and Transistors," McGraw-Hill Book Company, New York, 1959.
5. Valdes, L. B.: *Proc. IRE,* **42**:420 (1952).

3

Bulk-material Characterization

3-1. INTRODUCTION

In reviewing the development of semiconductor devices in Chap. 1, it becomes apparent that the rock on which so many efforts foundered was poor material. The erratic and frustrating course of many experiments can, with the usual advantage of hindsight, be attributed directly to either impure material or polycrystallinity or both. In the early 1940s, the importance of impurities in the raw material began to be appreciated, and this dates the beginning of real progress in semiconductors. Today, highly purified material, containing less than a few parts per billion of the electrically active elements, is routinely used as starting material. It presents problems to the analyst which are solved by classical methods only with considerable difficulty—if at all. Consequently, physical methods, many of them unfamiliar to the chemist, have been widely applied to the evaluation of semiconductor materials.

This chapter describes the methods used for preparing and characterizing bulk materials. In some cases, single-crystal material results directly in this production, but usually polycrystalline material is obtained. Methods for growing single crystals from this bulk material will be described in Chap. 4. The recovery and preparation of the purified elements have been included since this provides useful background, indicating likely impurities and providing a framework for interpretation of results.

3-2. HISTORY OF GERMANIUM

The element germanium was predicted by Mendeleev in 1871 from his periodic table and discovered by Winkler in 1886 in the mineral argyrodite. For half a century after that it remained very much a scientific curiosity, although its properties were studied extensively. In 1935, studies were begun at the Eagle-Picher plant in Henryetta, Oklahoma, on the recovery of the element from its zinc smelting operations, and in 1941 a pilot-plant production of 99.9% germanium oxide was achieved. The process has been described by Thompson and Musgrave.[1,†] At that

†Superscript numbers indicate References listed at the end of the chapter.

time, there was no commercial application of this material; but in 1942, in view of the importance of the rectifier in radar, the National Defense Research Council initiated a search for a substitute for silicon. A group at Purdue University began a study of germanium, and this was greatly facilitated by the availability of the relatively pure oxide from Eagle-Picher. Further work at Bell Telephone Laboratories,[2] Du Pont,[3] and General Electric[4] resulted in the preparation of a high-purity germanium with a resistivity of the order of 10 to 20 ohm-cm. This quality material was used to produce high-inverse-voltage germanium rectifiers and in 1948 led to the transistor.

Germanium transistors are still significant in the entertainment field, and they are likely to hold an important segment of the market for some time to come. However, they are being replaced steadily even in this field; and in the more sophisticated devices, including integrated circuits, germanium is only rarely used.

3-3. OCCURRENCE OF GERMANIUM

Germanium occurs as a minor constituent of a number of ores.[5,6] Argyrodite ($4Ag_2S$–GeS_2) occurs in Germany and contains about 6–7% of the metal. Germanite ($7CuS$–FeS–GeS_2) containing 5–10% Ge and renierite (a complex sulfide of Cu, Fe, Ge, and As) are both found in Africa, associated with zinc ores. None of these minerals has been found in sufficient quantity to be of significance commercially, and most of the supply of germanium is obtained as a by-product from other smelting operations. Many zinc ores contain amounts of germanium up to about 0.01%, and this can be recovered during the treatment of the mineral. The first commercial source was developed by Eagle-Picher as a by-product from their zinc smelting operations at Henryetta, Oklahoma, using Tri-State sphalerite; this is a zinc sulfide found in Missouri, Oklahoma, and Kansas and contains 0.005–0.015% Ge. The development of this production method has been described by Thompson and Musgrave.[1] Germanium has been found in trace amounts, not more than 0.003% Ge, in some coals, and a process for its recovery from flue dusts is described by Powell et al.[7] Currently, the most important source is from the zinc mining operations in Africa.[8] The zinc-copper ores at Tsumeb, Southwest Africa, contain 0.015% Ge, and oxide recovered from this is marketed by the Tsumeb Corporation. The zinc-copper deposits in Katanga, Democratic Republic of the Congo, contain about 0.01% Ge, and germanium oxide from this source is marketed by the Union Miniere de Haut-Katanga.

3-4. DETERMINATION OF GERMANIUM IN MINERALS

The literature relating to the determination of germanium in ores, coals, flue dusts, etc., is extensive, reflecting the considerable interest in this element in recent years. In general, the approach to the problem has been along two lines: spectrographic and colorimetric.

The most authoritative procedure for the emission spectrographic determination is given by Musgrave.[6] The powdered sample is mixed with a buffer of 2 parts lithium carbonate and 1 part graphite and burned completely in the arc. Lithium is the internal standard, and the matrix for the standards is a mixture of oxides and sulfides chosen to be applicable to both minerals and coal ashes. The range of this method is 0.002 to 0.1 percent with an accuracy of 10 percent. For coal, the sample is ashed prior to analysis. Waring and Tucker[9] have shown no detectable loss of germanium when samples were ashed at temperatures as high as 1000°C. However, germanous oxide does seem a possible product during combustion, and since this sublimes at 710°C, a temperature below this would seem preferable. Musgrave[6] prefers to ash at considerably lower temperatures; his experience leads him to the conclusion that the temperature should not exceed 800°F (427°C). An ashing procedure of 2 hr at 200°F, 2 hr at 400°F, 2 hr at 600°F, and 12 hr at 800°F is recommended. The result is a carbonaceous ash which can be used for spectrographic analysis.

More recent work by Menkovskii and Aleksandrova[10] on several varieties of coal suggests that germanous oxide can be oxidized at the lower temperature, after which no germanium is lost. They recommend an increase in temperature of 3.5°C/min over 3 hr to 500°C, followed by 30 to 60 min at 700 to 800°C. For most work, this is probably preferable since it is reasonably fast and gives a noncarbonaceous ash.

A variety of reagents have been used for the colorimetric determination of germanium, including molybdenum blue, hematoxylin, gallein, and quinalizarin; but the most widely used, because of its selectivity and high sensitivity, is phenylfluorone (2,3,7-trihydroxy-9-phenyl-6-fluorone). It was first applied to the analysis of flue dusts, coal, and coke by Cluley.[11] For all these colorimetric procedures, the sample must, of course, be rendered soluble; and, usually, a preliminary separation of the germanium is necessary to remove possible interferences. Some sulfide minerals may be treated with acids to extract the germanium; Schoeller and Powell[12] recommend the use of nitric acid for germanite and blende, and Strickland[13] used phosphoric acid for a variety of germaniferous ones. However, the presence of even small amounts of chloride can lead to the loss of volatile germanic chloride, and the safest method is to use a fusion, which is required in any case for flue dusts and coal ash. Alkali fusions with sodium carbonate, hydroxide, or peroxide have been variously recommended. In general, sodium peroxide fusion can be used for all these materials with no loss of germanium.[6] For coal, ignition mixed with sodium carbonate was recommended by Cluley[11] and Schoeller and Powell;[12] however, it seems simpler to ignite first, as suggested by Musgrave,[6] and then fuse with a more alkaline mixture. After fusion is complete, the cooled melt is leached with hot water, then made about 6 *N* in hydrochloric acid. Germanium chloride is separated either by distillation or by extraction with an organic solvent, usually carbon tetrachloride. Arsenic is the only element that will accompany germanium; but since this does not interfere in the phenylfluorone method,[11] its presence is not significant. Detailed procedures for this determination are given by Musgrave[6] and also by Sandell.[14] Its sensitivity is given as 1 ppm.

3-5. THE EAGLE-PICHER PROCESS

In the process described by Thompson and Musgrave,[1] the ore is concentrated by a flotation process, and the resulting zinc sulfide contains about 0.01% Ge. This is roasted to give a crude zinc oxide with evolution of sulfur dioxide. The roasted ore is then sintered with a mixture of coal and salt; at this stage the germanium, together with cadmium, lead, and some other metals, is volatilized as the chloride and condensed and collected in an electrostatic precipitator. The fume is leached with sulfuric acid and the lead filtered off as the sulfate. Zinc dust is added to the filtrate in sufficient amounts to precipitate germanium and copper without precipitating cadmium; arsenic and some other metals are also precipitated. The copper-germanium sludge is filtered off, redissolved in sulfuric acid, and reprecipitated to concentrate the germanium. The filtrates are retained for cadmium recovery. The germanium is roasted and dissolved in hydrochloric acid; distillation yields a crude germanium tetrachloride containing some hydrochloric acid and some arsenic trichloride. This crude chloride is purified by repeated distillation, the final distillations being carried out in the presence of chlorine and hydrochloric acid. Germanium tetrachloride is immiscible with water, and the acid floats on the top. Arsenic trichloride, on the other hand, is miscible and is extracted into the aqueous layer, where it is oxidized by the chlorine to the nonvolatile arsenic acid. Distillation thus yields an arsenic-free product. This pure germanium tetrachloride is hydrolyzed by water to yield germanium oxide.

3-6. GERMANIUM FROM COAL

In England, germanium has been recovered from the flue dusts obtained in the combustion of coal. The process has been described by Powell et al.[7] Certain coals from the Northumberland and Durham area contain as much as 0.003% Ge, and the flue dusts from producer-gas plants using these coals may contain up to 2% Ge. The dust is a mixture of oxides of iron, aluminum, silicon, zinc, and a number of other elements. It is smelted with soda and lime to flux the silica and with coal to reduce the iron oxide; iron acts as a good collector for germanium. Since gallium is also present, copper oxide is added since copper will collect this element. The result is a copper-iron regulus containing 3–4% Ge and 1.5–2% Ga. This is treated in a ferric chloride solution with a stream of chlorine. Distillation of this mixture gives a two-phase distillate; the upper layer is hydrochloric acid, the lower crude germanium chloride. The gallium remains in the still residue. The germanium chloride is distilled adiabatically to yield a product containing about 20 ppm arsenic which is refluxed with copper and redistilled to yield arsenic-free chloride. Hydrolysis with water forms germanium oxide.

3-7. BELGIAN PROCESSES[8]

The Tsumeb ores are treated at the mine by a selective flotation method, and a germanium concentrate is obtained which contains about 0.25% Ge in a mixture of lead and copper sulfides. This is treated in Belgium by roasting in a vertical retort in a stream of charcoal producer gas (30% CO, 1 to 2% H_2, remainder N_2) at about 900°C. Under these conditions, germanium and arsenic are volatilized, while most of the lead and all of the copper are not. The fume is condensed and then roasted at 550°C in air. Volatile arsenic trioxide is driven off and collected; germanium dioxide remains behind, together with some lead sulfate. This mixture contains 15 to 20% Ge.

In the Katanga mines, the germanium accompanies the copper during the flotation separation from zinc and concentrates in the dust in the waste gases from the smelter. This dust contains about 0.3 to 0.4% Ge in a mixture of crude lead and zinc sulfides. It is mixed with sulfuric acid and baked to remove most of the arsenic. The sulfated dust is leached with sulfuric acid and filtered; the insoluble material is chiefly lead sulfate. The filtrate contains the germanium along with considerable arsenic. It is oxidized with potassium permanganate and the pH adjusted to 2 to 2.2 to precipitate arsenic. After filtration, the filtrate is further neutralized with magnesia in two steps: first to pH 4.9 to precipitate a germanium cake, then to pH 5.5 to 5.7 to strip the solution of any residual germanium; the second precipitate is returned to the arsenic precipitation step. The germanium cake contains 8–10% Ge in a mixture of predominantly zinc and copper oxides.

The germanium-enriched material from either process is dissolved in concentrated hydrochloric acid and submitted to a series of distillations essentially similar to that described for the Eagle-Picher process. The purified chloride is hydrolyzed with water to form germanium dioxide.

3-8. DETERMINATION OF GERMANIUM IN THE RECOVERY PROCESS

For the various residues, concentrates, and other fractions from the germanium recovery process which contain more than 0.1% Ge, both gravimetric and titrimetric methods have been used. The usual gravimetric method uses tannin as the precipitating agent,[15] but cinchonine[12] and hydrogen sulfide[6] have also been used. None is entirely satisfactory, and all of them are extremely tedious. Cluley[11] suggested a simple titrimetric method in which germanate can be titrated with alkali in the presence of mannitol; the method is analogous to that for borates. However, it does require a preliminary separation by sulfide precipitation which takes about 12 hr for satisfactory coagulation. A somewhat faster method is due to Abel,[16] based on the hypophosphite method of Ivanov-Emin.[17] After fusion, the germanium is separated by distillation, and the germanium (IV) reduced to germanium (II) by sodium hypophosphite. It is then reoxidized by standard potassium iodate. A modification of this procedure is recommended by Musgrave.[6]

3-9. PRODUCTION OF ELECTRONIC-GRADE GERMANIUM

The germanium dioxide obtained in the recovery processes is reduced to the metal in a stream of hydrogen.[1,7] The oxide is loaded into graphite boats and heated in a stream of hydrogen at 650°C and then, when reduction is complete, to 1000°C to melt the metal to an ingot. The hydrogen is replaced by nitrogen and the ingot allowed to cool. The Belgian process[8] uses cracked ammonia for the reducing atmosphere but is otherwise similar.

This metal must be purified further, and as a first step is submitted to a gradient freeze. This process depends for its action on the phenomenon of segregation, the difference in solubility of an impurity in the liquidus and solidus. The distribution coefficient (or segregation coefficient) is defined as

$$K = \frac{C_s}{C_l}$$

where C_s = concentration of solute in solid phase
C_l = concentration of solute in liquid phase

For the most common impurities in germanium, K is less than unity. As the melt solidifies, impurities concentrate in the liquid phase, and the last-to-freeze portion contains most of the impurities. This last-to-freeze portion is cut off and returned to the next billet of metal.

In practice, this gradient freeze can be combined with the melting[18] by withdrawing the graphite boat from the 1000°C zone of the furnace over a period of about 3 hr, cooling finally to 200°C. The material is further purified by zone refining;[19] in this procedure a molten zone is moved through the length of the bar of germanium. Again, since the distribution coefficient is less than 1, impurities concentrate in the liquidus and are moved to the end of the bar. The process can be repeated until the material is of sufficient purity.

It is possible to produce single crystals by either of these treatments, although in many cases this is not the primary object. For this reason, both these procedures are dealt with in considerably more detail in Chap. 4.

3-10. RESISTIVITY OF HIGH-PURITY GERMANIUM

The real criteria for good electronic-grade germanium are, of course, its electrical properties, and of these resistivity is the most important since this will indicate the number of free carriers.

As mentioned in Sec. 2-13, the commonest method for determining resistivity is the four-point-probe method. However, for polycrystalline material, as the germanium frequently is at this stage, the two-point-probe method is more accurate. The procedures for both these determinations have recently been the subject of a tentative ASTM specification, F43-67T,[20] which should be the basis for acceptance tests.

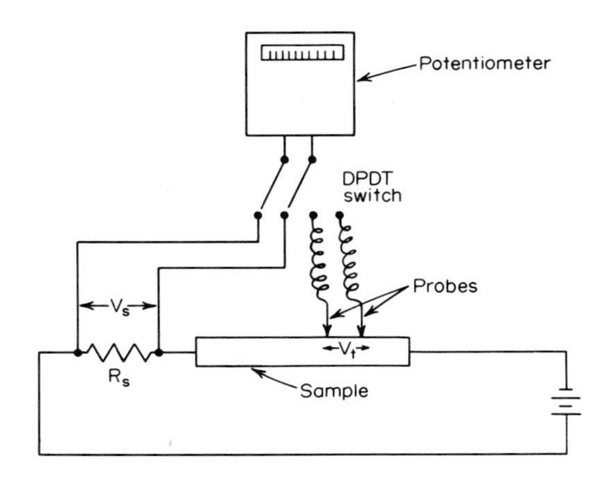

Fig. 3-1. The two-point-probe method for resistivity.

The two-point probe is shown in Fig. 3-1. It consists essentially of two spring-loaded steel points; nominally these tips are of 0.025 mm radius. The test specimen must be in the form of a strip or bar with its length at least three times its largest cross-sectional dimension. The bar must be as uniform as possible, and its two ends must be electroplated with nickel or silver to allow ohmic contacts to be made either by pressure or by soldering. An ohmic contact, frequently referred to in semiconductor-device preparation, is simply one that obeys Ohm's law, i.e., is free from rectification. The surface to be measured must be abraded by lapping with No. 600 alumina or sandblasting with No. 280 Carborundum. The two contacts to the ends of the bar are connected in series with a standard resistor R_s. A voltage is impressed across this network, and the potentials V_s, V_T measured across the standard resistor and across the test probes. The potential V_s is then remeasured; it should check the previous reading. The temperature of the bar (usually ambient) is noted and its dimensions measured. The resistivity is calculated as

$$\rho = \frac{V_T A}{IL} \qquad \text{ohm-cm}$$

where A = cross-sectional area, cm^2
L = length, cm
I = current, amp
$= V_s/R_s$

This method is applicable to material in the range 0.01 to 10,000 ohm-cm.

The four-point probe is dealt with fully in Sec. 4-16. It consists of four probes in line, equidistant from each other. Its advantage over the two-point probe is that it does not need a special bar but can be applied to thin slices or irregularly shaped pieces. However, it is not as accurate, particularly for polycrystalline material, and is not recommended as an acceptance test. The distance between the probes, a cm, is usually 0.05 in. A flat face must be prepared large enough to accommodate the probes and allow a distance of $4a$ from any probe to the nearest edge. This surface is prepared either by lapping or by sandblasting as before. The four spring-loaded probes are pressed against the surface and a voltage impressed on the two outside probes while the potential V_T is measured across the inside probes. The current I passing through the outer probes is also measured. It can be shown[21] that

$$\rho = 2\pi a \frac{V_T}{I} \qquad \text{ohm-cm}$$

3-11. CONDUCTIVITY TYPE OF HIGH-PURITY GERMANIUM

As well as the resistivity, it is usual to specify the type of germanium required, that is, whether it is n or p type. As was pointed out in Sec. 2-12, this can be found by determining the sign of the Hall voltage, and this method is preferred for germanium having a resistivity over 20 ohm-cm. Since the determination of Hall coefficient is an important one for crystalline material, it will be dealt with in detail in Sec. 4-17.

A simpler procedure for typing depends on the thermoelectric effect; both this and

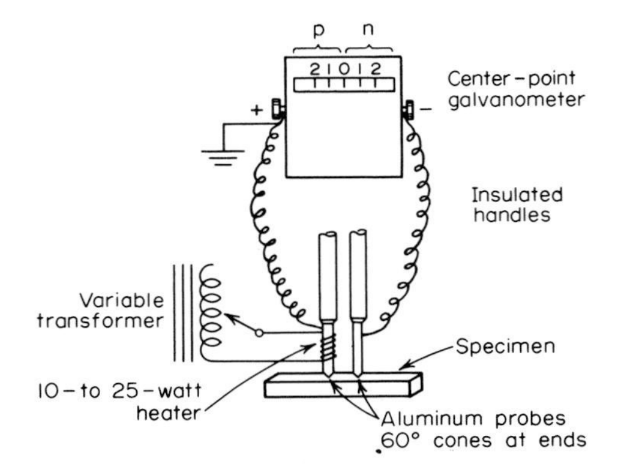

Fig. 3-2. Thermal emf method for typing by ASTM Method F42-64T. (*After ASTM Standards.*[20])

the Hall effect method are the subject of an ASTM specification, F42-64T.[20] The apparatus is shown in Fig. 3-2 and is really very simple, consisting of two probes, one heated, connected through a center zero galvanometer. If the probes are placed on a piece of n-type germanium, then the hot probe will generate more carriers and a space charge will result. To compensate for this, a potential difference is set up in the external circuit so that an equilibrium occurs. This requires that the hot probe be at a positive potential compared with the cold probe, and the galvanometer will deflect accordingly. Conversely, if the material is p type, then the hot probe will be the negative electrode.

It will be remembered that the higher the temperature, the more carriers are generated, so that, in general, the larger the difference in temperature between the electrodes, the greater the effect. However, as the resistivity increases and the material becomes more nearly intrinsic, the mobilities of the carriers become important and, with large temperature differences, will always type n. A difference of 40°C is therefore recommended, and even then this method is not used for material with a resistivity above 20 ohm-cm.

The test specimen should be cleaned prior to probing by the procedure used for the resistivity tests. The probes should be as close together as possible. The sample should be probed at several points since it is not unusual to find type variations, particularly in polycrystalline material.

3-12. EVALUATION OF GERMANIUM DIOXIDE

As a test for the suitability of germanium dioxide for subsequent treatment, a tentative ASTM method, F27-63T,[20] has been suggested for its reduction to the metal. A furnace is used which can be driven at a constant speed, as shown in Fig. 3-3. The tube, of fused quartz, is 3 cm diameter and 76 cm long and is mounted at a 1 in 42 incline. Fifteen grams of material is weighed into a quartz boat and placed in the cold tube, as shown. The hydrogen flow is started and the furnace raised to 650°C in 30 min. After 2 hr, the temperature is raised to 1000°C and the hydrogen replaced by argon or helium. After 15 min at this temperature, the drive mechanism for the furnace is started; this translates the furnace at 0.212 cm/min.

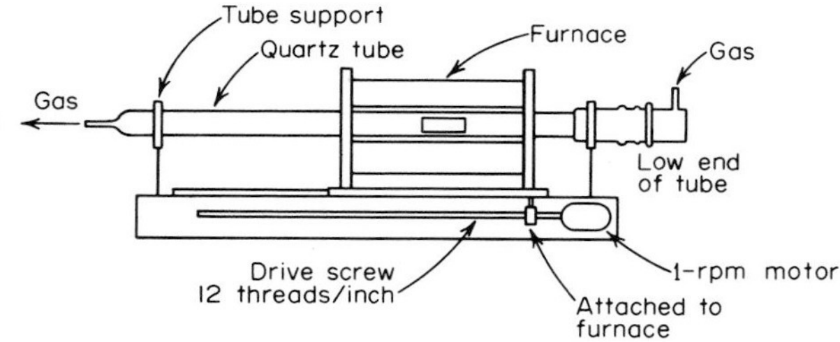

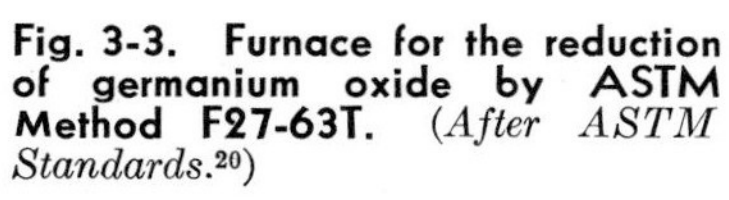
Fig. 3-3. Furnace for the reduction of germanium oxide by ASTM Method F27-63T. (*After ASTM Standards.*[20])

As the furnace slowly leaves the boat area, the metal is subjected to a gradient freeze. A bar of 3.20 mm^2 cross section is cut from the finished ingot for resistivity and typing measurements.

3-13. EVALUATION OF GERMANIUM

Germanium dioxide which is suitable for further refining will typically yield metal, when reduced by the procedure of Sec. 3-12, of better than 0.1 ohm-cm resistivity. Differences in values above 5 ohm-cm are not significant, and this is meant merely as a screening test.

Germanium metal should approach the intrinsic resistivity. Values for varying temperatures are given in Fig. 3-4. Acceptance tests usually will require values in excess of 40 ohm-cm at room temperature. The relationship between resistivity and carrier concentration (and, in effect, purity) is given in Sec. 4-5.

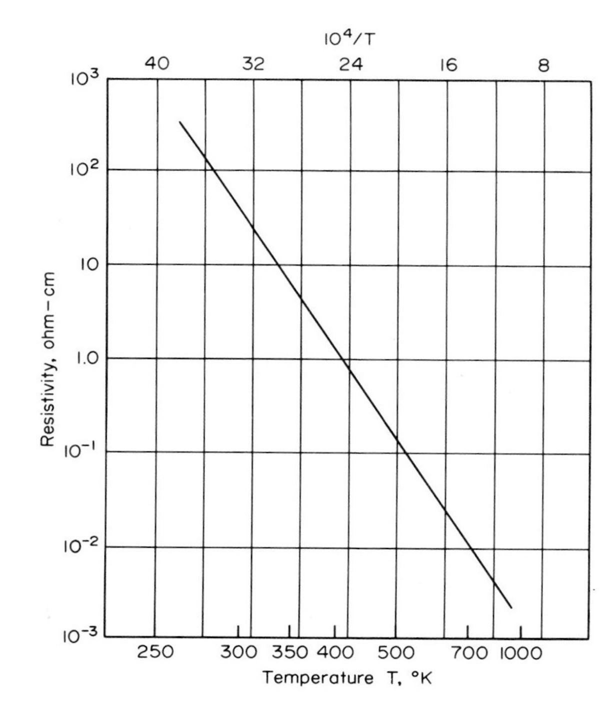

Fig. 3-4. Resistivity of intrinsic germanium. (*Adapted from Morin and Maita.*[22])

3-14. ANALYSIS OF HIGH-PURITY GERMANIUM AND GERMANIUM DIOXIDE

While the electrical properties of the material are extremely useful in assessing its quality, they do not identify the impurities present nor the dopant which may have been added. For this, analysis of a high order of sensitivity is required.

To this point, the analysis of the various refining fractions, including the original ores, has followed conventional lines. The methods, while possibly unfamiliar, present no great problem to the analyst. However, with the production of the high-purity germanium dioxide and its subsequent refinement to ultrapure metal we enter on a new concept of sensitivity, one measured in fractional parts per billion rather than parts per million. Since the methods applied to polycrystalline material are the same as those for crystalline material, we shall defer further description of these procedures to Chap. 5. It should be noted that the same methods can be applied to germanium dioxide with little or no change.

3-15. HISTORY OF SILICON

In contrast to germanium, silicon is one of the most abundant elements, second only to oxygen, with which it is usually associated. Forms of silica have been recognized from earliest times, and glass manufacture has been depicted in Egyptian frescoes of about 2000 B.C. However, the quartz form of silica was long felt to be elemental, and it was not until comparatively modern times that the element silicon was discovered.

Lavoisier first suggested that silica was an oxide as early as 1787, and in 1808 Berzelius claimed to have discovered the element silicon in an analysis of cast iron. However, this material was in fact ferrosilicon. The first isolation was by Gay-Lussac and Thenard in 1809, who heated potassium in silicon tetrafluoride, although they did not recognize it as such. In 1823, Berzelius succeeded in preparing the element by heating potassium fluosilicate with potassium. These and subsequent preparations were of amorphous silicon, but in 1854 Deville prepared, by accident from a fused mixture of aluminum and sodium chlorides, a crystalline form.

The early interest in the element was generated by the use of silicon as an alloying element in the steel industry. Originally, the element was prepared as ferrosilicon by heating a mixture of silica, carbon, and ferric oxide in a blast furnace. Later, electric-arc furnaces were used, and contents of silicon up to 99.8% were obtained. The subsequent efforts to upgrade this material were mentioned in Sec. 1-1.

A fuller description of the history of silicon is given in another volume of this series by Runyan.[23]

3-16. PRODUCTION OF ELECTRONIC-GRADE SILICON

Silicon has been obtained by a number of methods, but, for semiconductor grade, the usual starting material is a halide. Various methods have been proposed using a reduction by zinc, cadmium, or hydrogen, and these are reviewed by Runyan.[23] However, the commonest procedure uses either silicon tetrachloride or trichlorosilane, both of which are readily available in large quantities as starting materials for the silicone industry, and hydrogen as the reducing agent.

The chlorides are prepared by the reaction of chlorine with silicon or ferrosilicon or by the reaction of hydrochloric acid with a mixture of silica and carbon, both at elevated temperatures. The latter reaction tends to yield more trichlorosilane.

The processes using either halide are essentially the same. The raw material is received in a high state of purity and is further refined by distillation. The distillate is vaporized, mixed with hydrogen, and passed to a reactor at about 1000°C. The reactions are

$$SiCl_4 + 2H_2 = Si + 4HCl$$

and

$$SiHCl_3 + H_2 = Si + 3HCl$$

The reactor can be of various design; for example, quartz tubes, tantalum rods, and silicon rods have all been used for deposition. The quartz tube can be easily heated, but the silicon bonds so strongly to the quartz that it must be broken away and any residual pieces etched off with hydrofluoric acid. The tantalum rod can be resistance heated while the tube walls are kept cool. In this case, the tantalum rod must be dissolved from the center with a nitric acid etch. The most satisfactory reactor uses a high-purity silicon rod as the heater; this also gives the purest material.

Trichlorosilane is preferred as the starting material since it reacts somewhat faster and is easier to purify. However, recycling in the plant leads to a buildup of silicon tetrachloride by the side reaction

$$2SiHCl_3 = SiCl_4 + H_2 + Si$$

Distillation leads to an overhead of trichlorosilane, boiling point 33°C atmospheric, and a bottom of silicon tetrachloride, boiling point 57.6°C. Some tetrachloride can be tolerated, but it is necessary to remove some continuously from the plant.

The product is a polycrystalline dense mass of lustrous appearance. It is treated with suitable etchants to clean the surface and may be further purified by zone refining. In many cases, however, it is already sufficiently pure for crystal pulling.

3-17. ANALYSIS OF SILICON CHLORIDES

The trichlorosilane used for feedstock must be of high purity; particularly must it be free of volatile chlorides of doping elements such as phosphorus, arsenic, and boron. Organic derivatives should also be absent since these will form silicon carbide in the reactor and will deposit in the silicon. Heavy-metal chlorides will not in general distill over, but they may be carried over by entrainment. They should, therefore, not be present in appreciable amounts. The methods for silicon tetrachloride are rather more numerous in the literature than those for trichlorosilane, but generally the procedures are interchangeable.

The first attempts to analyze the halides for impurities were by emission spectrography. Two approaches are possible; one can hydrolyze the halide to silica and analyze this material, or one can evaporate the sample and analyze the residue. Generally speaking, the first approach is not sensitive enough, and more recent efforts have been aimed at the latter procedure. However, difficulties arise inasmuch as boron halides are volatile, and some method of holding them back must be devised. Veleker and Mehalchick[24] hydrolyzed a portion of the sample with methyl cyanide. The silica so formed adsorbed the boric acid, and the remaining silicon tetrachloride could be evaporated. The residue was then spectrographed by using a Stallwood jet with argon gas† to determine boron down to 0.8 ppb. Vecsernyes and Zombori[25] hydrolyzed the sample to silica with aqueous ammonia and then spectrographed. However, their sensitivity, as pointed out above, was only 0.9 ppm for boron; arsenic, which was also determined, gave even less sensitivity at 6 ppm. Neither of these values is of much use in semiconductor work. Pchelintseva et al.[26] suggested complexing the boron halide with chlorotriphenylmethane; this forms an insoluble complex after which silicon tetrachloride can be distilled off. The residue is submitted to emission spectrography. Using an 8-ml sample, as little as 4 ppb boron could be detected. Vecsernyes and Hangos[27] used the same method for boron contents of both halides, claiming a sensitivity of 1 ppb. Kawasaki and Higo[28] improved the sensitivity to 0.06 ppb of boron in both halides by introducing a hydrolysis step. Carbon interferes during arcing if the sample is large; they removed the organic residue by hydrolyzing the complex with sodium hydroxide and spectrographing the boric acid. This procedure is probably the best for determining boron.

Some other elements have been determined by emission spectrography. Usually these do not present any problem if they are nonvolatile. In fact, the previous methods for boron could probably also be extended to other impurities, although no attempt was made to do so. Martynov et al.[29] evaporated silicon tetrachloride on carbon powder and examined the residue by emission spectrography for Al, Fe, Mg, Mn, Cu, Ti, and Ca. No sensitivities were given, but values of 50 to 100 ppb were quoted. In another publication, Martynov et al.[30] suggested preconcentrating the impurities by passage of a large sample over silica gel and spectrographing the gel. Again, only the same elements, plus Pb and Ni, were mentioned. Tarasevich and Zheleznova[31] described a method specifically for Ta and Mn in trichlorosilane in which the sample was heated in moist air with silver chloride and the resulting silicic acid heated with hydrofluoric acid. The concentrate left was spectrographed. The sensitivity, for a 65-g sample, was given as 0.2 ppb for Mn and 6 ppb for Ta.

An activation analysis has been reported by Miyakawa and Kamemoto[32] in which sensitivities of 0.2 to 0.9 ppb were obtained for Na, As, Ga, and Cu and 5 ppb for Mn in a 10-ml sample of trichlorosilane. The sample must, however, be hydrolyzed first since no reactor authority will allow such a corrosive sample to be irradiated. Consequently, volatile chlorides can also be lost by this procedure.

Stripping (or amalgam) polarography was used by Vinogradova and Kamenev[33] for determining Bi, Pb, and Tl in trichlorosilane, but there again a preliminary evaporation was made. A more attractive approach was taken by Karbainov and Stromberg,[34] where 0.5 ml silicon tetrachloride was mixed with 4.5 ml of *n*-propanol and electrolyzed for 10 min. Subsequent anodic polarography separated Sb, Bi, and Sn at sensitivities of 0.04, 0.06, and 0.3 ppm, respectively.

A few colorimetric methods have been devised for specific elements. Alimarin et al.[35] determined tantalum by the fluorescent complex with Rhodamine 6G. A sensitivity of about 0.2 ppb in trichlorosilane was obtained following an evaporation

†This is to reduce interference from oxide bands and is discussed further in Sec. 5-3.

with hydrofluoric acid. Martynov et al.[36] evaporated silicon tetrachloride with carbon tetrachloride (to reduce the hydrolysis) and reacted the dissolved residue with *p*-dimethylaminobenzylidene-benzoylaminoacetic acid† to give a fluorescent compound of copper; the sensitivity was 0.1 ppb using a 50-g sample. Miyamoto,[37] in a method for boron in silicon tetrachloride, evaporated the sample with dimethylaniline, which also retains boron as a nonvolatile complex. A colorimetric finish used curcumin, and a sensitivity of 0.4 ppb was obtained (but on a 500-g sample!). Haas et al.[38] have used a method for boron in silicon tetrachloride in which the sample is extracted with a quinalizarin–sulfuric acid reagent. The color change in the reagent layer is proportional to the boron content; and, by using a 20-ml sample, a sensitivity of about 10 ppb seems possible. In another paper from the same laboratory, Lancaster and Everingham[39] described a procedure for the determination of phosphorus in silicon tetrachloride. This is a particularly difficult analysis since phosphorus is notoriously insensitive in emission spectrography and is analytically very similar to silicon. Lancaster and Everingham extracted the phosphorus compounds by shaking a 25-ml sample with 0.5 ml concentrated sulfuric acid. The sulfuric acid layer was oxidized with perchloric acid and the resulting phosphate reacted to form the yellow vanadophosphomolybdate. The sensitivity claimed is 15 ppb. This appears to be the only method published for the trace determination of phosphorus in silicon halides.

The foregoing analyses are concerned with trace elements, that is, in the sub-ppm range, and for the raw material these are of course extremely important. In addition to these metallic impurities, however, the halides may contain some organic compounds or, either before or during use, may become contaminated with hydrolysis products or products of side reactions. Infrared absorptiometry was used by Rakov[40] to determine the hydrochloric acid content of silicon tetrachloride and trichlorosilane; the measurements were made in the gas phase with hydrogen as the diluent. Tsekhovol'skaya and Zavaritskaya[41] described a method for determining $COCl_2$, CS_2, SO_2Cl_2, $POCl_3$, and COS in silicon tetrachloride by infrared absorption; and these, together with several more compounds, were determined by Rand.[42] He used a 10-cm path length and a liquid sample; Table 3-1 shows the absorption bands used.

The most convenient method of analysis of the halides, especially for plant control, is gas chromatography. Abe[43] examined a number of stationary phases for the separation of trichlorosilane, silicon tetrachloride, phosphorus trichloride, and boron trichloride and recommended a silicone, Dow-Corning DC-703. Turkel'taub et al.[44,45] used the discrimination of the flame ionization detector for organic compounds and described a procedure for determining benzene in silicon tetrachloride and trichlorosilane with a sensitivity down to 3 ppm. The stationary phase was petroleum oil on firebrick. Palamarchuk et al.[46] studied the separation of mixtures containing $(CH_3)_2SiCl_2$, CH_3SiCl_3, $(CH_3)_3SiCl$, CH_3HSiCl_2, $(CH_3)_2HSiCl$, $SiCl_4$, $SiHCl_3$, SiH_2Cl_2, and CH_3Cl. The stationary phases were benzyl benzoate, dibutyl phthalate, or diethyl phthalate, and a conventional thermal-conductivity detector was used. Bersadschi et al.[47] analyzed mixtures of silicon tetrachloride and trichlo-

†The translation from Consultants Bureau gives "lyumocupferron" as the trivial name, but it is not a cupferron derivative.

Table 3-1. Absorption Bands in Silicon Halides[a]

Measurements made with 10-cm liquid path

Impurity	Boiling point, °C	Absorption[b] maximum, μ	Absorptivity, wt. $\%^{-1}$ cm^{-1}	Detection limit, ppm
—OH	....	2.70–2.80[c]		
C—H in haloforms	....	3.28–2.35[d]		
HCl	−84	3.53	15	2
		3.41	10	3
HBr	−67	3.99		
CO_2	−78	4.27[e]	80	1
P—H	....	(4.3)		
$SiHCl_3$	33	4.43		
Si—H	....	(4.6)		
COS	−48	4.89	200	0.2
$COCl_2$	8	5.51	50	1
		6.05	5	10
$COBr_2$	65	(5.48)		
CCl_3COCl	118	5.54	40	1
CCl_4	77	6.44	0.5	100
CS_2	46	6.57	500	0.05
SO_2Cl_2	69	7.04	75	2
C_6Cl_6	325	7.69	10	5
CH_2Cl_2	40	7.93	20	5
$POCl_3$	105	7.95	80	2
		8.21	45	3
$SOCl_2$	79	8.08	100	0.5
		4.06	0.5	25
$SiCl_4$	57	8.18	0.5	100
$CHCl_3$	61	8.24	30	3
Si_2OCl_6	137	8.98	60	1
		5.43	0.8	40
		6.29	0.2	150
SiO_2	....	9.2		
$VOCl_3$	127	9.66	100	0.5
		4.84	0.5	50

[a]Adapted from Rand.[42]

[b]Values in parentheses are for the pure materials; it is not known whether any shift occurs in solution.

[c]The addition of water gives a broad, shallow band near 3 μ, attributed to hydrogen-bonded hydroxyl.

[d]For most organic compounds the C—H absorption is usually given as 3.3 to 3.4 μ.

[e]Doublet under high resolution.

rosilane for process control by using a stationary phase of transformer oil activated with glycerol. The sample was first dissolved in carbon tetrachloride to minimize hydrolysis.

Procedures due to Burson[48] are employed in the Texas Instruments laboratories. For inorganic impurities, a column of 20% SF-96 silicone fluid (General Electric

Co.) on deactivated Chromosorb P is used. Hydrogen chloride tails badly on the active support, and the column must be treated as follows prior to coating. To 40 g of support, add 100 ml saturated copper sulfate solution and 200 ml concentrated sulfuric acid. Reflux for 8 hr, rinse, and dry. With this SF-96 column at 40°C, good separation over a 6-ft length is obtained for air, hydrogen chloride, dichlorosilane, trichlorosilane, and silicon tetrachloride in this order, with a helium flow rate of 60 ml/min and a thermistor detector. Sensitivities are in the low-ppm range for a 10-μl sample.

For organic impurities, the stationary phase is 20% DC-200 silicone oil (Dow-Corning) on deactivated Chromosorb P. The 6-ft column is temperature programmed from 25°C at the start, rising at 10°/min to a final 150°C. A flame ionization detector is used to avoid interference from the silanes. The C_5 to C_8 hydrocarbons are eluted in the order of their boiling points. Again, sensitivities are in the low-ppm range for a 10-μl sample.

3-18. EVALUATION OF SILICON

Electronic-grade silicon, which is usually polycrystalline when sold, is characterized by its resistivity and type, and the methods are identical to those described for germanium in Secs. 3-10 and 3-11. The applicability range of the resistivity test is the same as for germanium, 0.01 to 10,000 ohm-cm. For the thermal type of test, the range of application is up to 1,000 ohm-cm for silicon; above this, the Hall effect method is recommended.

Silicon should approach the intrinsic resistivity. Values for varying temperatures are given in Fig. 3-5. Acceptance tests may require values in excess of 1,000 ohm-cm at room temperature. The relationship between resistivity and carrier concentration is given in Sec. 4-15.

The analysis of polycrystalline material and of silicon dioxide is identical to that of crystalline silicon and will be described in Chap. 5.

3-19. THE III-V COMPOUNDS

The III-V compounds are stoichiometric compounds prepared from elements of group IIIA in combination with elements of group VA. Although most of the combinations have been examined, only the gallium and indium compounds with arsenic and antimony are currently of any commercial significance, and consideration will be restricted to these.

The III-V compounds are always grown as single crystals, so that their preparation will be described in Chap. 4. However, highly purified elements must be used, and their isolation and analysis will be dealt with here.

3-20. HISTORY OF GALLIUM

Like germanium, the element gallium was predicted in 1871 by Mendeleev from his periodic table. It was also predicted about the same time by deBoisbaudran from a study of the spectral lines of the elements and was detected by him spectro-

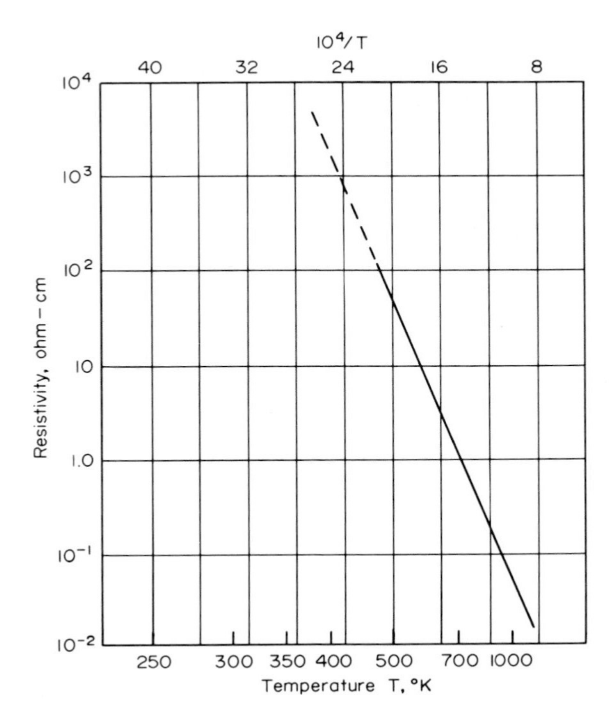

Fig. 3-5. Resistivity of intrinsic silicon. (*Adapted from Morin and Maita.*[49])

scopically in a Pyrenees zinc blende. In 1875, he isolated a small amount of the element, and it was shown to have properties very similar to those predicted by Mendeleev.

3-21. OCCURRENCE OF GALLIUM

Gallium occurs widely distributed in nature but in small amounts. Only one very rare ore, gallite, $CuGaS_2$, is known, and this is associated with the zinc ores of Tsumeb in Southwest Africa and of Katanga. It occurs[50] in the Tri-State zinc ores in a range about 55 ppm and is generally associated with aluminum wherever this element occurs. Bauxite, for example, contains about 20 ppm gallium, as do some coals.

3-22. DETERMINATION OF GALLIUM IN MINERALS

The determination of gallium in minerals was originally accomplished by gravimetric methods, being precipitated as the hydroxide or by cupferron or tannin. These complexes are nonstoichiometric and are ignited to the hydroxide by weighing. These classical methods, which include a preliminary concentration in which the chloride is extracted into ether, are described by Schoeller and Powell.[12]

More conveniently, ores can be examined by emission spectrography. This was, as pointed out in Sec. 3-20, the way in which gallium was originally discovered. The line usually employed[51] is 2943.6 Å, but even with a dc arc the sensitivity is limited to about 10 ppm. By using a chloride buffer, this has been increased to 1 ppm for silicate rocks.[52] Preconcentration techniques have been used by some workers; for example, Minczewski et al.[53] dissolved the ore in a suitable solvent, e.g., hydrofluoric acid, and extracted the 8-hydroxyquinolinate from a buffered solution into chloroform. Sensitivities approaching 0.1 ppm are possible by this method. Similar methods can be employed for coal. Bronshtein et al.[54] describe a procedure in which the coal is ashed at 600°C. A buffer mixture of coal and sodium chloride (1:1) is mixed with the ash in equal parts and arced. Ten parts per million of gallium can be determined in the ash.

An almost bewildering number of reagents have been suggested for the colorimetric determination of gallium, but probably the most generally used is Rhodamine B. Culkin and Riley[55] described a method using this reagent which is applicable to silicate, sulfide, oxide, and carbonate minerals. After solution, the gallium is extracted as the chloride into isopropyl ether and evaporated to dryness. The residue is dissolved in 6.5 N hydrochloric acid and extracted with a chlorobenzene–carbon tetrachloride mixture containing Rhodamine B. The absorption of the organic phase is measured. The sensitivity is about 1 ppm, but it can be increased about tenfold by using a fluorimetric finish. Such a procedure is given by Knipovich and Krasikova,[56] although 8-hydroxyquinoline has perhaps been more widely used for fluorescent procedures.[51] In general, they follow the same lines as the colorimetric method.

Activation analysis has been applied to the determination in rocks. Morris and Chambers[57] described a procedure in granite in which results around 20 ppm were quoted. However, this was by no means the lower limit of sensitivity. By using a flux of 10^{12} neutrons/(cm^2)(sec), it was calculated that about 0.1 ng should be detectable, or in the 0.5-g samples used, about 0.2 ppb. Jaskolska and Minczewski[58,59] describe a procedure using a somewhat lower flux of 3×10^{11} neutrons/(cm^2)(sec). The ore is dissolved and an aliquot of solution irradiated. A radiochemical separation finally yields an 8-hydroquinolate which is counted. A sensitivity of better than 1 ppm is attained.

A flame photometric method for gallium in zinc and aluminum ores has been described by Bode and Fabian.[60]

3-23. RECOVERY OF GALLIUM

Gallium, like germanium, is recovered as a by-product from other smelting operations. A review of the methods used has been given by Sheka et al.[61] Its close association with germanium in the zinc ores and coal suggests that it will probably be recovered in the same processes, and this is in fact the case.

In the Eagle-Picher process, the roasted zinc ore is leached with sulfuric acid and filtered. The filtrate is neutralized with the object of removing iron and aluminum. A precipitate, termed "iron mud," is filtered off, and this contains about 0.07% Ga. A sodium hydroxide leach dissolves aluminum and gallium along with some silica. Neutralization precipitates the hydroxides, which are ignited to render the silica insoluble. Leaching with hydrochloric acid gives a crude solution of aluminum and gallium chlorides. The gallium is separated by extracting with ether; the resulting gallium trichloride contains some iron, which is removed as the hydroxide by strong alkaline solution. This caustic solution is then electrolyzed to obtain gallium metal.

In the recovery of germanium from coal,[8] described in Sec. 3-6, the copper-iron regulus containing germanium and gallium is treated in a ferric chloride solution with a stream of chlorine. The germanium is distilled from this mixture, and gallium remains in the still. The still residue is then cooled to allow some of the copper salts to crystallize out. The mother liquor is diluted and treated with scrap aluminum to precipitate copper, arsenic, and some other metals; the ferric chloride is reduced to ferrous. After acidifying with hydrochloric acid, the gallium chloride is extracted with isopropyl ether. The ether solution is mixed with dilute hydrochloric acid and the ether distilled off. Heavy metals are removed by a hydrogen sulfide precipitation and the filtrate oxidized with nitric acid. Sodium hydroxide is added to first precipitate and then just redissolve gallium hydroxide; precipitated ferric hydroxide is filtered off, after which gallium metal is obtained by electrolysis.

The production of aluminum is by alkaline reactions; the Bayer process uses an alkaline solution, the dry process a soda-lime fusion and subsequent extraction. In either case, the hot solution is allowed to cool to deposit alumina which is filtered off. The gallium does coprecipitate to some extent, but most of it remains in solution. Since the alkaline solutions are used again in a cyclic process, it tends to concentrate in this mother liquor. If the solution is carbonated, more alumina is precipitated until, toward the end of the carbonation, gallium hydroxide also precipitates. This alumina containing gallium is dissolved again in alkali and the metallic gallium obtained by electrolysis.

3-24. DETERMINATION OF GALLIUM IN THE RECOVERY PROCESS

For the somewhat higher levels of gallium that are found in process liquors, either gravimetric or volumetric methods may be used. Generally, the latter are faster and more convenient. The usual procedure involves titration with EDTA, using morin as indicator. Gregory and Jeffery[62] devised a procedure for Bayer liquor in which interfering elements were removed by ion exchange prior to titration. Mizuno[63] has used this titration for the determination in red mud, another fraction in the Bayer process. He separates the chloride into isopropyl ether, evaporates the solvent, and treats the residue with excess standard EDTA. The excess is back-titrated with bismuth using xylenol orange as indicator.

3-25. PRODUCTION OF HIGH-PURITY GALLIUM

The gallium metal obtained by electrolysis is somewhat impure, containing several tenths of a percent of zinc together with several other metals. Since it is a liquid at room temperature, it can be purified by methods similar to those used for mercury, viz., filtration through glass wool followed by an acid wash. The material resulting is of good quality but can be further purified by either fractional crystallization[64] or zone refining.[65]

3-26. ANALYSIS OF HIGH-PURITY GALLIUM

Almost all the methods used for examining gallium arsenide can be used for the analysis of gallium. Since gallium arsenide is always prepared as single crystal, discussion of these methods will be deferred to Chap. 5.

A few absorptiometric methods for specific impurities have been described. Nazarenko et al.[66] determined arsenic by extracting as the diethyldithiocarbamate into chloroform, evaporating, and, after dissolving the residue, applying the molybdenum-blue method. Nazarenko and Flyantikova[67] evolved a method for silicon in which the gallium is first removed by volatilizing as the 8-hydroxyquinolate and then determined as the molybdenum-blue complex. The same workers[68] separate iron into a chloroform solution of hydroxyiminophenylhydroxylamine and, after evaporating and ashing, determine it by extracting the thiocyanate into isoamyl alcohol. A somewhat simpler procedure for iron is given by Knizek and Galik[69] in which the iron is reduced and the bathophenanthroline complex extracted into chloroform for measurement. Roberts et al.[70] removed the gallium by extraction of the chloride with isopropyl ether and determined copper in the aqueous residue using 1,5-diphenylcarbohydrazide. Knizek and Pecenkova[71] reduced the copper with hydroxylamine and extracted the neocuproine complex into chloroform for measurement. Antimony is determined by Biryuk[72] by separating into chloroform as the diethyldithiocarbamate, evaporating to fumes with sulfuric acid, extracting as the pyridine-iodide complex into ether, returning to dilute acid solution, and finally completing the determination with phenylfluorone. Monnier and Prod'hom[73] separated the gallium from zinc by extracting it into ether as the chloro compound and determined the zinc in the aqueous fraction with dithizone. Sulfur is determined by Goryushina and Biryukova[74] by reducing to sulfide, distilling off as hydrogen sulfide, and determining this as lead sulfide colorimetrically.

All these colorimetric methods have sensitivity levels in the 0.1- to 1-ppm range. A fluorimetric method for selenium, due to Vladimirova and Kuchmistaya,[75] has a similar sensitivity; after precipitation of metallic selenium from the sample solution, the redissolved element is reacted with 3,3′-diaminobenzidine and the fluorescent complex extracted into toluene. Sensitivities down to 1 ppb have been obtained by Lysenko and Kim;[76] in their procedure, 5 g of sample is dissolved in a hydrochloric-nitric acid mixture, and the gallium chloride extracted with butyl acetate. Photometric methods are described for copper using dithizone, nickel using α-dioxime, and cobalt using nitroso-R salt. Titration techniques using an absorptiometric end point are also given for silver and platinum using dithizone, bismuth using thiourea, and manganese using persulfate.

A preliminary separation of lead and zinc is made by Steffek,[77] using filter-paper chromatography. An ethyl acetate–nitric acid eluent is employed, and the colorimetric finish is with dithizone.

Several polarographic methods have been published. Pohl[78] described a procedure for copper and cadmium in gold which he claimed could also be applied to gallium; the gold was removed by extraction of the bromide into isopropyl ether prior to polarographing the aqueous residue. Sinyakova et al.[79] obtained 10 ppb sensitivity for indium by dissolving the gallium sample in aqua regia, adding cobalt, and coprecipitating the indium as the sulfide. This precipitate was then redissolved and the indium extracted as the dithizonate prior to polarography. Stripping polarography was used by Provaznik and Mojzis[80] to determine lead in gallium; a preliminary electrolysis was performed on a solution of the sample using a sessile mercury drop, followed by a rapid anodic scan. Miklos[81] described a procedure for zinc in which the gallium was complexed as the tartrate; however, for contents less than 0.1%, the zinc had to be separated first as the dithizone. A square-wave polarograph (Mervyn-Harwell) was used by Kaplan et al.[82] to determine tellurium. After solution of the sample, tellurium was reduced to the element by hydroxylamine and coprecipitated with sulfur prior to polarography. A variation of this method[83] uses a carbon tetrachloride extraction of the diethyldithiocarbamate to isolate the tellurium. Kaplan and Sorokovskaya[84] have used the same instrument to determine selenium, using the same coprecipitation with sulfur to separate it from gallium. The sensitivity for both these elements was 0.2 ppm. Lysenko and Kim[76] used the same concentration step described above for colorimetric analysis as a preliminary to polarography. After removal of the gallium chloride by butyl acetate extraction, the aqueous fraction is evaporated and then dissolved in either a bromide or an acetate electrolyte. Copper, zinc, cadmium, indium, and lead can be determined at sensitivities of 10 ppb or better.

Activation analysis has not been applied to this problem to any great extent. The high level of activity induced in gallium itself tends to increase the background and lower the sensitivity. Moreover, gallium metal is not allowed in the higher-flux reactors, so that a preliminary treatment to form oxide is necessary. Hoste and Van den Berghe[85] determined indium in gallium using a radium-beryllium source, but the sensitivity of 40 ppm is somewhat unrealistic for high-purity metal. Lerch and Kreienbuhl[86] described a procedure for calcium which had a sensitivity of 1 ppm and, in the same paper, were able to obtain a sensitivity of 50 ppb for zinc using an isotope dilution method. This method was later applied[87] to calcium to follow its distribution during the electrolytic separation of gallium; a sensitivity of 1 ppb could be obtained. A procedure for copper has been devised by Krivanek et al.[88] using the substoichiometric procedure. This uses a known, but insufficient, amount of diethyldithiocarbamate reagent to extract the carrier and active copper into chloroform for gamma-ray spectroscopy. The substoichiometric extraction is more selective, and a sensitivity of better than 0.1 ppm is attainable. More comprehensive procedures have been described by Alimarin et al.[89] A solution of the irradiated sample in hydrochloric acid was extracted with ether to remove gallium and gold; the latter was determined by evaporating, redissolving, and reducing with hydrogen peroxide. The aqueous solution was treated with hydrogen sulfide to precipitate arsenic and copper, phosphorus precipitated as bismuth phosphate, and zinc determined in the final solution. Values quoted for a high-purity sample varied from 5 ppb for arsenic up to 25% (sic) for phosphorus. Nagy et al.[90] used a half-life determination after 17 days' decay to determine zinc in gallium, although the values quoted were 10 ppm or higher. They also claimed to have determined iron and mercury by gamma-ray spectroscopy, but no level of sensitivity was given.

The determination of oxygen, hydrogen, and nitrogen in gallium is carried out by vacuum fusion. Both Wilson et al.[91] and Vasil'eva et al.[92] use a dry bath; the sensitivities are in the few-ppm range.

An indirect coulometric titration has been proposed by Kostromin and Anisimova[93] for the determination of beryllium, with a probable sensitivity of less than 1 ppm. After dissolution of the sample, the acetylacetonate is extracted into chloroform, the solution evaporated to dryness, and the complex dissolved in sulfuric acid; the liberated acetylacetone is titrated potentiometrically with electrolytically generated bromine.

While the above methods for specific elements undoubtedly have their uses, most of them do not have the general survey feature that an evaluation of a high-purity metal should have. This feature is available in the spectrographic methods, and these are by far the most useful, particularly since the sensitivity is quite high for emission as well as mass spectrography.

Most of the published emission spectrographic procedures call for a preliminary concentration step. Owens[94] removes the gallium as the chloro complex by extracting with isopropyl ether. Several metals are detected down to 0.02 ppm. An almost identical method is used by Oldfield and Bridge[95] with very similar sensitivity levels reported. Neeb[96] was able to achieve a sensitivity of 4 ppb for zinc by vaporizing the sample and condensing the impurity on a cold finger prior to sparking. Lysenko and Kim[76] used their butyl acetate extraction of the chloro compound to effect a preconcentration and achieved sensitivities of as much as 0.1 ppb.

Undoubtedly these preconcentration procedures can enhance the sensitivity, but they are tedious to carry out, and contamination from the reagents is always a possibility. Moreover, some elements of interest may be removed with the gallium. It is preferable to use a direct method. Massengale et al.[97] first applied the split-burn technique to gallium arsenide, and this will be dealt with more fully in Sec. 5-3. Essentially, it consists in arcing the sample in the conventional way but splitting the burn into three consecutive periods of time. This has the effect of reducing the background to one-third in the first period while the volatile elements are mostly evolved in this same period; i.e., the signal-to-noise ratio is increased by 3. Similarly, in the last period the nonvolatiles are enhanced. The method, including that for gallium, has been given in detail by Kane.[98] The sensitivities are the same as those given in Sec. 5-3 for gallium arsenide and are about 1 ppb for copper and magnesium and about 1 ppm for many others. This compares quite favorably with the preconcentration methods and gives a more comprehensive and reliable evaluation of the metal.

Mass spectrography gives even more complete coverage and, in general, better sensitivity than emission spectrography. This technique is very important in the analysis of semiconductor materials and will be dealt with in somewhat more detail in Sec. 5-4. Briefly, self-electrodes of a solid sample are sparked in vacuum at radiofrequencies, and the ions generated are analyzed by a double-focusing spectrometer system. The readout is a photographic plate. Wolstenholme[99] described an attachment, shown in Fig. 3-6, for maintaining the sample as a solid during analysis. A similar attachment has also been used by Nalbantoglu.[100,101] It consists of a glass tube, closed at its lower end and carrying two copper leads which clamp

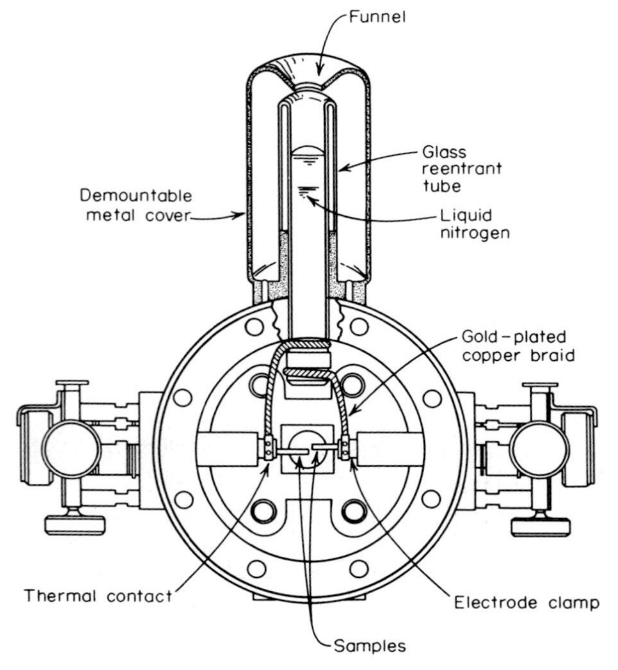

Fig. 3-6. Cold-finger attachment for the analysis of gallium. (*Adapted from Wolstenholme.*[99])

to the electrode holders. The tube is surrounded by a jacket which exhausts into the source to form a vacuum insulator when the unit is mounted in place of the shutter mechanism of the mass spectrograph. The gallium sample is frozen, cut into small bars, and clamped in the holders. The tube is kept filled with liquid nitrogen during the sparking period to conduct away heat from the electrodes. The method has been given in detail by Kane.[98] The sensitivities are the same as those given for gallium arsenide in Sec. 5-4 and range from 1 ppb up to about 50 ppb for most elements. Fitzner[102] has described his experiences with this same procedure, and his analyses for three grades of material are given in Table 3-2; note that these values are in ppm by weight.

3-27. HISTORY OF INDIUM

The discovery of indium preceded that of gallium; and, like gallium, indium was found spectrographically. Reich and Richter, in 1863, were examining a residue from the treatment of zinc blende for thallium, which had been found 2 years earlier by Crookes in another sulfide mineral, when they noted, instead, two blue lines due to a new element. They named their element indium after the indigo color of the emission. They isolated the metal the same year from this same ore.

The metal remained a curiosity until well into the 1930s. It is reported[103] that the world's supply was only 1 g in 1924. During World War II, production was considerably expanded because of its use in aircraft engine bearings.

Table 3-2. Mass Spectrographic Analysis of Gallium†
All values ppm by weight

Element	Highest-purity material	Average material	Comparatively impure material
C	<0.2	<0.2	<0.2
F	0.06	0.05	0.1
Mg	0.03	0.04	0.3
Al	0.04	0.04	0.25
Si	<0.3	<0.3	0.7
P	0.02	0.02	0.1
S	0.05	0.05	0.15
Cl	0.05	0.07	0.3
K	0.03	0.06	0.061
Ca	0.05	0.07	0.71
Ti	.....	0.03	0.04
V	.....	0.03	0.03
Cr	.....	0.02	0.04
Mn	.....	0.02	0.08
Fe	.....	0.04	0.07
Ni	.....	.....	0.05
Cu	.....	0.06	0.4
Zn	.....	0.03	0.6
In	.....	.....	3
Sn	.....	.....	0.15
Hg	.....	.....	1.5
Pb	.....	.....	0.8

†From Fitzner.[102]

3-28. OCCURRENCE OF INDIUM

Like gallium, indium is found in many minerals throughout the world but in very minute amounts. It is associated generally with zinc and, to a lesser extent, lead. Amounts, even in ores more abundant in the element, are less than 0.1%.

3-29. DETERMINATION OF INDIUM IN MINERALS

Although indium was first detected spectrographically, the classical methods for its determination in minerals are gravimetric. Schoeller and Powell[12] describe a method in which the ore is decomposed and treated to obtain the indium in a hydrochloric acid solution. Zinc is added and the resulting sponge dissolved in nitric acid. Indium hydroxide is precipitated from homogeneous solution using hexamethylenetetramine and reprecipitated before weighing as the oxide. A few other precipitants have been used and are reviewed by Onishi.[51] For example, Patrovsky[104] has applied diethyldithiocarbamate to the determination in zinc and iron ores. A comprehensive survey of the analytical chemistry of this element has been made by Busev,[105] and this should be consulted for work prior to 1957.

Spectrographic analysis has been extensively applied to the determination of indium in minerals, and in general the same comments can be made that were made for gallium. By direct methods, e.g., that of Morris and Brewer[106] or Raikhbaum and Kostynkova,[107] a sensitivity limit of 10 ppm is usual, but with the addition of a chloride[108,109] this can be increased to 1 ppm. Preconcentration can be employed to improve on this. Minczewski et al.[53] applied the method described for gallium to the simultaneous determination of indium and obtained a similar sensitivity (0.1 ppm). Brooks[110] has described a method in which the iodine complex is extracted into ether, which is then distilled off and the residue submitted to emission spectrography. By application of this method to silicate rocks, a sensitivity of about 0.01 ppm was obtained.

Of the colorimetric reagents, the most popular, as for gallium, seem to be the Rhodamine dyestuffs. Blyum and Dushina[111] describe a method in which the ore is dissolved in nitric acid–hydrochloric acid mixture and evaporated to dryness, the residue treated with hydrochloric acid, and the bromo-indium complex extracted into butyl acetate. The indium is reextracted into hydrochloric acid and then precipitated as the hydroxide with ferric iron as the collector. The precipitate is dissolved in dilute hydrobromic acid, Rhodamine 3B is added, and the complex extracted into benzene where it is determined fluorimetrically. By comparing with standards visually, a sensitivity of 0.2 ppm is obtained on a 0.1-g sample. An essentially similar method using Rhodamine 6G was given by Blyum et al.;[112] using Rhodamine B, by Knipovich et al.;[113] both used fluorimetric finishes. Rozbianskaya[114] determined indium in cassiterite by both colorimetry and fluorimetry, depending on the level, using Rhodamine B; a preliminary extraction of the bromo compound into ether was made. Levin and Azarenko,[115] on the other hand, devised a method of extraction into alkyl hydrogen phosphates which eliminated the necessity of the final extraction of a Rhodamine G complex into benzene; their finish was colorimetric. Other colorimetric reagents have included arsenazo,[116] 5,7-dibromo-8-hydroxyquinoline,[117] and bromopyrogallol red.[118] 8-hydroxyquinoline has been used[119] as a fluorimetric reagent.

Activation analysis was applied by Smales et al.[120] to a number of rocks including granite and diabase, and a similar procedure by Irving et al.[121] to cylindrite. The indium is precipitated several times as the hydroxide or sulfide and finally recovered as the oxinate. Sensitivities of 2 ppb were obtained on a 400-mg sample if ^{114}In were counted or on a 20-mg sample if ^{116}In were determined by using a flux of 10^{12} neutrons/(cm^2)(sec). Abdullaev et al.,[122] using a polonium-beryllium source, determined 10 ppm by gamma-ray spectroscopy, using the ^{116}In, in sphalerite. The procedure of Jaskolska and Minczewski[58,59] described earlier (Sec. 3-22) for gallium also includes an additional step for indium. The sensitivity is about the same. Pierce and Peck[123] introduced a method of separating ^{116}In by passage of a sample solution of rock through a cellulose column impregnated with dithizone. Interfering elements were removed by the column, although an additional extraction step was recommended by Mapper and Fryer[124] to remove gallium. Two analyses of ores were made by Okada and Kamemoto[125] using very short (2.5 sec) irradiations and gamma-ray spectroscopy; the metastable ^{116m}In was used. Down to 1 ppm indium was determined by Tomov et al.[126] in sphalerite and lead-zinc ores, also by using gamma-ray spectroscopy.

Polarography is a particularly useful technique for determining indium in ores since the wave is well separated from those of both zinc and gallium. Weiss[127] has described a method for ores using chloride as the base in which vanadium is masked with fluoride and cadmium removed as the ammonia complex; a sensitivity of 30 ppm is obtained. Kvacek and Kuhn[128] obtained a similar sensitivity in the presence of lead, cadmium, and tin by using a chloride-bromide electrolyte and applied this to the determination in sphalerite and zinc ores. These same workers[129,130] subsequently published modifications for use in the presence of large amounts of lead or tin. Kaplan[131] applied a pulse polarograph to the determination of indium in acid-soluble ores, using a hydrochloric acid base, but no sensitivities were given.

The flame photometric method of Bode and Fabian,[60] mentioned in Sec. 3-22 for gallium, has also been applied to indium.

Patrovsky[132] applied the EDTA titration using morin as indicator to the determination of indium in sphalerite, and Tsyvina and Vladimirova[133] titrated amperometrically using EDTA.

3-30. RECOVERY OF INDIUM

Since indium is present only at very low levels in minerals, it is recovered as a by-product in other metallurgical operations. The commonest source is the electrolytic production of zinc.[134]

Sulfide ores of zinc are first roasted to form the oxide and then leached with sulfuric acid. The solution is neutralized to precipitate ferric hydroxide, which carries down with it many other metals, some of which may be worth recovering. In some of the larger operations, e.g., the Cominco plant at Trail, British Columbia, the ore is essentially a lead-zinc ore and the leach residue is high in lead. It is therefore treated in a blast furnace to recover the lead. The slag from this furnace contains zinc which was carried down with the precipitate; it is fumed in another furnace and additional zinc separated as the oxide. This is leached and precipitated as before and the residue returned to the lead furnace. The solutions from these leach operations proceed to the electrolysis step for zinc recovery.

The lead passes to a drossing stage in which the lead bullion is melted prior to casting into ingots which are subsequently purified by electrolysis. In this stage, a slag is separated which contains most of the indium. The treatment of this slag has been described by Mills et al.[135] The flow of the indium through the plant is shown in Fig. 3-7. The bulk of the indium ends up in the dross slag, which is treated by the process shown in Fig. 3-8. The slag, which contains about 2.5% indium, is ground, and copper, which is present in significant amounts, is separated by flotation. The tailings, mostly lead, are sintered and reduced in an electric furnace with coke and limestone. The speiss is returned to the smelter; and the metal, a mixture of lead, tin, indium, and antimony, is cast to form anodes for the electrolysis. The electrolyte is lead fluosilicate, and a lead-tin alloy (about 10% tin) is deposited on the cathode. The indium forms an indium antimonide slime at the anode, and only relatively small amounts of these two elements go into solution. The slime is heated to 300°C with sulfuric acid and the product leached with water.

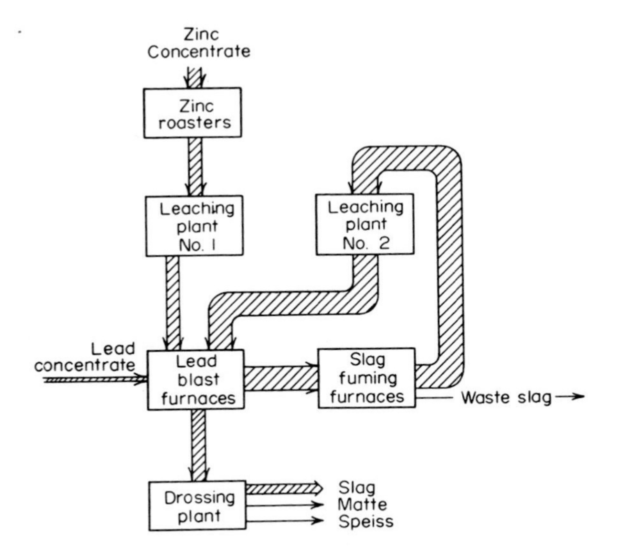

Fig. 3-7. Indium flow through the Trail lead-zinc plant. (*After Mills et al.*[135])

Antimony, tin, and lead are insoluble and can be filtered off. The solution, mostly indium sulfate and a little copper sulfate, is adjusted to pH 1.0 and sodium chloride added. Sheet indium is introduced, and copper separates by electrodeposition. When complete, the sheets are removed, the pH adjusted to 1.5, and zinc or aluminum sheets introduced. Indium deposits electrolytically in the form of a sponge, which is cast. This metal is 99.5%. The spent electrolyte from the first electrolysis is treated as shown in Fig. 3-8 to recover lead and electrolyte.

Variations on this method are used by other producers, dependent on the method used for zinc refining. Several of these are reported by Mills et al.[103]

3-31. DETERMINATION OF INDIUM IN THE RECOVERY PROCESS

In industrial processes, gravimetry and volumetry have been used occasionally. Zettler,[136] for example, determined indium in smelter fume by a series of precipitations of the hydroxide and sulfide. Indium was determined in the hydrochloric acid solution from the germanium recovery process by Kalina and Baburina[137] by extracting the bromo complex into ether and precipitating finally as the hydroxide. Lead fractions containing indium were examined by Sayun and Tikhanina[138] by extracting the iodo complex into ether and then titrating with EDTA with 4-(2-pyridylazo) resorcinol as indicator.

Spectrographic methods have been devised by Yudelevich and his coworkers[139–142] for the control of the lead-zinc process, and these methods include determinations for indium. The samples are arced, and for powders, the sensitivity is 10 ppm. For solutions, the sample is atomized into the arc, or carbon powder is impregnated with the sample. Sensitivities are about 10 μg/ml.

Colorimetric and fluorimetric methods have also been applied to the process control of lead-zinc by-products. Ginzberg and Shkrobot[143] removed interfering elements from zinc and lead dusts by ion exchange followed by a fluorimetric

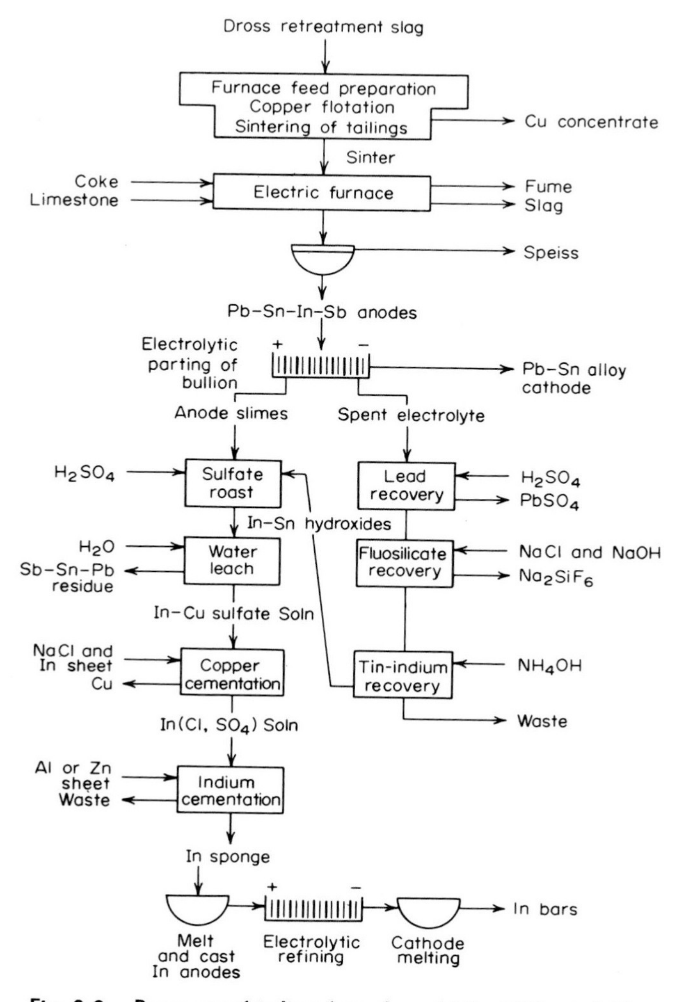

Fig. 3-8. Recovery of indium from slag. (*After Mills et al.*[135])

determination using 8-hydroxyquinoline, and the same reagent was applied to the analysis of flue dusts by Gurev et al.[144] using a preliminary extraction of the bromo complex into ether. A colorimetric method was used by Collins and Kanzelmeyer[145] for determinations in various zinc fractions; after a separation of the bromo compound into isopropyl ether, the indium was returned to aqueous solution and then extracted into a chloroform solution of dithizone, where its color was measured. The alkyl phosphate extraction and Rhodamine 6G colorimetric procedure of Levin and Azarenko,[115] previously mentioned for minerals (Sec. 3-29), has also been applied to fractions in the lead-zinc industry. Other reagents used for this purpose include arsenazo,[116] phenylfluorone,[146] and xylenol orange.[147]

Activation analysis has been used for analyzing final products such as lead and zinc for indium, but needless to say it has not been applied as a process-control method. Hoste and van den Berghe[85] applied their method for indium in gallium to its determination in zinc also, but again the sensitivity was poor. Kusaka[148] used the same type of radium-beryllium source for determinations of 0.05% or more in zinc. Kosaric and Leliaert[149] determined about 5 ppb of indium in zinc by irradiating at 8×10^{11} neutrons/(cm^2)(sec) and separating the bromo indium complex into isopropyl ether. Jaskolska and Minczewski[58,59] used the same method described earlier for gallium (Sec. 3-22) and for indium in ores (Sec. 3-29) for its determination in metallic lead. Gibbons and Lawson[150] described a gamma-ray spectroscopic method for zinc samples with a sensitivity of 10 ppb. An irradiation of 30 sec was given at a flux of 10^{12} neutrons/(cm^2)(sec), and the ^{116}In peak was measured.

Although polarography is stated by Mills et al.[135] to be the technique used in process control in the lead-zinc industry's recovery of indium, information on the procedures is not given. Busev[105] reports several methods, and Moeller and Hopkins[151] refer to one or two, pointing out the interference by cadmium. This was avoided by Treindl[152] by using a potassium iodide electrolyte and applied to the determination in zinc. Nizhnik and Chaus[153] treated the zinc-lead concentrate with sulfuric acid and added zinc amalgam to remove interfering ions by electrodeposition; alkali chloride was added to form the base electrolyte. Pats and Tsfasman[154] described methods for lead concentrates in which the lead is removed as sulfate and the indium coprecipitated with iron as the hydroxide. This precipitate is dissolved in hydrochloric acid, phosphoric acid is added to prevent interference by tin, and the solution is examined on a square-wave polarograph. Banks et al.[155,156] made studies of the partition of indium between molten lead and zinc and obtained good waves by using a tartrate electrolyte after removal of the lead as sulfate. By using a sine-wave polarograph, they were able to avoid the lead precipitation and to use a chloride base. Kovalenko and Musaelyants[157] polarographed the zinc sulfate electrolyte from the process directly; the wave at -0.56 volt in this medium is due to cadmium. On addition of potassium chloride, a combined wave due to both this and indium is obtained, and indium is determined by difference.

3-32. PRODUCTION OF HIGH-PURITY INDIUM

The 99.5% metal obtained above is further refined electrolytically; the process has been described by Mills et al.[135] A sodium chloride–indium chloride electrolyte is used, and by careful electrolysis a product of 99.999% purity can be obtained.

3-33. ANALYSIS OF HIGH-PURITY INDIUM

Since the metals indium and gallium are so similar in their chemical properties, it is not perhaps surprising that many of the methods referred to in the discussion on gallium (Sec. 3-26) occur again here. In addition, many of the methods for indium antimonide and indium arsenide, dealt with in Chap. 5, will also be applicable to the metal.

Several absorptiometric methods have been described for specific impurities. Nazarenko et al.[66] applied their diethyldithiocarbamate separation and molybdenum-blue finish to the determination of arsenic in indium. In fact, Nazarenko and his coworkers have been quite active in this field, having published methods for cadmium[158] using dithizone, silicon[67] using molybdenum blue, zinc[159] using dithizone, iron[68] as the thiocyanate, and, more recently, arsenic[160] by a rapid method in which the element is first separated as arsine. Nishimura and Imai have also published for several elements. Lead[161] is determined by extraction into benzene of the dithizonate after a preliminary removal of indium as the bromo complex by isopropyl ether; however, results tend to be low. Gallium[162] is separated from indium by extracting the chloride into ether, after which the gallium is determined with Rhodamine B. Iron[163] was also extracted as the chloride into ether and determined by 1:10-phenanthroline. Copper[163] is extracted into chloroform as the diethyldithiocarbamate and measured. Another procedure for copper is given by Busev and Bozenkova[164] in which a sulfuric acid solution of the sample to which nickel diethyldithiophosphate has been added is shaken with carbon tetrachloride; the intense orange-yellow copper salt is extracted and measured. Peshkova et al.[165,166] have determined nickel by extracting with benzene the complex with furil α-dioxime or, with better sensitivity, benzil α-dioxime. The methods applied to gallium (Sec. 3-26) by Biryuk[72] for antimony and by Goryushina and Biryukova[74] for sulfur have also been applied to indium.

With the exception of Peshkova et al.'s methods for nickel, which are in the 5 to 50-ppb range, all these colorimetric procedures give sensitivities between 0.1 and 1 ppm.

The fluorimetric method of Vladimirova and Kuchmistaya[75] for selenium (Sec. 3-26) using 3,3′-diaminobenzidine has also been applied to indium, as have the procedures of Steffek[77] (Sec. 3-26) in which lead and zinc are separated by filter-paper chromatography prior to determination as the dithizonates. A turbidimetric method has been used by Vydra and Stulik[167] for cadmium in which a precipitate is formed with iodide and 1:10-phenanthroline.

Polarography has been extensively applied to the examination of high-purity indium. Dolezal[168] used an ethylenediamine–potassium hydroxide base electrolyte for the simultaneous determination of copper, thallium, lead, and cadmium, but the sensitivity was not high; a lower limit of 0.02% was quoted for copper. Subsequently,[169] the base electrolyte was changed to bis-(2-hydroxybutyl)-2-hydroxyethylamine and sodium hydroxide for separation of the last three impurities.

The method of Pohl[78] for copper and cadmium in gold (Sec. 3-26) was claimed to be applicable to indium. This procedure, in which the gold was extracted as the bromide by diisopropyl ether, was adapted by Pohl and Bonsels[170] to a more comprehensive analysis of indium. Three groups of analyses were carried out: (1) a hydrobromic acid solution of the sample was extracted with diisopropyl ether to remove indium and the aqueous solution treated to obtain a tartrate base in which bismuth, copper, and lead were determined; (2) the same extraction was made but the aqueous layer treated to yield an ammonia base in which copper, cadmium, and zinc were determined; and (3) a hydrochloric acid solution was extracted with diisopropyl ether (indium chloride is not extracted) and dissolved in tartrate base in which iron and thallium were determined. Since 5-g samples can be used, sensitivities down to 10 ppb can be achieved. Towndrow et al.[171] separated zinc from indium in hydrochloric acid solution on a cellulose column;

the eluate containing zinc was evaporated and then dissolved in ammonia for polarography. The sensitivity was about 10 ppm. Kopanica and Pribil[172] masked the indium with 1,2-diaminocyclohexane-*NNN'N'*-tetraacetic acid in ammonia buffered to pH 2.7 to 3.5. Under these conditions, cadmium is displaced from its complex by thorium and can be determined. By using triethylene tetramine-hexaacetic acid (TTHA), Conradi and Kopanica[173] were able to determine copper, lead, cadmium, and bismuth in the presence of indium; they recommend removing the bulk of the indium by an ether extraction of the bromo complex before this determination, and thallium can be determined after stripping from this organic phase. Musil and Kopanica[174] later found that the more readily available ascorbic acid could be used in place of TTHA; the sensitivity in both was about 10 ppm. Molybdenum was determined by Bikbulatova and Sinyakova[175] in a sulfuric-nitric acid base electrolyte from its catalytic nitrate wave at a sensitivity of 20 ppb.

These conventional polarographic techniques require some preconcentration step to achieve a useful sensitivity. More sensitive instrumentation has been applied to the problem. Shirai[176] used an alternating-current polarograph to determine cadmium, lead, and zinc by using a phosphoric-nitric acid base electrolyte, and Ishibashi et al.[177] used a similar instrument to determine cadmium and lead but with a perchloric-nitric base. In both cases, indium can be tolerated in the same solution. The square-wave polarograph has been applied by Kaplan and his coworkers to a number of determinations in indium. For copper,[178] as little as 0.1 ppm can be determined in a phosphoric-nitric acid base without separating the indium. For tellurium[82] and selenium,[84] the methods applied to gallium (Sec. 3-26) were used also for indium; the impurity was coprecipitated with sulfur prior to polarography in acid potassium chloride. For thallium,[179] a direct method using an ammoniacal EDTA base electrolyte was sensitive to 2 ppm; a tenfold increase can be achieved by a preliminary extraction of the chloride into ether and the use of a phosphoric-sulfuric acid base electrolyte. Nishimura and Imai[161] extracted the indium as its bromo complex into isopropyl ether prior to determining lead, cadmium, and zinc in a phosphoric acid medium by square-wave polarography. The sensitivities were about 1 ppm.

Stripping or amalgam polarography is attractive in that the preconcentration step is carried out in the same vessel and medium as the polarography. Sinyakova et al.[180] used a preliminary isopropyl ether extraction of indium bromide before electrolyzing copper, lead, cadmium, and zinc into a hanging-mercury-drop electrode (HMDE). The subsequent anodic wave was capable of detecting as little as 10 ppb in the sample. A similar sensitivity was obtained by Stepanova et al.[181] for germanium by a preliminary extraction of the chloride into carbon tetrachloride followed by stripping polarography on an HMDE. Mesyats et al. have described methods for thallium,[182] in which 20 ppb can be determined with a preliminary concentration of thallium chloride into ether, and for copper;[183] this latter is a direct procedure in phosphoric acid with a sensitivity of 40 ppb. Detailed methods for the determination of zinc, cadmium, tin, lead, copper, thallium, and bismuth using stripping polarography are given by Kane.[98] The sensitivities vary from 50 ppb upward.

No methods have been published for the activation analysis of indium presumably because of its high capture cross section.

Oxygen, nitrogen, and hydrogen were determined by Vasil'eva et al.[92] in indium by vacuum fusion using a dry bath; the sensitivity was 10 ppm for oxygen.

For a general survey of the purity of this metal, the broad coverage of the spectrographic methods is preferred. Hyman has described two solution methods: the first,[184] using the porous-cup technique, was intended for alloying constituents in the 0.10% range or higher, but the second[185] has sensitivities as low as 5 ppm. A solution of the sample is evaporated on a graphite electrode and excited by a high-voltage ac arc. Lead, tin, silver, copper, zinc, iron, gallium, and nickel were determined quantitatively by using bismuth as the internal standard. Mercury was determined at a level of 0.1 ppm by Porkhunova et al.[186] by evaporating in an ac arc, essentially a fractional volatilization. To determine lead, Nazarenko et al.[159] dissolved the sample and coprecipitated the lead with strontium sulfate; the precipitate was mixed with carbon and arced using bismuth as an internal standard. Neeb[96] applied his distillation procedure for zinc, described earlier as applied to gallium (Sec. 3-26), to indium with a comparable sensitivity. Caldararu[187] devised a method for several impurities at sensitivities as low as 3 ppb. A preconcentration was carried out by dissolving the sample in hydrochloric acid, extracting the indium bromide into ether, and concentrating the aqueous solution to dryness on a carbon electrode. The split-burn technique has been applied to indium as well as gallium, and the sensitivity levels are comparable. Full details are given by Kane.[98]

Mass spectrography, as might be expected, can be applied to this metal rather more easily than to gallium since it is a solid. Full details of the procedure are given by Kane[98] and are essentially the same as those given in Sec. 5-4.

3-34. PRODUCTION OF HIGH-PURITY ANTIMONY

Unlike germanium, gallium, and indium, antimony has been known for many centuries and was identified as a metal in the sixteenth century.[188] It is widely distributed in nature and commonly occurs as stibnite, the sulfide Sb_2S_3. Since its preparation is so well documented, it is not intended to deal with it here in any great detail nor to dwell at any length on the analytical methods used in determining the element in ores and concentrates. However, some background will be useful in assessing possible contaminants in the element. Fuller descriptions can be found in standard works such as Kirk-Othmer.[189]

The low-grade (5–25% Sb) ores are concentrated by roasting. The sulfide is oxidized by heating in a furnace with coke to form the volatile trioxide, which is collected in a condenser or precipitator. Arsenic trioxide, being more volatile, can be removed in this stage. For more concentrated ores, the sulfide can be separated from the gangue by liquation. The antimony trisulfide liquefies at 500 to 600° and can be run off from the bottom of a reverberatory furnace. The residue, containing 12 to 30% Sb, can be volatilized as any other low-grade ore.

The concentrated oxide or sulfide is converted to the metal by smelting. The richer, liquated ores can be heated with iron, which displaces antimony to form a matte of iron sulfide which can be separated from the metallic antimony. The intermediate-grade ores are smelted by techniques similar to those for lead; that is, the roasted sulfide ores or oxides are reduced in water-jacketed blast furnaces with coke to form the metal, which is separated from the slag in a heated forehearth.

This crude antimony typically contains about 95% Sb with considerable amounts of iron, sulfur, and arsenic. It is refined by slagging, in which, for example, a mixture of sodium sulfate and charcoal is added to the molten metal. The iron (and any copper) forms the sulfide, and arsenic is converted to arsenate; both are carried off in the slag formed. Sulfur can be removed by adding antimony oxysulfide. Lead is difficult to remove and is usually avoided by choosing suitable ores. The metal which results is typically 99.1 to 99.9% pure with arsenic contents less than 0.1%. Other principal impurities will be lead and sulfur.

Purification of this commercial metal to the high purity required by the semiconductor industry has been reviewed by Haberecht.[190] Several procedures are mentioned, but the preferred technique is zone refining. Tanenbaum et al.[191] started with commercial 99.8% metal and, after seven passes in an atmosphere of nitrogen, significantly reduced the spectrographic impurities to nondetectable levels. Arsenic was not affected; its segregation coefficient is close to unity. By starting with antimony trichloride, distilling from hydrochloric acid, and then reducing with carbonyl iron, they were able to obtain a substantially arsenic-free metal. After 10 passes in a zone refiner, the only impurities detected were zinc and arsenic at the 0.1-ppm level. An alternative method of removing arsenic is described by Haberecht.[190] Aluminum is added as a scavenger in the zone-refining process and, after six passes, carries the arsenic to the back of the ingot. By removing the last-to-freeze portion and subjecting the remainder to six further passes, a material was obtained with less than 1 ppm total spectrographic impurities. As Haberecht points out, this does not take into account such nonmetallic impurities as carbon, which may have significant effects on the final III-V product.

3-35. ANALYSIS OF HIGH-PURITY ANTIMONY

Many colorimetric methods have been devised for specific impurities. Nazarenko et al.,[66] after dissolving the sample, reduced it with zinc in a stream of hydrogen and separated the arsenic as arsine; the evolved arsenic was absorbed in a mercuric chloride solution and determined by the molybdenum-blue method. Maekawa et al.[192] retained arsenic in solution during the solution of antimony in aqua regia by adding metallic copper; the arsenic is then coprecipitated with iron as the hydroxide, reduced to the arsenous form, extracted into chloroform, and determined as molybdenum blue. Kowalczyk[193] also used a molybdenum-blue finish after separating the arsenic as arsine; he used hydroxylamine as the reductant. Nazarenko and Flyantikova[67] removed the matrix as antimony tribromide prior to determining silicon by the molybdenum-blue procedure. Klein and Skrivanek[194] determined as little as 1 ppm gallium by removing the antimony by sulfide precipitation and reacting the filtrate with malachite green. Copper was determined by Provaznik and Knizek[195] with a sensitivity of 0.5 ppm by extracting the diethyldithiocarbamate into chloroform, and Ishihara and Koga[196] determined it with neocuproine after first removing antimony as the bromide or masking it as the citrate. Steffek[197] separated copper and iron from antimony by paper chromatography and determined copper as the diethyldithiocarbamate and iron with salicylaldoxime or salicylic acid. Iron was determined by Lipshits et al.[198] as the 2,2′-bipyridyl complex. An indirect method for lead was used by Zaglodina;[199] after volatilization of the antimony as the bromide, lead was precipitated as the chromate and this latter determined with diphenylcarbazide. A similar method was applied by Nazarenko et al.[200] for the determination of chromium. A test for the heavy-metal content was devised by Haberli,[201] in which a succession of aliquots of dithizone in carbon tetrachloride was shaken with the sample solution until one remained colorless; the volume was a measure of the contamination and was calibrated against a known addition of zinc. Goryushina and Biryukova's[74] method for sulfur, which was mentioned for both gallium (Sec. 3-26) and indium (Sec. 3-33), was applied also to antimony after its removal as the bromide.

The fluorimetric procedure[75] for selenium (Secs. 3-26 and 3-33), in which the 3,3′-diaminobenzidine is extracted into toluene, has been applied to antimony. A fluorimetric method for aluminum is described by Nazarenko et al.[200] Great care is taken to remove antimony prior to reaction with 1:10-phenanthroline by evaporating with hydrobromic acid, extracting the chloride with an organic phase, and finally masking any remaining traces with iodide. A sensitivity of 50 ppb is claimed for this procedure, an order of magnitude better than the preceding colorimetric procedures.

A few polarographic methods have been described. Conventional polarography was applied by Aref'eva and Pats[202] to the determination of copper, cadmium zinc, and nickel in an ammonia–ammonium chloride base after removal of the antimony as the bromide; cobalt could also be determined by isolating it first as the 1-nitroso-2-naphthol complex and lead by removing other interfering elements with iron. Pohl[78] claimed that his method for copper and cadmium in gold (Secs. 3-26 and 3-33) could also be applied to antimony. A rapid method for tin is described by Sulcek et al.,[203] in which a preliminary separation is made on a silica gel column; the base electrolyte is acid ammonium chloride, and a sensitivity of 5 ppm is claimed. The same procedures described earlier for gallium (Sec. 3-26) and indium (Sec. 3-33) were applied by Kaplan and his coworkers to the determination of tellurium[82] and selenium[84] in antimony; the impurities were coprecipitated with sulfur and examined by using the square-wave polarograph. Lysenko[204] extracted the antimony from solution as the acid chloride into butyl acetate and determined bismuth, copper, lead, cadmium, indium, and zinc on the aqueous residue in an acid potassium chloride base using an ac polarograph; sensitivities between 1 and 5 ppb were possible. Stripping polarography was used for the determination of lead by Provaznik and Mojzis.[80] The antimony was first fumed off as the bromide, and the residue submitted to a preliminary electrolysis using a sessile mercury cathode followed by a fast anodic scan. A sensitivity of 0.6 ppm was obtained. Although not stated, the method given by Kane[98] for indium antimonide (see Sec. 5-8) could also be applied to antimony; this is a stripping polarographic method for zinc, indium, cadmium, tin, lead, copper, thallium, and bismuth using a hanging-mercury-drop electrode.

A coulometric method for determining microgram amounts of aluminum has been applied by Kostromin and Akhmadeev;[205] the impurity is precipitated as the 8-hydroxyquinolate, and this organic fraction titrated with electrolytically generated permanganate.

Activation analysis has been used to determine very-low-level impurities. Kulak[206] determined nickel, copper, cobalt, tellurium, and arsenic in the low-ppb range by a radiochemical separation. Zone-refined antimony was examined by Rakovskii et al.[207] for phosphorus, chromium, manganese, copper, zinc, gallium, and arsenic using chromatographic separations as well as the more conventional chemical separations. The radiochemical separation of these elements is described also by Alimarin et al.[89] Iron was determined at the 0.5-ppm level by Simkova et al.[208,209] by precipitation as the sulfide and subsequent ion exchange to achieve radiochemical purity; only 0.5 ppm sensitivity was claimed with a flux of 7.5×10^{12} neutrons/(cm^2)(sec). Artyukhin et al.[210] eliminated antimony from solution following irradiation by extracting the pentachloride with β,β'-dichlordiethyl ether. The impurities remaining in the aqueous phase were separated by ion exchange. Copper, cobalt, zinc, indium, arsenic, tin, and tellurium were determined at levels as low as 10 ppb in a high-purity sample.

In common with all other high-purity materials, the most useful techniques are those which give a review of the impurities. Emission spectrography was applied by, among others, Yudelevich et al.[211] and Knipe[212] to the analysis of antimony, but their methods were aimed at the fractional percentage ranges. Jones[213] analyzed high-purity material using arc excitation and was able to achieve better than 10 ppb sensitivity for some impurities without pretreatment. Lysenko[204] used the same extraction with butyl acetate described above for the polarographic determination as a preconcentration step in emission spectrography; using carbon powder as a buffer, he obtained a sensitivity of 0.5 ppb for 15 metallic impurities. Copper was determined by Kowalczyk[214] by depositing it electrolytically on the tip of a graphite electrode prior to arcing. In common with most of the other semiconductor metals, the split-burn technique has been applied to antimony and is described by Kane.[98] Further details will be found in Sec. 5-3.

A flame photometric method was used by Nazarenko et al.[200] to determine 1 ppm calcium; the antimony was removed as the tribromide and the ignited residue dissolved in water before being aspirated into the burner. Calcium has also been determined, along with the alkali metals, by this same technique by Neeb,[215] who removed the antimony by distillation in a stream of chlorine. His sensitivity was about 0.1 ppm.

Hannay and Ahearn's original paper[216] on the application of the solids mass spectrograph included results on a sample of antimony containing 100 ppm arsenic, but, rather surprisingly, there appears to be no subsequent reference to this application. Undoubtedly it has been used, and the method described in Sec. 5-4 would be suitable.

3-36. PRODUCTION OF HIGH-PURITY ARSENIC

Like antimony, arsenic has a history dating back many centuries. The oxide was known medicinally even before Albertus Magnus separated the metal in the thirteenth century.[188] It is wisely distributed in nature and is obtained commercially as a by-product in the smelting of ores for copper, silver, gold, lead, nickel, and cobalt. The crude oxide is recovered in the flue dusts. Details will be found in Kirk-Othmer[217] and other standard works. Metallic arsenic is obtained from this oxide by reduction with charcoal.

The purification of the metal has been reviewed by Blum,[218] who lists the following methods which have been used:

1. Vacuum sublimation.
2. Sublimation in hydrogen at elevated temperatures.
3. Distillation from lead arsenic solutions.
4. Growth of arsenic single crystals by Bridgman technique.
5. Reduction of arsenic trioxide.
6. Reduction of arsenic trichloride.
7. Thermal decomposition of arsine.
8. Electrodeposition.
9. Vapor zone refining.
10. Zone refining of arsenides.

Of these, Beau[219] recommends method 5, carrying out the reduction in a stream of hydrogen in the vapor phase. A second sublimation yields material of 99.999%; the main impurities are silica and carbon with less than 20 ppb of sulfur. Brau[220] obtained very-high-purity material by what was essentially a combination of methods 5 and 3. Redistilled arsenic trichloride was reduced in the vapor phase in a stream of hydrogen. The condensed arsenic was sublimed in the apparatus shown in Fig. 3-9. The metal was heated at 600°C and the lead held at 400°C. The purified arsenic solidified in the upper bulb. By using radioactive tracers, it was shown that an initial concentration of sulfur of 3.6×10^{17} atoms/cm^3 could be reduced to 3.9×10^{14} by this treatment and carbon from 4×10^{16} to 1.7×10^{15}.

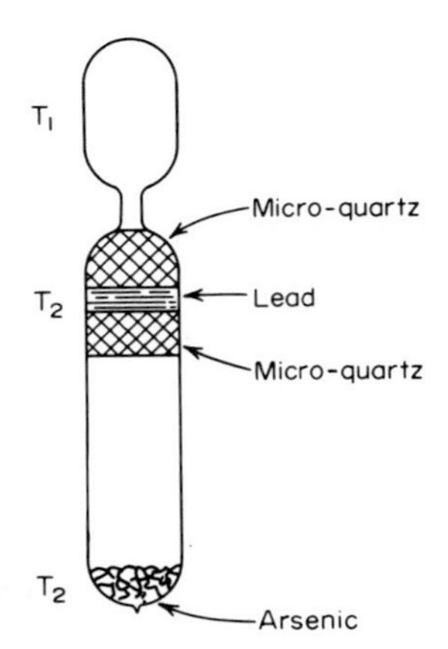

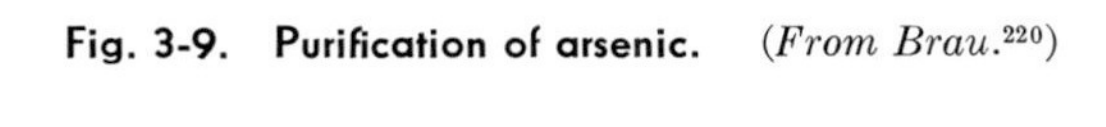
Fig. 3-9. Purification of arsenic. (*From Brau.*[220])

3-37. ANALYSIS OF HIGH-PURITY ARSENIC

The published methods for analysis of high-purity arsenic are all comparatively recent and reflect the growing importance of gallium arsenide.

A number of colorimetric methods have been devised by Kristaleva and her coworkers. Antimony[221] was complexed in hydrochloric acid solution with Brilliant Green and the complex extracted into benzene, toluene, or xylene for measurement. Iron[221] was determined with *o*-phenanthroline and copper[221] with diethyldithiocarbamate, both in aqueous solution. Bismuth[222] was determined after the removal of arsenic by distillation as arsenious chloride in the presence of hydroxylamine; the bismuth was complexed with thionalide, extracted into chloroform to remove interferences, and then returned to aqueous solution for measurement. Phosphorus[223] was determined as molybdenum blue after removal of the arsenic as the bromide and extraction of the phosphomolybdate into ether. Goryushina and Esenina[224] criticized this procedure and modified it to ensure essentially complete removal of the arsenic before reaction with the molybdate. Goryushina and Biryukova[74] also applied their sulfur procedure (Secs. 3-26, 3-33, and 3-35) to arsenic after its removal as the bromide. The methods that Knizek et al. used for gallium (Sec. 3-26) were also applied to arsenic; iron[69] was reduced and extracted into chloroform as the bathophenanthroline complex, copper[71] as the neocuproine complex. Roberts et al.[70] removed the arsenic by heating the sample with hydrochloric acid and bromine dissolved in carbon tetrachloride (to moderate the reaction); the residue is reacted with molybdate and extracted into *n*-pentanol to determine silicon. As usual, these methods have sensitivities in the 0.1- to 1-ppm range generally.

The fluorimetric method of Vladimirova and Kuchmistaya[75] (Secs. 3-26, 3-33, and 3-35) for selenium, in which the 3,3′-diaminobenzidine is extracted into toluene, has also been applied to arsenic. Tellurium was also determined by Vladimirova et al.[225] by extracting the diethyldithiocarbamate into carbon tetrachloride, evaporating, redissolving, reacting with butylrhodamine, and determining its fluorescence as a benzene–butyl acetate extract.

The polarographic method for copper and cadmium suggested by Pohl[78] as applicable to gallium (Sec. 3-26) was also stated to be applicable to arsenic. Copper and iron were separated on an ion-exchange column, eluted in a tartrate electrolyte, and polarographed by Rozanova and Kamaev.[226] The stripping polarographic method of Provaznik and Mojzis,[80] using a sessile-drop electrode and described for the determination of lead in gallium (Sec. 3-26) and antimony (Sec. 3-35), was applied also to arsenic. The same technique, using a hanging-mercury-drop electrode, was used by Kataev et al.[227,228] for determining copper and lead. After solution in potassium hydroxide, the impurities were plated into the drop at 1.0 volt for 30 min and the cell then scanned anodically. Sensitivities of 0.2 ppm for lead and 0.08 ppm for copper were obtained.

Activation analysis for chlorine, phosphorus, and sulfur was carried out by Niese[229] by radiochemical separation. The effect of interfering reactions was determined by duplicating the analysis in thermal-neutron and fast-neutron fluxes. A method for copper has been devised by Nizet et al.[230] in which the irradiated sample was precipitated as barium arsenate, the filtrate passed over an anion-exchange resin, and the copper precipitated as the sulfide and electrolyzed. About 0.1 ppm was determined in a 140-mg sample.

An emission spectrographic procedure was devised by Mack[231] in which the arsenic was first removed by the method used above by Roberts et al.[70] for their silicon determination. Gallium was added as a carrier during evaporation, and zinc, cadmium, lead, aluminum, nickel, vanadium, titanium, and cobalt were all determined at the 0.5-ppm level. Kataev and Otmakhova[232] separated copper, aluminum, iron, zinc, and magnesium on a cation-exchange resin and determined the concentration spectrographically in the eluate. Sensitivities from 1 ppb for copper to 100 ppb for iron were obtained. A detailed procedure for determining 22 elements in arsenic is described by Kane.[98]

Neeb's flame photometric method[215] for calcium and the alkali metals which was used for antimony (Sec. 3-35) has also been applied to arsenic.

The solids mass spectrograph has been used by Brown et al.[233] for the analysis of arsenic.

REFERENCES

1. Thompson, A. P., and J. R. Musgrave: *J. Metals,* **4**:1132 (1952).
2. Theurer, H. C., and H. J. Scaff: *NDRC Interim Rep.* No. 1, Contract OEMsr-1408, 1944.
3. Rick, C. E., and T. D. McKinley: *OSRD Progr. Rep.* NWP-P-44-3K, Contract OEMsr-1139, 1944.
4. North, H. Q.: *NDRC* 14–328, 1945.
5. Anon.: *Eng. Mining J.,* **157**:75, 204 (May, 1956).
6. Musgrave, J. R.: in I. M. Kolthoff and P. J. Elving (eds.), "Treatise on Analytical Chemistry," pt. II, vol. 2, p. 207, Interscience Publishers, Inc., New York, 1962.
7. Powell, A. R., F. M. Lever, and R. E. Walpole: *J. Appl. Chem. (London),* **1**:541 (1951).
8. Anon.: *Eng. Mining J.,* **157**:78 (May, 1956).
9. Waring, C. L., and W. P. Tucker: *Anal. Chem.,* **26**:1198 (1954).
10. Mekovskii, M. A., and A. N. Aleksandrova: *Zavodsk. Lab.,* **29**:797 (1963).
11. Cluley, H. J.: *Analyst,* **76**:517 (1951).
12. Schoeller, W. R., and A. R. Powell: "The Analysis of Minerals and Ores of the Rarer Elements," 3d ed., Hafner Publishing Company, Inc., New York, 1955.
13. Strickland, E. H.: *Analyst,* **80**:548 (1955).
14. Sandell, E. B.: "Colorimetric Determination of Traces of Metals," 3d ed., Interscience Publishers, Inc., New York, 1959.
15. Davies, G. R., and G. T. Morgan: *Analyst,* **63**:388 (1938).
16. Abel, A. J.: *Anal. Chem.,* **32**:1886 (1960).
17. Ivanov-Emin, B. N.: *Zavodsk. Lab.,* **13**:161 (1947).
18. Wilson, J. M.: in A. F. Gibson, R. E. Burgess, and P. Aigrain (eds.), vol. 3, "Progress in Semiconductors," p. 27, John Wiley & Sons, Inc., New York, 1958.
19. Pfann, W. G.: "Zone Melting," 2d ed., John Wiley & Sons, Inc., New York, 1966.
20. "1967 Book of ASTM Standards," ASTM, Philadelphia, Pa., 1967.
21. Valdes, L. B.: *Proc. IRE,* **42**:420 (1952).
22. Morin, F. J., and J. P. Maita: *Phys. Rev.,* **94**:1525 (1954).
23. Runyan, W. R.: "Silicon Semiconductor Technology," McGraw-Hill Book Company, New York, 1965.
24. Veleker, T. J., and E. J. Mehalchick: *Anal. Chem.,* **33**:767 (1961).

25. Vecsernyes, L., and V. Zombori: *Chem. Anal.* (*Warsaw*), **7**:429 (1962).
26. Pchelintseva, A. F., N. A. Rakov, and L. P. Slyusareva: *Zavodsk. Lab.*, **28**:677 (1962).
27. Vecsernyes, L., and I. Hangos: *Z. Anal. Chem.*, **208**:407 (1965).
28. Kawasaki, K., and M. Higo: *Anal. Chim. Acta*, **33**:497 (1965).
29. Martynov, Yu. M., I. I. Kornblit, N. P. Smirnova, and R. V. Dzhagatspanyan: *Zavodsk. Lab.*, **27**:839 (1961).
30. Martynov, Yu. M., I. I. Kornblit, A. A. Sytnik, I. G. Syrkina, and N. P. Smirnova: USSR Patent 137918, May 10, 1961.
31. Tarasevich, N. I., and A. A. Zheleznova: *Tr. Kom. Anal. Khim.*, **15**:121 (1965).
32. Miyakawa, Y., and Y. Kamemoto: *J. Chem. Soc. Japan, Pure Chem. Sect.*, **83**:1029 (1962).
33. Vinogradova, E. N., and A. I. Kamenev: *Tr. Kom. Anal. Khim.*, **15**:175 (1965).
34. Karbainov, Yu. A., and A. G. Stromberg: *Zh. Anal. Khim.*, **19**:1341 (1964).
35. Alimarin, I. P., A. P. Golovina, I. M. Gibalo, and Yu. A. Mittsel: *Zh. Anal. Khim.*, **20**:339 (1965).
36. Martynov, Yu. M., E. A. Kreingold, and B. M. Maevskaya: *Zavodsk. Lab.*, **31**:1447 (1965).
37. Miyamoto, M.: *Japan Analyst*, **12**:233 (1963).
38. Haas, C. S., R. A. Pellin, and M. R. Everingham: *Anal. Chem.*, **36**:245 (1964).
39. Lancaster, W. A., and M. R. Everingham: *Anal. Chem.*, **36**:246 (1964).
40. Rakov, N. A.: *Zavodsk. Lab.*, **29**:437 (1963).
41. Tsekhovol'skaya, D. I., and T. A. Zavaritskaya: *Tr. Kom. Anal. Khim., Akad. Nauk SSSR*, **13**:399 (1963).
42. Rand, M. J.: *Anal. Chem.*, **35**:2126 (1963).
43. Abe, Y.: *Japan Analyst*, **9**:795 (1960).
44. Turkel'taub, N. M., S. A. Ainshtein, and B. V. Kuznetsov: *Khim. i Tekhnol. Topliv i Masel*, **1961**(12):44.
45. Turkel'taub, N. M., V. T. Shemyatenkova, S. A. Ainshtein, and S. V. Syavtsillo: *Tr. Kom. Anal. Khim., Akad. Nauk SSSR*, **13**:284 (1963).
46. Palamarchuk, N. A., S. V. Syavtsillo, N. M. Turkel'taub, and V. T. Shemyatenkova: *Tr. Kom. Anal. Khim., Akad. Nauk SSSR*, **13**:277 (1963).
47. Bersadschi, D., V. Stefan, and S. Petroianu: *Rev. Chim.* (*Bucharest*), **15**:224 (1964).
48. Burson, K. R.: Unpublished results, Texas Instruments Incorporated, 1966.
49. Morin, F. J., and J. P. Maita: *Phys. Rev.*, **96**:28 (1956).
50. Thompson, A. P.: in C. A. Hampel (ed.), "Rare Metals Handbook," p. 147, Reinhold Publishing Corporation, New York, 1954.
51. Onishi, H.: in I. M. Kolthoff and P. J. Elving (eds.), "Treatise on Analytical Chemistry," pt. II, vol. 2, p. 1, Interscience Publishers, Inc., New York, 1962.
52. Borisenok, L. A.: *Zh. Anal. Khim.*, **12**:704 (1957).
53. Minczewski, J., H. Maleszenska, and T. Steciak: *Chem. Anal.* (*Warsaw*), **7**:791 (1962).
54. Bronshtein, A. N., T. I. Sendul'skaya, and M. Ya. Shpirt: *Zavodsk. Lab.*, **26**:973 (1960).
55. Culkin, F., and J. P. Riley: *Analyst*, **83**:208 (1958).
56. Knipovich, Yu. N., and V. M. Krasikova: *Tr. Vses. Nauch. Issled, Geol. Inst.*, **117**:105 (1964).
57. Morris, D. F. C., and M. E. Chambers: *Talanta*, **5**:147 (1960).
58. Jaskolska, H., and J. Minczewski: *Chem. Anal.* (*Warsaw*), **6**:149 (1961).
59. Jaskolska, H., and J. Minczewski: *Acta Chim. Acad. Sci. Hung.*, **32**:9 (1962).
60. Bode, H., and H. Fabian: *Z. Anal. Chem.*, **170**:387 (1959).
61. Sheka, I. A., I. S. Chaus, and T. T. Mityureva: "The Chemistry of Gallium," Elsevier Publishing Company, Amsterdam, 1966.

62. Gregory, G. R. E. C., and P. G. Jeffery: *Talanta,* **9**:800 (1962).
63. Mizuno, K.: *Japan Analyst,* **14**:410 (1965).
64. Hoffman, J. I., and B. F. Scribner: *J. Res. Nat. Bur. Stand.,* **15**:205 (1935).
65. Detwiler, D. P., and W. M. Fox: *J. Metals,* **7**:205 (1955).
66. Nazarenko, V. A., G. V. Flyantikova, and N. V. Lebedeva: *Zavodsk. Lab.,* **23**:891 (1957).
67. Nazarenko, V. A., and G. V. Flyantikova: *Zavodsk. Lab.,* **24**:663 (1958).
68. Nazarenko, V. A., and G. V. Flyantikova: *Zavodsk. Lab.,* **27**:1339 (1961).
69. Knizek, M., and A. Galik: *Z. Anal. Chem.,* **213**:254 (1965).
70. Roberts, J. A., J. Winwood, and E. J. Millett: "Proceedings of the SAC Conference, Nottingham, 1965," p. 528, W. Heffer & Sons, Ltd., Cambridge, England, 1965.
71. Knizek, M., and V. Pecenkova: *Zh. Anal. Khim.,* **21**:260 (1966).
72. Biryuk, E. A.: *Zavodsk. Lab.,* **30**:651 (1964).
73. Monnier, D., and G. Prod'hom: *Anal. Chim. Acta,* **31**:101 (1964).
74. Goryushina, V. G., and E. Ya. Biryukova: *Zavodsk. Lab.,* **31**:1303 (1965).
75. Vladimirova, V. M., and G. I. Kuchmistaya: *Zavodsk. Lab.,* **30**:528 (1964).
76. Lysenko, V. I., and A. G. Kim: *Tr. Kom. Anal. Khim.,* **15**:200 (1965).
77. Steffek, M.: *Chem. Listy,* **58**:957 (1964).
78. Pohl, F. A.: *J. Polarogr. Soc.,* **4**:8 (1958).
79. Sinyakova, S. I., N. A. Rudnev, Yu-Ch'ih Shen, and R. Dzhumaev: *Zh. Anal. Khim.,* **16**:32 (1961).
80. Provaznik, J., and J. Mojzis: *Chem. Listy,* **55**:1299 (1961).
81. Miklos, I.: *Magy. Kem. Folyoirdt,* **69**:66 (1963).
82. Kaplan, B. Ya., I. A. Sorokovskaya, and O. A. Shiryaeva: *Zavodsk. Lab.,* **30**:659 (1964).
83. Kaplan, B. Ya., and O. A. Shiryaeva: *Zavodsk. Lab.,* **31**:39 (1965).
84. Kaplan, B. Ya., and I. A. Sorokovskaya: *Zavodsk. Lab.,* **30**:783 (1964).
85. Hoste, J., and H. van den Berghe: *Mikrochim. Acta,* **1956**:797.
86. Lerch, P., and L. Kreienbuhl: *Chimia,* **15**:519 (1961).
87. Lerch, P., and C. Vuilleumier: *Chimia,* **16**:414 (1962).
88. Krivanek, M., F. Kukula, and J. Shunecko: *Talanta,* **12**:721 (1965).
89. Alimarin, I. P., Yu. V. Yakovlev, M. N. Shulepnikov, and G. P. Peregozhin: *Radioakt. Izotopy i Yadernye Izlucheniya v Nar. Khoz. SSSR,* **1**:293 (1961).
90. Nagy, L. G., J. Bodnar, Z. Demjen, J. Sandor, and T. Szekrenyesy: *Period. Polytech.,* **7**:147 (1963).
91. Wielson, C. M., D. Hazelby, M. L. Aspinal, and J. A. James: *CVD Res. Rep. RP* 4/3, *R.p.* G1374, Associated Electrical Industries (Rugby) Ltd., 1962.
92. Vasil'eva, N. M., N. F. Litvinova, and Z. M. Turovtseva: *Zh. Anal. Khim.,* **18**:250 (1963).
93. Kostromin, A. I., and L. A. Anisimova: *Uch. Zap. Kazansk. Gos. Univ.,* **124**:179 (1965).
94. Owens, E. B.: *Appl. Spectry.,* **13**:105 (1959).
95. Oldfield, J. H., and E. P. Bridge: *Analyst,* **86**:267 (1961).
96. Neeb, K. H.: *Z. Anal. Chem.,* **194**:255 (1963).
97. Massengale, J. F., D. Andrychuk, and C. E. Jones: Symposium on Impurities in III-V Elements and Compounds, Battelle Memorial Institute, June 2, 1960.
98. Kane, P. F.: in F. J. Welcher (ed.), "Standard Methods of Chemical Analysis," vol. III B., p. 1764, D. Van Nostrand Company, Inc., Princeton, N.J., 1966.
99. Wolstenholme, W. A.: *Appl. Spectry.,* **17**:51 (1963).
100. Nalbantoglu, M.: in "Advances in Mass Spectrometry," vol. 3, in press.
101. Nalbantoglu, M.: *Chim. Anal. (Paris),* **48**:85, 148 (1966).
102. Fitzner, E.: *Chem. Rundsch.,* **18**:389 (1965).

103. Mills, J. R., R. C. Bell, and R. A. King: in C. A. Hampel (ed.), "Rare Metals Handbook," p. 191, Reinhold Publishing Corporation, New York, 1954.
104. Patrovsky, V.: *Chem. Listy,* **48**:1047 (1954).
105. Busev, A. I.: "The Analytical Chemistry of Indium," The Macmillan Company, New York, 1962.
106. Morris, D. F. C., and F. M. Brewer: *Anal. Chim. Acta,* **14**:183 (1956).
107. Raikhbaum, Ya. D., and E. S. Kostynkova: *Zavodsk. Lab.,* **25**:961 (1959).
108. Ravlenko, L. I., and Z. M. Davydova: *Zh. Anal. Khim.,* **17**:199 (1962).
109. Steciak, T.: *Chem. Anal. (Warsaw),* **7**:503 (1962).
110. Brooks, R. R.: *Anal. Chim. Acta,* **24**:456 (1961).
111. Blyum, I. A., and T. K. Dushina: *Zavodsk. Lab.,* **25**:137 (1959).
112. Blyum, I. A., I. T. Solov'yan, and G. N. Shebalkova: *Zavodsk. Lab.,* **27**:950 (1961).
113. Knipovich, Yu. N., V. M. Krasikova, and V. G. Zherekhov: *Tr. Vses. Issled. Geol. Inst.,* **117**:63 (1964).
114. Rozbianskaya, A. A.: *Tr. Inst. Mineral., Geokhim. i Kristallokhim. Redkikh Elementov, Akad. Nauk SSSR,* **1961** (6):138.
115. Levin, I. S., and T. G. Azarenko: *Zh. Anal. Khim.,* **20**:452 (1965).
116. Matsumae, T.: *Japan Analyst,* **8**:167 (1959).
117. Minczewski, J., U. Stolarczyk, and Z. Marczenko: *Chem. Anal. (Warsaw),* **6**:51 (1961).
118. Talipov, Sh. T., Kh. S. Abdullaeva, and G. P. Gor'kovaya: *Uzblksk. Khim. Zh.,* **1962** (5):16.
119. Kuznetsova, L. N., and B. L. Serebryanyi: *Izv. Vyssh. Ucheb. Zavedenii, Tsvetn. Met.,* **1962** (3):107.
120. Smales, A. A., J. van R. Smit, and H. Irving: *Analyst,* **82**:539 (1957).
121. Irving, H., J. van R. Smit, and L. Salmon: *Analyst,* **82**:549 (1957).
122. Abdullaev, A. A., E. M. Lobanov, A. V. Novikov, M. M. Romanov, and A. A. Khaidarov: *Zh. Anal. Khim.,* **15**:701 (1960).
123. Pierce, T. B., and P. F. Peck: *Analyst,* **86**:580 (1961).
124. Mapper, D., and J. R. Fryer: *Analyst,* **87**:297 (1962).
125. Okada, M., and Y. Kamemoto: *Nature,* **197**:279 (1963).
126. Tomov, T., C. Popov, G. Stefanov, and J. Tolgyessy: *Chem. Zvesti,* **18**:705 (1964).
127. Weiss, D.: *Hutnicke Listy,* **13**:641 (1958).
128. Kvacek, M., and P. Kuhn: *Chem. Listy,* **55**:1296 (1961).
129. Kuhn, P., and M. Kvacek: *Chem. Listy,* **57**:62 (1963).
130. Kvacek, M., and P. Kuhn: *Chem. Listy,* **58**:584 (1964).
131. Kaplan, B. Ya.: *Zavodsk. Lab.,* **25**:1168 (1959).
132. Patrovsky, V.: *Chem. Listy,* **47**:1338 (1953).
133. Tsyvina, B. S., and V. M. Vladimirova: *Zavodsk. Lab.,* **24**:278 (1958).
134. Weimer, F. S., G. T. Wever, and R. J. Lapee: in C. H. Mathewson (ed.), "Zinc: The Science and Technology of the Metal, Its Alloys and Compounds," p. 174, Reinhold Publishing Corporation, New York, 1959.
135. Mills, J. R., B. G. Hunt, and G. H. Turner: *J. Electrochem. Soc.,* **100**:136 (1953).
136. Zettler, H.: *Z. Erzberg. Metallhuettenw.,* **16**:350 (1963).
137. Kalina, Yu. P., and V. V. Baburina: *Zavodsk. Lab.,* **31**:946 (1965).
138. Sayun, M. G., and S. P. Tikhanina: *Zavodsk. Lab.,* **28**:544 (1962).
139. Naimark, L. E., and I. G. Yudelevich: *Izv. Akad. Nauk Kaz. SSR, Ser. Met. Obogashch. i Ogneuporov,* **1959** (1):90.
140. Yudelevich, I. G., I. R. Shelpakova, T. I. Sosnovskaya, and L. S. Bortnik: *Zavodsk. Lab.,* **25**:959 (1959).

141. Bortnik, L. S.: *Sb. Nauch Tr. Vses. Nauch Issled. Gorno-Met. Inst. Tsvetn. Metal.*, **1959**:196.
142. Yudelevich, I. G., and I. R. Shelpakova: *Zh. Anal. Khim.*, **17**:174 (1962).
143. Ginzberg, L. B., and E. P. Shkrobot: *Zavodsk. Lab.*, **21**:1289 (1955).
144. Gurev, S. D., L. B. Ginzberg, and A. P. Shibarenkova: *Sb. Nauchn. Tr., Gos. Nauchn.-Issled. Inst. Tsvetn. Metal.*, **1955**:387.
145. Collins, T. A., and J. H. Kanzelmeyer: *Anal. Chem.*, **22**:245 (1961).
146. Stolarczykowa, U.: *Chem. Anal. (Warsaw)*, **9**:161 (1964).
147. Orlovskii, S. T., and P. O. Kish: *Ukr. Khim. Zh.*, **29**:209 (1963).
148. Kusaka, Y.: *J. Chem. Soc. Japan, Pure Chem. Sect.*, **80**:1419 (1959).
149. Kosaric, N., and G. Leliaert: *Nature*, **191**:703 (1961).
150. Gibbons, D., and D. Lawson: "Proceedings of the 1965 International Conference on Modern Trends in Activation Analysis," p. 172, Texas A&M University Press, 1966.
151. Moeller, T., and B. S. Hopkins: *Trans. Electrochem. Soc.*, **93**:84 (1948).
152. Treindl, L.: *Chem. Listy*, **50**:534 (1956).
153. Nizhnik, A. T., and I. S. Chaus: *Zh. Anal. Khim.*, **14**:37 (1959).
154. Pats, R. G., and S. B. Tsfasman: *Zavodsk. Lab.*, **27**:266 (1961).
155. Hofer, R. J., R. Z. Bachman, and C. V. Banks: *Anal. Chim. Acta*, **29**:61 (1963).
156. Kasagi, M., and C. V. Banks: *Anal. Chim. Acta*, **30**:248 (1964).
157. Kovalenko, P. N., and L. N. Musaelyants: *Ukr. Khim. Zh.*, **30**:753 (1964).
158. Nazarenko, V. A., and G. V. Flyantikova: *Zavodsk. Lab.*, **24**:801 (1959).
159. Nazarenko, V. A., N. A. Fuga, G. V. Flyantikova, and K. A. Esterlis: *Zavodsk. Lab.*, **26**:131 (1960).
160. Flyantikova, G. V.: *Zavodsk. Iab.*, **32**:529 (1966).
161. Nishimura, K., and T. Imai: *Japan Analyst*, **13**:423 (1964).
162. Nishimura, K., and T. Imai: *Japan Analyst*, **13**:518 (1964).
163. Nishimura, K., and T. Imai: *Japan Analyst*, **13**:713 (1964).
164. Busev, A. I., and N. P. Bozenkova: *Zavodsk. Lab.*, **27**:13 (1961).
165. Peshkova, P. M., V. M. Bochkova, and L. I. Lazareva: *Zh. Anal. Khim.*, **15**:610 (1960).
166. Peshkova, P. M., V. M. Bochkova, and E. K. Astakhova: *Zh. Anal. Khim.*, **16**:596 (1961).
167. Vydra, F., and K. Stulik: *Chemist-Analyst*, **54**:77 (1965).
168. Dolezal, J.: *Chem. Listy*, **51**:1058 (1957).
169. Dolezal, J., V. Petrus, and J. Zyka: *J. Electroanal. Chem.*, **3**:274 (1962).
170. Pohl, F. A., and W. Bonsels: *Z. Anal. Chem.*, **161**:108 (1958).
171. Towndrow, E. G., R. Hutchinson, and H. W. Webb: *Analyst*, **85**:769 (1960).
172. Kopanica, M., and R. Pribil: *Collect. Czech. Chem. Commun.*, **26**:398 (1961).
173. Conradi, G., and M. Kopanica: *Chemist-Analyst*, **53**:4 (1964).
174. Musil, J., and M. Kopanica: *Chemist-Analyst*, **55**:9 (1966).
175. Bikbulatova, R. V., and S. I. Sinyakova: *Zh. Anal. Khim.*, **19**:1434 (1964).
176. Shirai, H.: *Japan Analyst*, **9**:206 (1960).
177. Ishibashi, M., T. Fujinaga, and A. Saito: *Collect. Czech. Chem. Commun.*, **25**:3387 (1960).
178. Kaplan, B. Ya., I. A. Sorokovskaya, and G. A. Smirnova: *Zavodsk. Lab.*, **28**:1188 (1962).
179. Kaplan, B. Ya., and O. A. Shiryaeva: *Zavodsk. Lab.*, **31**:658 (1965).
180. Sinyakova, S. I., A. G. Dudareva, I. V. Markova, and I. N. Talalaeva: *Zh. Anal. Khim.*, **18**:377 (1963).
181. Stepanova, O. S., M. S. Zhakarova, and L. F. Trushina: *Zavodsk. Lab.*, **30**:1180 (1964).
182. Mesyats, N. A., B. F. Nazarov, M. S. Zakharov, and A. G. Stromberg: *Zh. Anal. Khim.*, **19**:959 (1964).

183. Mesyats, N. A., A. A. Kaplin, M. S. Zakharov, and G. K. Tychkina: *Izv. Tomsk. Politekh. Inst.*, **128**:42 (1964).
184. Hyman, H. M.: *Appl. Spectry.*, **12**:95 (1958).
185. Hyman, H. M.: *Appl. Spectry.*, **14**:125 (1960).
186. Porkhunova, N. A., L. K. Larina, and N. S. Bakaldina: *Sb. Nauch. Tr. Vses. Nauch. Issled. Gorno-Met. Inst. Tsvetn. Metal.*, **1962**:389.
187. Caldararu, H.: *Rev. Chim. (Bucharest)*, **14**:39 (1963).
188. Aitchison, L.: "A History of Metals," vol. 2, p. 477, Interscience Publishers, Inc., New York, 1960.
189. Gonser, B. W., and E. M. Smith: Antimony, in R. E. Kirk and D. F. Othmer (eds.), "Encyclopedia of Chemical Technology," vol. 2, p. 50, Interscience Publishers, Inc., New York, 1948.
190. Haberecht, R. R.: in R. K. Willardson and H. L. Goering (eds.), "Compound Semiconductors," vol. 1, "Preparation of III-V Compounds," p. 92, Reinhold Publishing Corporation, New York, 1962.
191. Tanenbaum, M., A. J. Goss, and W. G. Pfann: *J. Metals*, **6**:762 (1954).
192. Maekawa, S., Y. Yoneyama, and E. Fujimori: *Japan Analyst*, **11**:497 (1962).
193. Kowalczyk, M.: *Chem. Anal. (Warsaw)*, **9**:331 (1964).
194. Klein, P., and V. Skrivanek: *Chem. Prumysl*, **13**:250 (1963).
195. Provaznik, J., and M. Knizek: *Chem. Listy*, **55**:79 (1961).
196. Ishihara, Y., and M. Koga: *Japan Analyst*, **11**:246 (1962).
197. Steffek, M.: *Chem. Listy*, **56**:951 (1962).
198. Lipshits, B. M., G. K. Smirnova, and F. S. Kulikov: *Zavodsk. Lab.*, **27**:1199 (1961).
199. Zaglodina, T. V.: *Sb. Nauch. Tr., Gos. Nauch. Issled. Inst. Tsvetn. Metal.*, **1962**:727.
200. Nazarenko, V. A., M. B. Shustova, R. V. Ravitskaya, and M. P. Nikonova: *Zavodsk. Lab.*, **28**:537 (1962).
201. Haberli, E.: *Z. Anal. Chem.*, **160**:15 (1958).
202. Aref'eva, T. V., and R. G. Pats: *Sb. Nauch. Tr., Gos. Nauch. Issled. Inst. Tsvetn. Metal.*, **1955**:353.
203. Sulcek, Z., J. Dolezal, J. Michal, and V. Sychra: *Talanta*, **10**:3 (1963).
204. Lysenko, V. I.: *Tr. Kom. Anal. Khim.*, **15**:195 (1965).
205. Kostromin, A. I., and M. Akhmadeev: *Uch. Zap. Kazansk. Gos. Univ.*, **124**:166 (1965).
206. Kulak, A. I.: *Zh. Anal. Khim.*, **1957**:727.
207. Rakovskii, E. E., L. A. Smakhtin, and Yu. V. Yakovlev: *Zavodsk. Lab.*, **26**:1199 (1960).
208. Simkova, M., F. Kukula, and R. Stejskal: *Jaderna Energie*, **9**:165 (1963).
209. Simkova, M., F. Kukula, and R. Stejskal: *Collect. Czech. Chem. Commun.*, **30**:3193 (1965).
210. Artyukhin, P. E., E. N. Gilbert, and V. A. Pronin: *Zh. Anal. Khim.*, **21**:504 (1966).
211. Yudelevich, I. G., F. P. Polatbekov, and V. G. Kovaleva: *Zavodsk. Lab.*, **25**:305 (1959).
212. Knipe, G. F. G.: *Spectrochim. Acta*, **15**:49 (1959).
213. Jones, C. E.: Unpublished results, Texas Instruments Incorporated, 1960.
214. Kowalczyk, M.: *Chemia Anal.*, **10**:395 (1966).
215. Neeb, K. H.: *Z. Anal. Chem.*, **200**:278 (1964).
216. Hannay, N. B., and A. J. Ahearn: *Anal. Chem.*, **26**:1056 (1954).
217. Roush, G. A.: Arsenic, in R. E. Kirk and D. F. Othmer (eds.), "Encyclopedia of Chemical Technology," vol. 2, p. 113, Interscience Publishers, Inc., New York, 1947.
218. Blum, S. E.: in R. K. Willardson and H. L. Goering (eds.), "Compound Semiconductors," vol. 1, "Preparation of III-V Compounds," p. 85, Reinhold Publishing Corporation, New York, 1962.

219. Beau, R.: in M. S. Brooks and J. K. Kennedy (eds.), "Ultrapurification of Semiconductor Materials," p. 173, The Macmillan Company, New York, 1962.
220. Brau, M. J.: 19th Southwest Regional Meeting, ACS, Houston, Texas, 1963.
221. Kristaleva, L. B.: *Zavodsk. Lab.*, **25**:1294 (1959).
222. Kristaleva, L. B., and P. V. Kristalev: *Sb. Nauch. Tr. Permsk. Politekh. Inst.*, **1963** (14):68.
223. Soldatova, L. A., and L. B. Kristaleva: *Tr. Tomsk. Univ.*, **157**:279 (1963).
224. Goryushina, V. G., and N. V. Esenina: *Zh. Anal. Khim.*, **21**:239 (1966).
225. Vladimirova, V. M., N. K. Davidovich, G. I. Kuchmistaya, and L. S. Razumova: *Zavodsk. Lab.*, **29**:1419 (1963).
226. Rozanova, L. N., and G. A. Kamaev: *Zh. Prikl. Khim.*, **32**:2574 (1960).
227. Kataev, G. A., and E. A. Zakharova: *Zavodsk. Lab.*, **29**:524 (1963).
228. Kataev, G. A., E. A. Zakharova, and L. I. Oleinik: *Tr. Tomsk. Univ.*, **157**:261 (1963).
229. Niese, S.: *Kernenergie*, **8**:499 (1965).
230. Nizet, G., J. Fouarge, and G. Duyckaerts: *Anal. Chim. Acta*, **35**:370 (1966).
231. Mack, D. L.: *Analyst*, **88**:481 (1963).
232. Kataev, G. A., and Z. I. Otmakhova: *Zavodsk. Lab.*, **30**:40 (1964).
233. Brown, R., R. D. Craig, and J. D. Waldron: in R. K. Willardson and H. L. Goering (eds.), "Compound Semiconductors," vol. 1, "Preparation of III-V Compounds," p. 106, Reinhold Publishing Corporation, New York, 1962.

4

Materials Characterization in Single-crystal Growth

4-1. INTRODUCTION

The concept of materials characterization begins to play a more important role when the analyst is dealing with single crystals. The chemical purity or perfection of the single-crystal semiconductor is only one consideration. A complete evaluation or characterization of a material includes structural properties such as crystallographic defects and physical properties such as mobility and lifetime. Electrical properties are especially important because these are the final criteria by which the crystal will be judged. Dopant solubility, distribution, and behavior during and after crystal growth are all important characteristics of the material. The analyst will find himself called upon to give analytical services both during the single-crystal growth and in the analysis of bulk materials, gases, and materials used in the single-crystal growth, such as graphite susceptors, boats, and furnace tubes and parts. It is obviously outside the scope of this book to discuss techniques for the analysis of those materials not directly used in the single crystal itself. It is apparent that these materials are of exceptionally high purity and will in themselves require sophisticated techniques for their analyses.

Crystal Quality. The single-crystal semiconductor must be of both high crystalline perfection and high chemical purity. These requirements are vital to the entire semiconductor industry. The finished crystal may be cut and used directly for the fabrication of discrete devices and integrated circuits. Imperfections of any type would be extremely detrimental to the overall quality of the finished devices. These finished crystals could also be cut, polished, and used directly for substrates for the growth of high-perfection epitaxial layers (Chap. 8). Here again any physical imperfections in the single-crystal substrate would be propagated from the bulk substrate directly into the epitaxial layer.

Since the electrical behavior of single crystals is so highly sensitive to the quality of the crystal structure and the chemical purity, the analytical chemist can expect stringent demands on his services. This is particularly true in the study of the intrinsic properties of semiconductors, where the crystal must be as nearly perfect, in all respects, as possible. There is good reason to believe that new devices based on excitons or electron-hole pairs are possible only in these extremely pure intrinsic single-crystal materials. This area of application of semiconductors is only now in its infancy but will grow rapidly, and the analyst must be ready.

4-2. CRYSTAL-GROWTH TECHNIQUES

The techniques for obtaining high-purity single crystals with high physical perfection have received a large amount of attention during the growth of the semiconductor industry. Since virtually every semiconductor device produced uses a single crystal, it is obvious that it is necessary to have rigorous analytical control and examination of both the crystal-growth process and the finished crystal itself.

Although there are many methods of growing crystals, they all fundamentally involve crystallization of the semiconductor in a very pure form and in a crystallographically perfect arrangement. This crystallization can take place from the melt or from the vapor. On examination of the various methods now used, it is generally found that each semiconductor (e.g., silicon, gallium arsenide, silicon carbide) is grown by a slightly different method to optimize crystalline perfection and ease of growth. It should be obvious that the compound semiconductors such as gallium arsenide, indium antimonide, and indium arsenide must be synthesized before single-crystal growth, while the elemental semiconductors can be grown directly.

4-3. ELEMENTAL SEMICONDUCTORS

Silicon and germanium are by far the most important semiconductors in use today. There is little doubt that this situation will stand for a very long time, and the analytical chemist can expect to be asked to analyze even higher-purity crystals with better crystalline perfection in the coming years. Since these two materials are now in volume production, their crystal-growing techniques have reached a high degree of perfection.

Vertical-pull Method. This crystal-growth technique is often referred to as the Czochralski method, even though Czochralski[1,†] originally developed the method to study the speed of crystallization of various metals. Teal and Little[2] modified this technique and applied it to the growth of silicon and germanium single crystals (see Sec. 1-2).

In general, this vertical-pull technique starts by preparing a melt of the semiconductor in a quartz or graphite crucible. A small, oriented single crystal of the semiconductor is introduced into the top of the melt, and then the seed crystal is rotated while being slowly withdrawn from the melt. The heat input to the melt and the rate of pull of the crystal from the melt are adjusted to yield the desired crystal shape and size. A schematic of a typical vertical-pull apparatus is shown in Fig. 4-1. As can be seen, the heating is usually carried out by a large radiofrequency generator, and the molten material is protected by maintaining a positive pressure of very-high-purity helium or argon over the melt. Figure 4-2 shows some typical semiconductor crystals obtained by this vertical-pull technique.

†Superscript numbers indicate References listed at the end of the chapter.

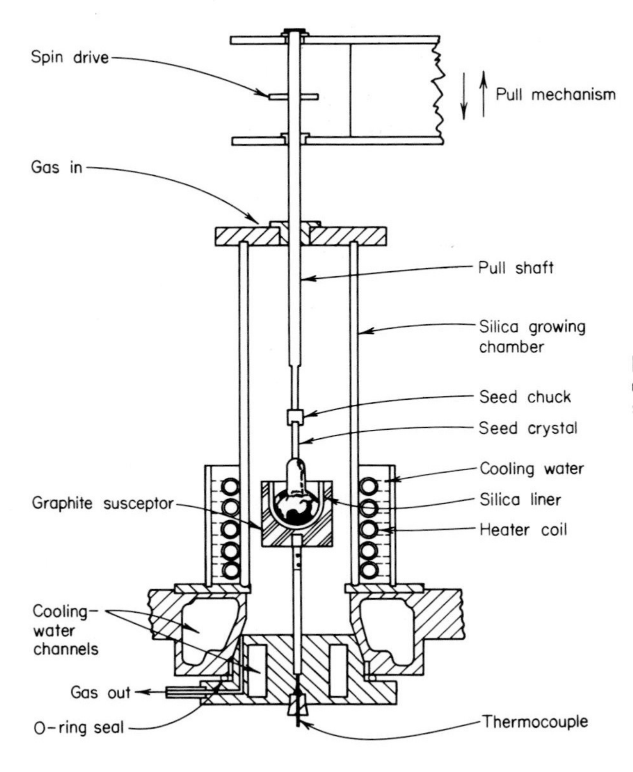

Fig. 4-1. Vertical-pull apparatus used in the growth of single-crystal silicon. (*From Runyan.*[3])

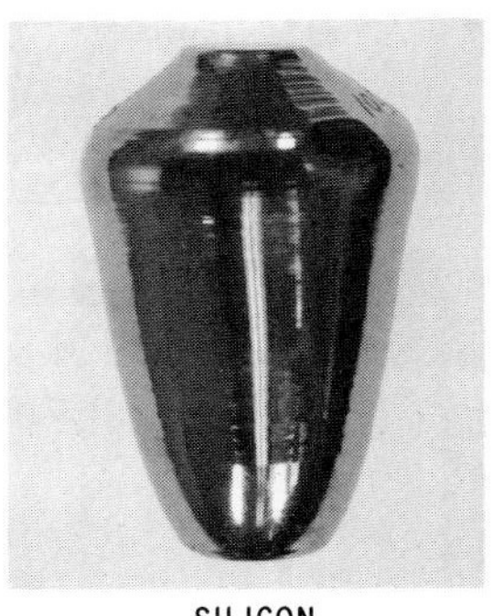

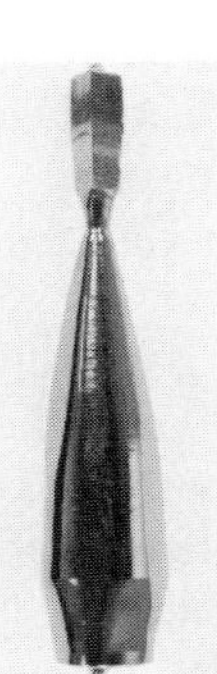

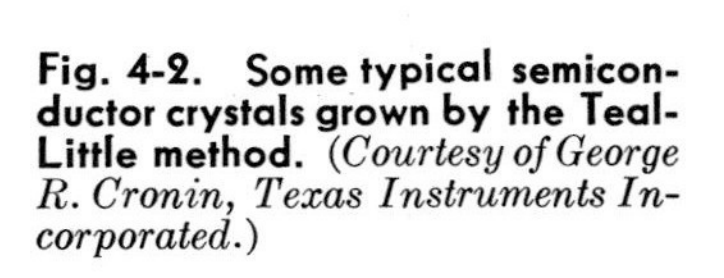
Fig. 4-2. Some typical semiconductor crystals grown by the Teal-Little method. (*Courtesy of George R. Cronin, Texas Instruments Incorporated.*)

The growth of single crystals by vertically pulling the crystal from the melt has several advantages over other techniques with regard to impurity distribution in the crystal. The continuous stirring by the rotating crystal maintains a fairly uniform distribution of impurity in the melt. Simultaneously the large temperature differential at the solid-liquid interface minimizes the movement or diffusion of impurities from the melt into the single crystal. This was the original method[4] used for the first commercial production of silicon devices.

The vertical puller does not lend itself to the growth of long, uniformly doped single crystals because the dopant or impurity concentration in the melt is constantly changing. As a result the crystal will show an impurity-concentration difference that will be graded from the top to the bottom of the crystal. The concentration of impurity over the length of the pulled crystal is described by the equation

$$C_x = kC_0(1 - x)^{(k-1)} \tag{4-1}$$

where C_0 = initial concentration in melt
C_x = concentration at any point x, where x is the fraction of the original volume which has solidified
k = effective segregation coefficient

The effective segregation coefficient is slightly different for each apparatus and for each set of pull conditions; and to overcome the difficulty of producing large, uniformly doped single crystals, the horizontal-zone-refining technique was developed.

Horizontal Zone Refining. This technique is sometimes referred to as zone leveling because of its ability to yield uniformly doped single crystals. In this method, shown schematically in Fig. 4-3, only a narrow band of the semiconductor is melted. This molten zone is then slowly moved down the length of the boat. If a single crystal seed were placed at the front end of the boat and its tip made part of the initial melted zone, then the entire length of the semiconductor in the boat would be grown single crystal. Similarly, if a known amount of dopant or impurity had been introduced into the initial melted zone, it would have been distributed through the length of the crystal according to Eq. (4-2), developed by Pfann.[5]

$$C = kC_0\, e^{-(kx/l)} \tag{4-2}$$

where C = concentration in solid
C_0 = initial concentration in solid
k = effective distribution coefficient
l = length of molten zone
x = distance zoned

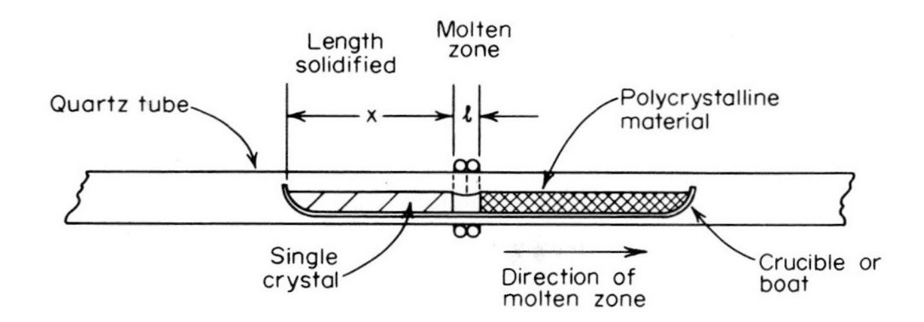

Fig. 4-3. Schematic illustrating horizontal zone refining

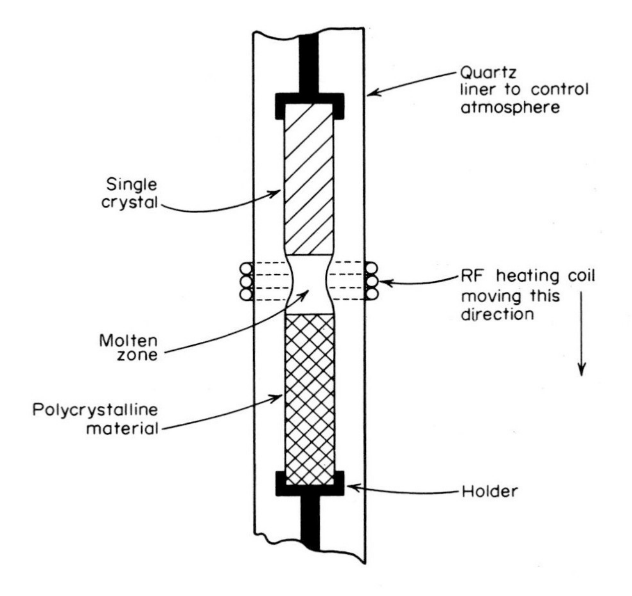

Fig. 4-4. Schematic illustrating vertical zone refining.

The horizontal-zone method has become the major technique used in the production of single-crystal germanium. One of the big advantages of this technique is the ease with which large crystals of uniform cross section can be produced. While there is a large risk of contamination from the large crucible or boat area used, it is relatively easy to dope the crystal and obtain a large amount of uniformly doped material from any one run.

Vertical Zone Refining. In cases in which crucible contamination is a problem, or in which highly reactive materials are being grown, it is possible to "float zone" the ingot. This is shown schematically in Fig. 4-4, where a small section of the ingot is melted and this melted zone slowly moved down the bar, thus effecting zone refining. The small melted zone is held in place by surface tension of the liquid. The impurities move either into the molten zone or the back, freezing section, depending on whether the segregation coefficient is greater than or less than 1. The distribution of impurities is also described by the same equation (4-2) as that given for horizontal zone leveling.

If the initial melt is seeded with a single crystal, this vertical-float-zone method will grow single crystals. This technique will yield high-quality materials, and large silicon crystals of 2 to 2.5 in. diameter are now being produced for device fabrication. Mechanical vibration or change in zone-leveling conditions can result in loss of the molten zone.

4-4. COMPOUND SEMICONDUCTORS

The growth of single-crystal compound semiconductors must of necessity include, or be preceded by, a synthesis step. Since every effort has been made to purify the elements prior to this synthesis, this extra handling step before single-crystal growth presents a potential source of contamination. For this reason, the preferred growth technique is in situ synthesis and single-crystal growth at the same time. Techniques used for the preparation of single-crystal III-V compounds (for example, AlP, GaAs, InSb) have been described by various authors,[6] and basically the growth techniques are similar to those used for silicon and germanium.

Vertical Pull. The Teal-Little or Czochralski technique is generally directly applicable to the growth of the antimonides AlSb, GaSb, and InSb, because they have low melting points. The compounds can be synthesized by weighing stoichiometric amounts of the elements into the crucible, melting, and then introducing a seed and pulling a crystal by using the germanium single-crystal-growth techniques.

The III-V compound arsenides and phosphides present other problems because of their high vapor pressure at the melting point. It is necessary to grow the crystal in a sealed chamber with the vapor pressure adjusted to the equilibrium vapor pressure (InAs 0.3 atm, GaAs 0.9 atm, InP 60 atm, and GaP 50 atm). The problems of sealing these chambers and providing means of heating all the walls to control the vapor pressure and prevent condensation make the crystal-growth apparatus very complex. However, good single crystals of InAs and GaAs are now routinely grown in production areas.

Horizontal Crystal Growth. The only direct application of the horizontal-zone-leveling technique to the III-V compounds, as developed by Pfann,[7] has been to the antimonides. As discussed by Richards,[8] a modified technique has been applied to gallium arsenide but has not been widely used. The more accepted horizontal-growth technique used for gallium arsenide is the Bridgman technique.[9] In this technique, illustrated in Fig. 4-5, the boat or the furnace is moved so that a temperature gradient exists across the crystal. This allows a gradual solidification of the molten crystal. Some of the best melt-grown gallium arsenide ever produced has been from the horizontal Bridgman technique (high 10^{15} to low 10^{16} carriers/cm^3).

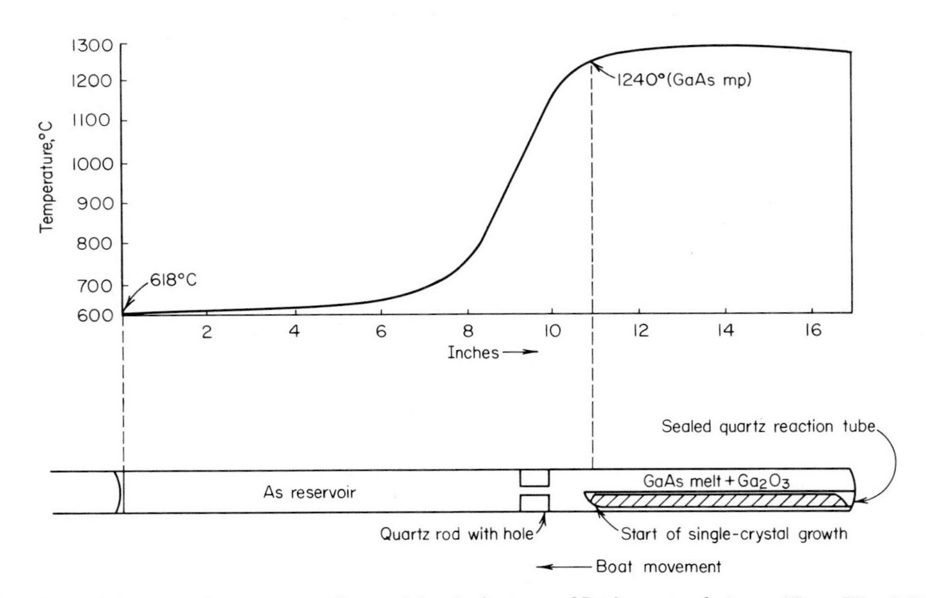

Fig. 4-5. Schematic showing crystal growth by the horizontal Bridgman technique. (*From Woodall.*[10])

Vapor Transport. This method of crystal growth is almost exclusively restricted to epitaxial-layer growth (Chap. 8). However, certain of the III-V compounds with extremely high vapor pressures, such as InP or GaP, lend themselves more to this type of crystal growth than to the normal Czochralski or horizontal Bridgman methods. Basically, the vapor-transport method makes use of the high volatility of the halides, the fact that chlorine is a neutral impurity, and the fact that lower chlorides and iodides of indium and gallium are not particularly stable compounds.

The general equation describing the process is

$$5M^{\mathrm{III}}X_3 + 2N^{\mathrm{V}} \rightleftharpoons 3M^{\mathrm{III}}X_5 + 2M^{\mathrm{III}}N^{\mathrm{V}}$$

where X = halide
M = In or Ga
N = P or As

The reactants are sealed in a quartz tube and placed in a furnace with a temperature gradient. The compound is deposited, in polycrystalline form, at the coolest spot of the reaction tube.

4-5. DOPANT OR IMPURITY BEHAVIOR

The materials scientist attempts to alter the electrical characteristics of the single crystal during growth to meet the needs of the user. The crystal may be intended for substrates for epitaxial deposition for use in the fabrication of integrated circuits. In this case the crystal would be required to be heavily doped (in the order of 10^{18} carriers/cm^3) and free of physical imperfections that would be propagated into the epitaxial layer. Another crystal might be intended directly for device fabrication,

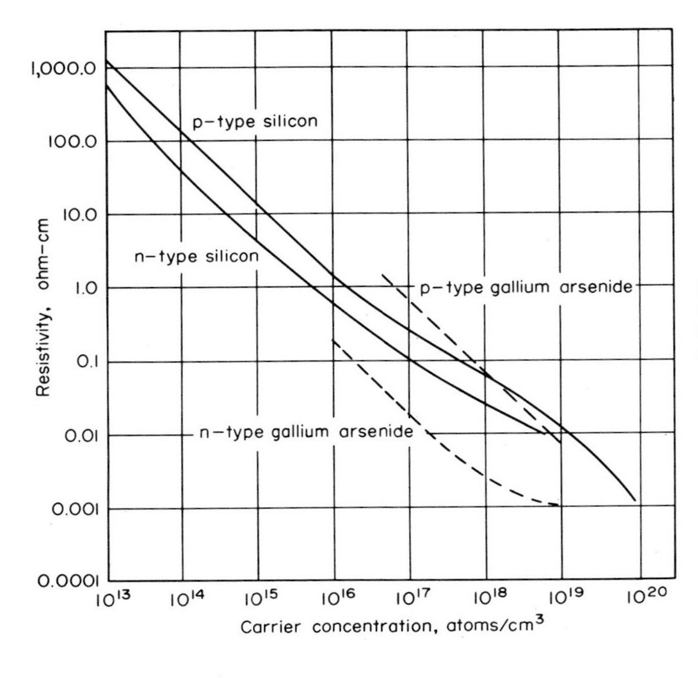

Fig. 4-6. Resistivity as a function of carrier concentration for silicon and gallium arsenide.

and the doping level would have to be adjusted to the 1 to 10 ohm-cm level for silicon and germanium (10^{14} to 10^{16} carriers/cm^3). If high-resistivity material were required, then a "trap" such as gold in silicon (in the order of 10^{13} atoms/cm^3) or chromium in gallium arsenide (10^{15} atoms/cm^3) would be introduced into the crystal during the growth.

Figure 4-6 shows a graph relating resistivity to carrier concentration that the materials scientist would use to correlate resistivity and doping concentration for a particular crystal. For any type of dopant, say p-type silicon, the choice of boron, aluminum, gallium, or indium would be up to the device engineer or the person requesting the crystal. Each of these doping elements has a different energy level and although all are acceptors, each dopes in a slightly different manner. For example, in silicon, the indium energy level is 0.16 ev above the valence band, while the boron level is only 0.045 ev above the band edge. It follows that the probability of an electron's moving up from the valence band to the boron level requires less energy than filling the indium level. As a result boron would, under most circumstances, have a higher degree of electrical activity. Similarly, for n-type dopants which lie below the conduction band, it requires more energy to ionize the electron from a deep level into the conduction band than electrons from a shallow level. When these deeper-lying impurities are involved, a discrepancy can be expected between electrical activity and chemical analysis. Surprisingly enough, this discrepancy becomes largest at the higher doping levels because of the interaction between the Fermi level and the doping concentration. The Fermi level is pulled toward the band edge with increasing dopant concentration. As the Fermi level approaches the band edge, the probability that the dopant states will be ionized decreases and the percentage of impurity atoms electrically active decreases. Figure 4-7 shows a computer-generated

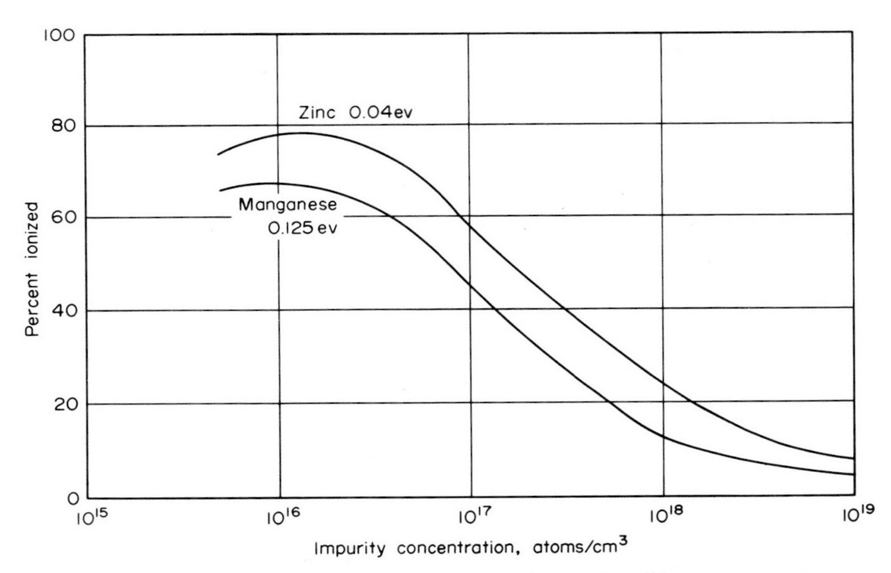

Fig. 4-7. Effect of ionization level on the amount of electrically active impurity for zinc and manganese in gallium arsenide. A base donor concentration of 1×10^{15} atoms/cm^3 was assumed.

curve showing percent ionized (amount that would be observed electrically) plotted versus actual concentration. In the case of gallium arsenide this discrepancy is enhanced by the fact that n-type gallium arsenide becomes degenerate when the doping level reaches 4.9×10^{17} donors/cm^3. (A degenerate semiconductor no longer obeys classical Boltzmann statistics and occurs when the Fermi level lies in the conduction band.)

The analytical chemist must be always cognizant of this expected variation or discrepancy between electrical and chemical analysis of a semiconductor material, particularly at higher concentrations.

4-6. DISTRIBUTION OF IMPURITIES IN GROWN CRYSTALS

During both routine production and research studies of single-crystal growth, the analyst is frequently called upon to determine the distribution of an impurity or dopant. If the analyst is working closely with the materials scientist, it is imperative that he have at least a working knowledge of impurity distributions. The materials scientist will rely heavily on the judgment of the analyst with regard to statistical analysis of the chemists' analyses and how this variation compares with that observed in analyses of samples from top, middle, and bottom of the crystal. Since the analyses will probably be in the 0.001- to 10-ppm range, the analyst must be able to recognize anomalous behavior of the dopant even at these concentration levels.

4-7. DISTRIBUTION COEFFICIENTS†

In any solid-liquid system, where slow freezing is occurring, the impurity distribution is described by K_0, the *equilibrium* distribution coefficient. In semiconductor crystal growth every effort is made to prevent slow cooling and, in effect, set up nonequilibrium conditions. As a result, if $K_0 < 1$, the solid rejects the impurity into the liquid more rapidly than the impurity can diffuse into the solid. The solid interface then advances an impurity-enriched layer which builds up ahead of the interface (behind if $K_0 > 1$). Because of this nonequilibrium buildup at the solid-liquid interface, the amount of impurity freezing out is controlled at the interface and not by the main liquidus body. Therefore, the distribution coefficient is controlled by the concentration in the solid C_s and in the liquid C_l and is now an effective segregation coefficient (k) or a nonequilibrium segregation coefficient. It is this k which is used in all semiconductor crystal-growth studies.

It should be obvious from the above discussion that, since k is controlled by the interface, it is very susceptible to changes in crystal-growth conditions. The value of k will be affected by the speed of crystal growth, crystal orientation, operator, and apparatus conditions. In any interpretation of analytical data, all these factors must be taken into account. Variations in the absolute value of the effective distribution coefficient between runs in a study of impurity behavior during crystal growth may well be explained by variations in growth parameters.

†For complete derivations and description of distribution coefficients the reader should consult Pfann.[7]

4-8. CHARACTERIZATION PROBLEMS ENCOUNTERED WITH GROWN SINGLE CRYSTALS

In the chemical analysis of the grown-single-crystal semiconductor, the analytical chemist will generally be expected to provide results that can be used both to understand and to control the distribution of impurities and dopants. As mentioned earlier, this can be a formidable task, particularly when dealing with ultrapure undoped semiconductors. Analyses of ultrapure silicon and germanium are particularly difficult because they are the most pure semiconductors currently in production. All n- and p-type impurities are less than 1 ppb, and lifetime killers such as gold and copper are a factor of 100 to 1,000 less. The levels of impurities in the melt-grown III-V semiconductors are in the 10- to 1,000-ppb range.

The analytical techniques usually available for analysis in these concentration ranges are emission spectroscopy, mass spectroscopy, and radiochemical techniques. As discussed in Sec. 5-3, emission spectrographic techniques have been particularly useful with the III-V intermetallic compound semiconductors. Solids mass spectroscopy has carried the brunt of the load in the analysis of all semiconductors because of its broad coverage of all elements in one analysis. Neutron-activation analysis is by far the most sensitive, but is limited mostly to silicon and, in certain instances, germanium. As will be seen later in this chapter, radiotracer techniques have supplied the bulk of knowledge on the behavior of dopants and impurities during the growth of single-crystal semiconductors.

4-9. SOURCES OF IMPURITY CONTAMINATION

Growth of single-crystal semiconductors is almost always carried out in quartz containers. As a result, the molten semiconductor comes in contact with the quartz and is contaminated by reaction with and/or dissolution of some of the quartz. Silicon contamination is obviously not a problem in silicon single-crystal growth but is a serious problem with the III-V intermetallic semiconductors since silicon is an electron donor.[11] Kern[12] at RCA synthesized gallium arsenide in neutron-activated boats of both natural and synthetic quartz. The concentration and distribution of silicon were determined by utilizing the 2.6-hr silicon-31 isotope. All crystals were found to have bulk concentrations of 1×10^{17} to 3×10^{18} atoms/cm^3 and were completely enclosed by a silicon-rich surface layer (0.02 mm) containing up to 1,500 ppm of silicon. Other impurities observed in the gallium arsenide included copper, sodium, antimony, gold, and phosphorus. Contamination of gallium arsenide by impurities from neutron-irradiated quartz has recently been reported by Gensauge and Hoffmeister[13] at much lower temperatures during diffusion studies. Similar diffusion studies with irradiated quartz in the Texas Instruments Incorporated laboratories[14] have yielded similar results. As a result, the quartz containers must be regarded as a large potential source of contamination during single-crystal growth.

Ekstrom and Weisberg,[15] in a study of sources of contamination in GaAs crystal growth, observed that several hours of vacuum baking of gallium at 650°C in a quartz boat increased only the copper content. Significant quantities of copper and silicon were introduced during vacuum sealing of the quartz ampules. Back diffusion of a contaminated high-vacuum pump was observed. The most serious contamina-

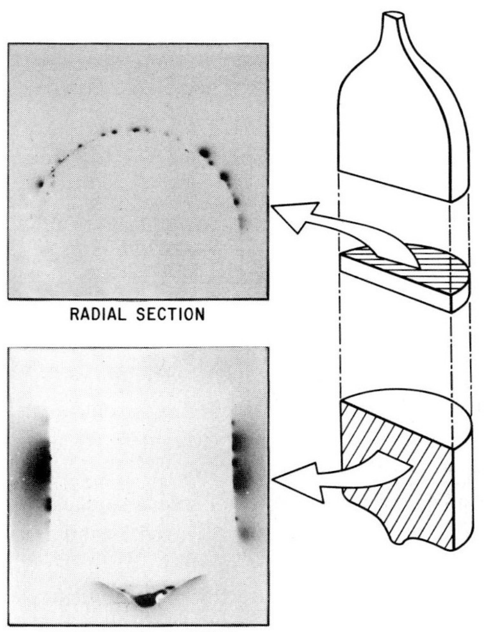

Fig. 4-8. Autoradiograms showing carbon-14 distribution in single-crystal silicon grown by the Teal-Little method. Shaded areas are those surfaces used for autoradiography.

tion was found to be reaction of the GaAs melt with the quartz boat, which agrees with Kern's work. Lithium in GaAs has also been reported to originate in the quartz boats.[16]

Silicon single crystals are sometimes grown from silicon carbide–coated boats. Scace and Slack[17] have shown carbon solubility in silicon at the melting point to be of the order of 3×10^{18} atoms/cm^3. Carbon at this level would be a serious problem and would preclude the use of any carbon-containing boat or crucible with silicon. In order to determine the behavior of carbon in silicon,[18] some carbon-14 labeled amorphous carbon was introduced into a silicon melt in a quartz crucible. The melt was stirred for about 1 hr, and then a single crystal was grown from the melt by using the Teal-Little method. The crystal was sectioned, and autoradiograms were run on the sections to determine the carbon-14 distribution. As can be seen in Fig. 4-8, the carbon was rejected into the melt and was precipitated at the outer edge of the crystal. An effective segregation coefficient of less than 1×10^{-4} was estimated in this work, which is close to the value of 0.005 reported by Newman and Willis.[19]

Ducret and Cornet[20] reported finding 500 to 2,000 ppm (1.2×10^{20} atoms/cm^3) carbon in different silicon semiconductor samples by utilizing their analytical method (see Sec. 5-6). Germanium samples were also found to contain 5 to 10 ppm carbon. This method was based on the conversion of the carbon to carbon disulfide followed by the colorimetric determination of the amount of carbon disulfide. Discussions following the paper indicated that other workers observed only 5 to 20 ppm carbon (1.2×10^{18} atoms/cm^3) by other analytical methods. Schink[21] subsequently reported an analytical method based on the oxidation of the carbon with a mixture of lead chromate and lead chloride. All silicon samples analyzed by Schink contained at least 25 ppm carbon (3×10^{18} atoms/cm^3). Single-crystal float-zoned silicon showed the least amount of carbon (3 to 4×10^{18} atoms/cm^3), while crucible-pulled crystals contained 6.5×10^{18} atoms/cm^3. Schink[21] feels that the carbon probably originates in the trichlorosilane and exists in the single-crystal silicon as elemental carbon.

4-10. DOPANT SOLUBILITY IN A SEMICONDUCTOR

Both electrical techniques and chemical analyses can be used to determine the solubility of an impurity in a semiconductor; Boltaks[22] has given an excellent review of this topic. Basically, the semiconductor is saturated with a given impurity at a given temperature for a long period of time; then the concentration of impurity is determined. Fuller et al.,[23] for example, used resistivity and Hall effect methods to study copper in germanium. The problem with the electrical techniques is that they are very susceptible to the presence of other electrically active impurities.

Chemical analysis appears to offer the best method for determining impurity solubility, but accurate analysis at 1- to 100-ppm levels is difficult. Many workers[24,25] have used radiotracer techniques and determined the impurity concentration, after equilibration, by counting the sample. It is, of course, possible to utilize activation analysis, but straightforward radiotracer techniques are more attractive where a useful radioisotope exists.

In the radiotracer technique for determining impurity solubility, the equilibrated sample is quenched to room temperature and the sample assayed by conventional counting techniques. It is of little interest if precipitation occurs during quenching, since the datum of interest is the solubility at the equilibration temperature. On the other hand, in single-crystal growth, it is important to be aware of any precipitation when determining maximum impurity solubility for a particular crystal-growth technique.

4-11. CONSTITUTIONAL SUPERCOOLING DURING CRYSTAL GROWTH

In studying the solubility of an impurity or dopant in a semiconductor it becomes apparent that there are two maximum solubilities to be considered. As discussed earlier, there is the maximum solubility for an impurity at each elevated temperature, and it is of no interest that precipitation occurs on cooling to room temperature. On the other hand, during semiconductor single-crystal growth, where very heavy doping is required (e.g., material for Esaki diodes or heavily doped substrates for epitaxial-film growth), constitutional supercooling can occur.

During single-crystal growth from an impure melt with a solute distribution coefficient less than 1, solute is being continually rejected into the melt. At some point, dependent upon growth parameters, the concentration of impurity at the interface will be high enough to cause the onset of constitutional supercooling. Hurle[26] and Bardsley et al.[27] reported an excellent study of constitutional supercooling and applied the theory to heavily gallium-doped germanium.

Some work in the Texas Instruments Incorporated laboratories [28] was carried out on the maximum amount of tin that could be incorporated into gallium arsenide as an efficient dopant using radioactive tin-113 tracer. Initially, two [115] oriented tin-doped crystals were grown by a vertical-pulling technique in the 10^{17} and 10^{18} concentration ranges. The rate of withdrawal of the crystal was 0.5 in./hr. The [115] orientation was chosen since previous experience in these laboratories indicated that the best uniformity of dopant distribution was associated with the [115] orientation. The elemental tin added to the melt was tagged with radioactive tin-113 to determine the actual tin concentration and its uniformity of distribution. Hall bars were cut from each slice and the tin concentration determined by radiochemical gamma-counting techniques. The net carrier concentration was then determined by conventional Hall coefficient measurements.

The crystal doped in the $10^{17}/cm^3$ range showed good correlation between electrical activity and radiochemical concentration, indicating normal donor behavior as high as 4.5×10^{17} atoms/cm^3. The $10^{18}/cm^3$ crystal showed 30 percent lower electrical activity than the radiochemical concentration would indicate. This electrical inactivity might be explained by onset of tin amphoteric behavior or inaccuracies inherent in electrical measurements on degenerate semiconductors.

Autoradiograms made from several representative slices from the $10^{17}/cm^3$ crystal, cut perpendicular to the growth direction, showed no irregularities; i.e., no evidence of precipitation, faceting, growth striations, or other indications of nonuniform tin distribution. Figure 4-9 shows three typical autoradiograms made from the second crystal (10^{18} range). There is evidence of tin precipitation starting at 2.6×10^{18} and pronounced precipitation when the tin concentration was 3.9×10^{18} atoms/cm^3. Close examination of the darkened area in the third autoradiogram, which has the outward appearance of a facet, discloses what is probably dendritic growth caused by constitutional supercooling in a tin-rich solution.

A third crystal was grown in the $10^{18}/cm^3$ concentration range on the [111] orientation, and autoradiograms again were made on each slice. Precipitation was observed to begin between the sixth and seventh slice, which again was about 2.9×10^{18} atoms/cm^3. As expected, all these autoradiograms showed nonuniform distribution of tin resulting from facets near the edge of the crystal.

It was concluded that the solubility limit for tin in gallium arsenide crystals grown from the melt by vertical-pulling techniques was about 2.5×10^{18} atoms/cm^3. This would seem to preclude the use of tin-doped gallium arsenide as n^+ substrates

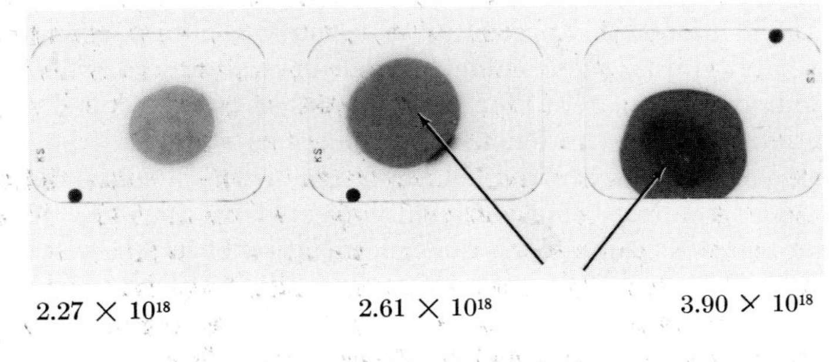

Fig. 4-9. Autoradiograms of slices from a pulled GaAs crystal showing onset of tin precipitation.

for epitaxial growth if electrical concentrations greater than $2 \times 10^{18}/cm^3$ in the substrate are desired. At lower concentrations, however, there was no evidence of tin nonuniformity in crystals grown on the [115] orientation.

4-12. RADIOCHEMICAL TECHNIQUES IN THE DETERMINATION OF SEGREGATION COEFFICIENTS

Pulled Crystals. To obtain an accurate distribution coefficient for an impurity in a semiconductor, it is essential that the impurity concentration in the starting material, C_0, be uniform and be known. This requirement is easily met with vertical-pulled crystals since they are grown from a melt. It is then only necessary to analyze for C (concentration in the solid) at any point in the crystal and know the mass fraction (x) of the original melt at that point. The effective segregation coefficient is then calculated from Eq. (4-1).

The concentration of impurity in each fraction can be determined by Hall measurements if the impurity is electrically active. This type of measurement can be very misleading unless the concentration of impurity is much higher than all other electrically active impurities in the crystal. This is particularly difficult with the III-V intermetallics such as GaAs, InAs, or GaSb, where the best pulled single crystals grown have residual donor concentrations of around 1×10^{16} electrons/cm³. This is equivalent to about 1 ppm of impurity, and for semiconductors, levels of impurities higher than this are normally deliberately added dopants and not trace impurities. For silicon and germanium, where residual impurity levels are from 10^{12} to 10^{13} carriers/cm³, this is rarely a problem.

Zone-refined Crystals. The methods commonly used to determine effective segregation coefficients for zone refining rely upon the ability of the analyst to obtain accurate and reproducible concentrations. Zone refining is restricted, for this discussion, to the purification of a uniform ingot by passing a molten zone down the crystal, as shown in Fig. 4-3. It can be shown[7,29] that Eq. (4-3) describes the zone-refining process, where

$$\frac{C}{C_0} = 1 - (1 - k)e^{-kx/l} \tag{4-3}$$

C is the concentration at a distance x from the starting end, and l is the length of the molten zone. Since the ingot was homogeneous at the start, C_0 is readily obtained by analysis before the zone-refining step. After a single zone pass, analyses for C at points x along the ingot are made. A semilog plot of C/C_0 versus x/l will yield results similar to those shown in Fig. 4-10. At the front end of the ingot $x/l = 0$, and k can be obtained by extrapolating (dashed line in Fig. 4-10) the curve back. The intersection of this line with the ordinate yields the value of the effective segregation coefficient k, since $C/C_0 = k$ when $x/l = 0$. A series of values of k can be obtained by solving Eq. (4-3) for k for each value of C and x/l.

The determination of k after multiple passes is mathematically considerably more difficult, and the reader is referred to Pfann.[7] From the analyst's viewpoint, the problem remains unchanged since C_0 and C, at known values of x/l, must be determined by some analytical method.

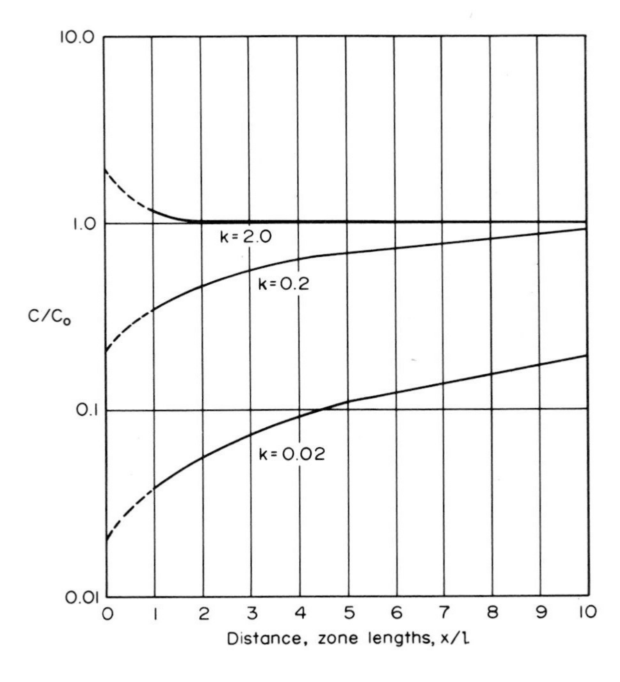

Fig. 4-10. Typical impurity profiles observed during zone refining of a uniform ingot. Single zone pass showing effect of effective segregation coefficient k.

Zone Leveling. In this technique, the impurity of interest is introduced into one end of an ingot of an ultrapure semiconductor, and a single zone-leveling pass is made over the length of the bar. In this case the concentration C at each point x on the ingot is

$$C = kC_i e^{-kx/l} \tag{4-4}$$

where C_i is the concentration in the first zone length x/l.

As in the case of zone refining, the problem of determining the effective segregation coefficient depends on the analyst's ability to determine C_i and C at known zone lengths x/l. A semilog plot of C/C_i versus x/l yields a straight line, as shown in Fig. 4-11. The effective segregation coefficient k can be determined by extrapolating to $x/l = 0$, where $C/C_i = k$. Here, as in zone refining, multiple passes complicate only the mathematical interpretation of the analyst's results.

While, generally, Hall measurements are an acceptable method of analysis, they can be misleading. In some of the earliest work on indium antimonide, Harman[30] determined the segregation coefficient of several impurities by using electrical measurements to determine electrically active impurity concentrations. Harman estimated the segregation coefficient for zinc to be 10, which would mean that zinc could be readily zone-refined out of indium antimonide. Mullin[31] and Strauss,[32] in subsequent studies using radiotracer techniques, determined the zinc segregation coefficient to be 2.3, which would mean that zinc was quite difficult to remove by zone refining. These comments are not directed in any way as criticism of Harman's excellent work (good agreement between both workers was obtained for selenium and tellurium), but only to point out the type of difficulty one can encounter when

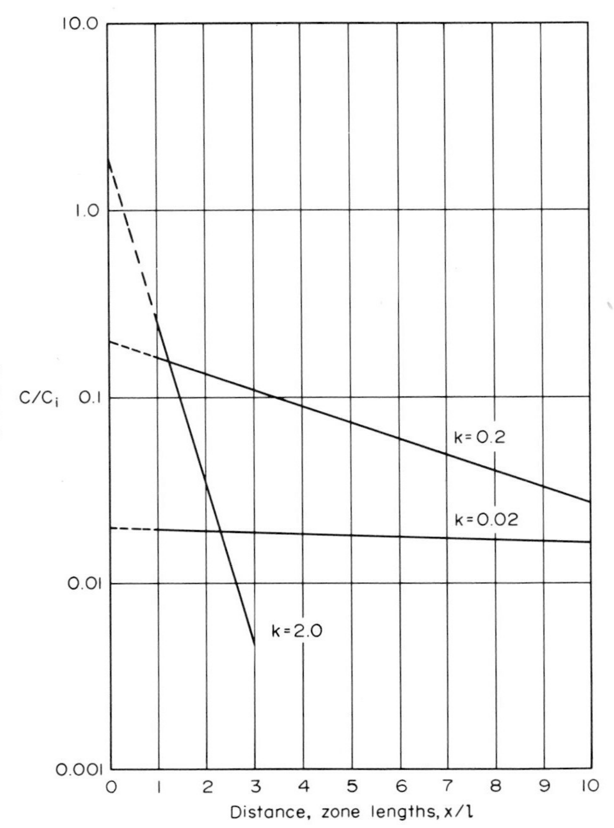

Fig. 4-11. Typical impurity profiles observed during zone leveling an impurity into an ingot. Single zone pass showing the effect of effective segregation coefficient k.

relying solely on electrical measurements. It can be stated, without reservation, that all the reliable values for segregation coefficients for impurities in semiconductors have been determined by using radiotracer techniques.

4-13. RADIOCHEMICAL TECHNIQUES IN THE STUDY OF ANISOTROPIC SEGREGATION

One of the distinct advantages of the use of radiotracer techniques to study the segregation of impurities in semiconductors is the use of autoradiography to determine the distribution of the radioactivity in the sample. It can be stated unequivocally that it is not enough just to analyze for the concentration of impurity in the crystal or slice and report a segregation coefficient from those data. If the impurity is not uniformly distributed, then the analyst's results are open to question.

In 1953, Burton et al.[33,34] reported on discontinuities or variations in the distribution of impurities in germanium crystals grown from the melt. They were able to correlate these variations with resistivity and lifetime by using antimony radiotracers and autoradiography. Figure 4-12 shows the variation of a densitometer trace of the autoradiogram and the corresponding variations in resistivity and lifetime. A similar autoradiogram, obtained in the Texas Instruments Incorporated

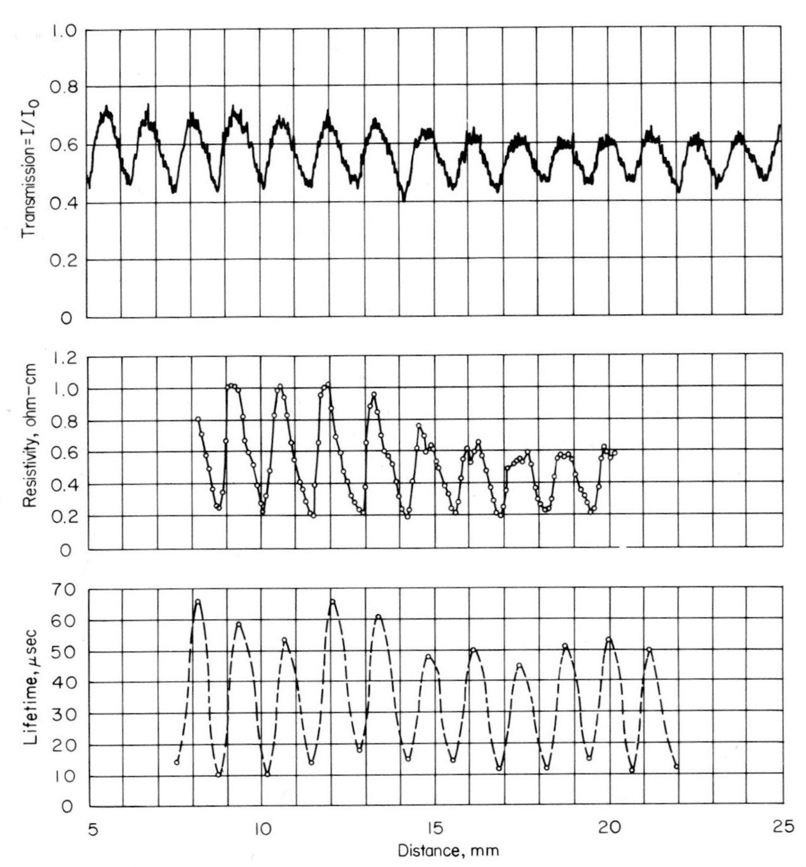

Fig. 4-12. Hole lifetime, electrical resistivity, and optical transmission of autoradiogram measured on a germanium crystal. *(From Slichter and Burton.[35])*

laboratories for a gold-doped (1×10^{13} atoms/cm^3) silicon crystal which had been tagged with gold-198, is shown in Fig. 4-13. Burton found that these striations could be removed, and a homogeneous crystal obtained with more intense stirring of the melt. Weisberg[36] presented evidence to show the adverse effect of inhomogeneous impurity distribution on Hall mobility. This reduction in mobility was attributed to the buildup of large space charge regions surrounding these local inhomogeneities. Dikhoff,[37] in further work on silicon and germanium, showed that the striations had become much smaller with faster stirring and could no longer be resolved in the autoradiogram. Dikhoff used a pulsed copper-plating technique developed by Camp[38] to delineate the striations. By using this pulsed technique, striations separated by as little as 10μ could be resolved. In heavily doped crystals, striations as small as 1 μ were made visible and resolved by etching the crystal in an HF–HNO_3–alcohol etch. Witt, Gatos, and Morizane [39–42] have studied impurity striations and their deliberate introduction into InSb as an aid to crystal-growth investigations. In the course of this work Witt[43] developed an excellent permanganate etch which was vastly superior to other delineation etches such as CP-4. With

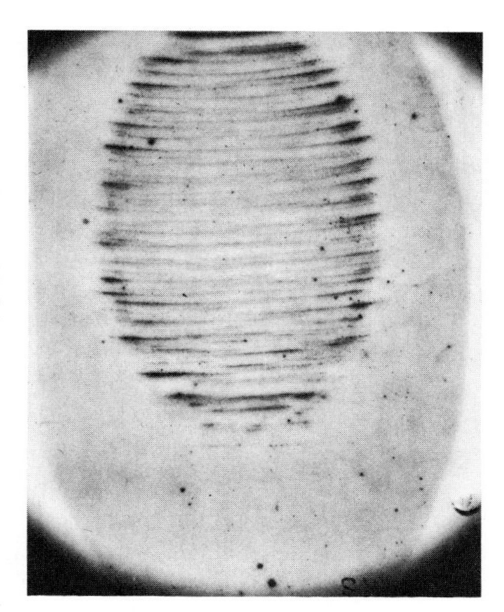

Fig. 4-13. Autoradiogram of a gold-doped silicon crystal after neutron activation to produce ^{198}Au, showing striations. The crystal was cut at 60° to the growth direction.

this etch,[44] it was possible to obtain clear micrographs of closely spaced minute impurity striations at ca. 1,320 magnifications.

Dikhoff[37] examined some silicon and germanium crystals by sawing a slice perpendicular to the crystal length and found a single spiral striation which extended from the center to the edge of the slice. Cronin et al.[45] observed similar striations in pulled tellurium-doped GaAs crystals, using both autoradiograms and etching techniques. Heinen[46] used activation analysis to observe striations in silicon. While it has been shown that these impurity striations are strongly dependent upon stirring of the melt during crystal growth, Ueda[47] found similar striations in horizontal-zone-melted crystals. Dikhoff feels that there is strong evidence for "fundamental" striations and was able to observe them in a crystal which was not rotated and was pulled rapidly from the melt.

These striations present a twofold challenge to the analyst. First, while working with the materials scientist, the analyst must be able to develop methods to determine or delineate these striations. Second, in the analysis of any semiconductor sample, the analyst must always be cognizant of the problem of nonuniform sampling through the use of a small specimen that may be striated. If the sampled area is small, e.g., in the solids spark-source mass spectrometer, the problem may be acute. On the other hand, if a 1-g sample is dissolved for analysis, then the striations are small compared with the overall size of the sample and will present little problem as long as the materials scientist is satisfied with an average value.

Cores, Facets, and Anisotropic Segregation. Another phenomenon frequently encountered in semiconductor single-crystal growth is the appearance of an impurity-rich center core down the length of the crystal. Extensive work has been carried out on these cores in germanium,[37,48] silicon,[49] indium antimonide,[50–52] and gallium arsenide,[45,53] and it is generally accepted that the cores are in fact due to anisotropic segregation of the impurity. The single crystals are crystallographically perfect; but during the crystal growth, a planar crystallographic (111) facet forms at the solid-liquid interface, and a new distribution coefficient k^* causes the impurity to segregate preferentially in the area of the (111) facet. Burton[54] proposed the use of k^* and defined it as the ratio of solute concentration in the solid and liquid *at the interface* under the growth conditions. The value of k^* will be different on facet than off facet, and a core of impurity will appear at the (111) facet. These cores of impurities are frequently and incorrectly referred to as "facets" but are in fact anisotropic segregations of the impurity at the crystallographic facet. Multiple facets, or annular facets, have been observed by Cronin et al.[45] and are believed to be caused by the presence of an irregularly curved solid-liquid interface resulting in several {111} facets present during growth.

Here, as with impurity striations, the analyst must watch carefully for anisotropic segregation both during crystal-growth studies and in the analysis of pulled single crystals. Autoradiography provides the most straightforward method of determining the presence of impurity cores in a crystal during crystal-growth studies. This requires that a radioactive tracer be added to the melt and autoradiograms run on slices of the pulled single crystal. Figure 4-14 shows some autoradiograms of anisotropic segregation of several impurities in GaAs and silicon. As can be seen, the segregation is not always restricted to the center of the crystal. This is probably caused by a slightly misoriented seed.

When a radioactive tracer is not used, it is still possible to determine the presence of impurity striations and cores by chemical staining. Banus and Gatos[52] used a 0.2 N Fe^{3+} in 6 N HCl etch for indium antimonide. Dikhoff[37] used pulsed copper-plating and etching in an HF–HNO_3–alcohol etch for germanium and silicon. Plaskett and Parsons[55] used a $3HNO_3$:$1HF$:$4H_2O$ etch with strong illumination for detecting impurity inhomogeneities in gallium arsenide.

Electron microscopy[56] and transmission of infrared radiation[57] viewed with an infrared image converter have been used to observe gallium arsenide inhomogeneities. Massengale and Klein[58] used emission spectroscopy to determine the presence of a germanium-enriched facet or core in a pulled indium antimonide crystal. The crystal slice was diced by scribing into 40 to 50 pieces, and each piece was analyzed. Then a topogram was constructed, as shown in Fig. 4-15. The presence of a germanium-enriched section is apparent at the top of the crystal.

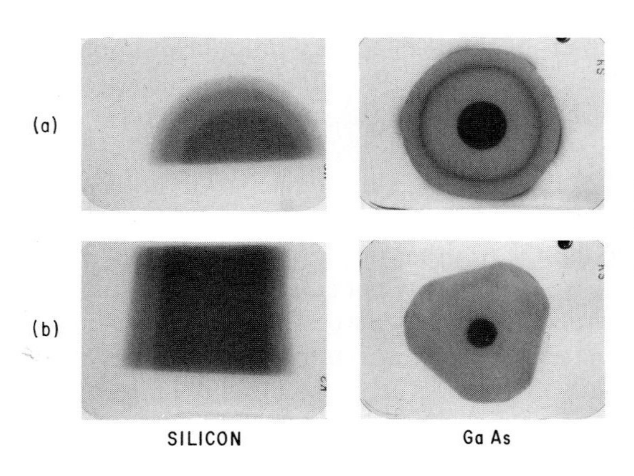

Fig. 4-14. Autoradiograms showing anisotropic segregation in GaAs and silicon. The (*a*) silicon is a radial section, and (*b*) an axial section of the same crystal.

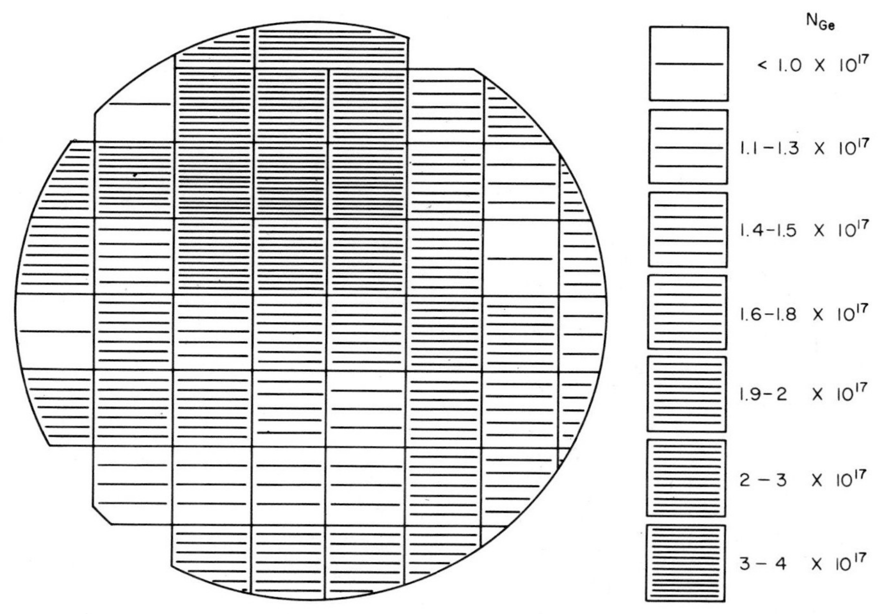

Fig. 4-15. Emission spectrographic determination of anisotropic segregation of germanium in indium antimonide. (*From Massengale and Klein.*[58])

In the analysis of samples of single-crystal semiconductors, sampling can be a serious problem if the impurity is segregated through coring or facet formation. Table 4-1 shows some reported values of the variation in impurity concentration. This difference is reported as a ratio of concentrations on and off the crystallographic facet. As can be seen, the choice of a small sample on a slice, which may be either on or off facet, can result in a serious analytical error. Conversely, if the entire slice is used in the analysis, an average value for the impurity content will be obtained, and comparison with electrical evaluation will be very difficult.

4-14. IMPURITY DISTRIBUTION AS A FUNCTION OF GROWTH ORIENTATION

Cronin et al.[28,45,59] have used radiotracers to study the effect of orientation during crystal growth on impurity distribution. In vertical-pulled crystals of gallium arsenide they observed that growth on the [115] orientation rather than the conventional [111] gave consistently more uniformly doped material. Impurity striations and central facets were not present in tellurium- and tin-doped crystals grown on this orientation.

In similar studies on horizontal leveling of indium antimonide[59] using ^{65}Zn, ^{113}Sn, and ^{110}Ag radiotracers, Cronin observed a pronounced orientation effect on impurity distribution. It was observed that crystals zone-leveled on the [113] orientation were significantly more uniform than those leveled on the [111] or [110] orientations. The segregation coefficients for each impurity were independent of growth orientation.

Table 4-1. Observed Variations in Impurity Concentrations on and off Facet in Single Crystals

Material	Dopant	Concentration ratio on/off facet	Reference
GaAs	Te	2.1–2.4	45
InSb	S	3.2–5.5	52
InSb	Te	1.7–15	50
Ge	Sb	1.5	37
Si	Sb	1.6	49

4-15. ELECTRICAL TECHNIQUES

In the final analysis, the ultimate measure of success of the growth of the semiconductor single crystal is the electrical properties. If the electrical properties do not satisfy the demands of the particular device that the crystal will be used to fabricate, then the semiconductor material must be rejected. As pointed out in Chap. 2, the electrical properties are a direct measure of the chemical and physical imperfections in the host crystal. Each type of electrical measurement provides a different type of information about the crystal. If the semiconductor type is known, the resistivity can be used along with Fig. 4-6 to yield an approximate net majority-carrier concentration. The Hall coefficient and resistivity yield the mobility and net majority-carrier concentration. The sign of the Hall coefficient gives the carrier type, positive for p type and negative for n type. Lifetime measurements give information on neutral impurities and imperfections which act as recombination centers for minority carriers.

4-16. RESISTIVITY

A number of methods not requiring the attachment of electrical contacts have been reported but have not received wide application. These methods include radio-frequency spreading resistance from a small probe on a flat surface,[60] eddy-current losses in a sphere set in an induction coil,[61] and microwave transmission through a thin sample of semiconductor material.[62]

Two methods have received wide acceptance through the efforts of the American Society for Testing and Materials.[63] These are the two-point and four-point probes for silicon rods, Method F43-67T, and the four-point probe for slices (proposed method). These procedures have evolved from those used by early workers in this field.[64–67]

Two-point Probe. ASTM[63] considers this method to be the most precise for determining the resistivity of single-crystal semiconductors. Basically, the procedure involves making ohmic contact to the ends of the bar, passing a known current through the bar, and then measuring the voltage drop across two probes applied to the surface. A schematic of a typical apparatus is shown in Fig. 3-1, where, during the measurement, the potential V_s across the standard resistor R_s is measured, and then

Fig. 4-16. A four-point probe. *(Courtesy of A & M Fell, Ltd., London, England.)*

V is measured across the two probes. The constant current through the specimen can then be calculated:

$$I = \frac{V_s}{R_s} \tag{4-5}$$

The resistivity of the sample is then calculated:

$$\rho = \frac{V}{I}\frac{A}{L} \tag{4-6}$$

where A = cross-sectional area normal to current
L = distance between two probes

For any given rod of material, the resistivity is measured at regular intervals along its length. The resistivity will be higher near the seed end of the single crystal.

Earleywine et al.[68] have automated the two-point-probe technique so that the instrument measures the diameter of the rod and the voltage drop across the probes, calculates the resistivity, and types it out along with the probe position on the rod.

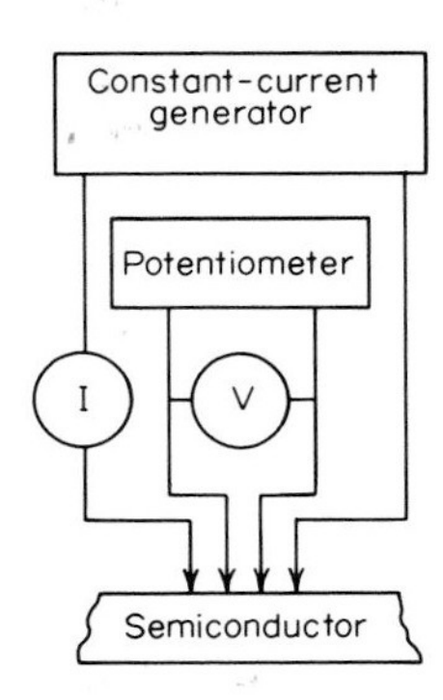

Fig. 4-17. Schematic showing the dc circuit used with a four-point probe.

This procedure rapidly provides a resistivity profile of the length of the rod. They report a precision of ± 1.5 percent for 95 percent of the time on resistivities between 0.1 and 1,000 ohm-cm. The ASTM procedure has a coefficient of variation of ± 6 percent at the 3-sigma confidence level.

Four-point Probe. The four-point probe has been discussed in Sec. 3-10, and a picture of such a probe is shown in Fig. 4-16. A typical circuit used with this probe is shown in Fig. 4-17, where it can be seen that a small constant current is applied through the sample by using the outside probes. A potentiometer is used to obtain a galvanometer null, and the voltage and current readings are recorded. The resistivity is calculated with the probe spacing a known:

$$\rho = 2\pi a \frac{V}{I} \tag{4-7}$$

The four-point-probe technique has the advantage that it is not necessary to make an electroplated or alloyed contact to obtain an ohmic contact. However, since an alloyed contact is not deliberately made, as in the case of the two-point probe, care must be taken to eliminate surface leakage and ensure that good ohmic contact is made with all four probes. The problem becomes more acute with higher-resistivity samples. These problems can be circumvented by using an ac system rather than the described dc system. The block diagram of a commercial instrument using a Fell's probe is shown in Fig. 4-18. It is a direct-reading instrument and covers a resistivity range from 0.001 to 300 ohm-cm.

The ASTM procedure has measured the precision of this technique on 5 to 20 ohm-cm material, but it undoubtedly has a much wider range of application. A precision of ± 18 percent expressed as the relative percentage error at the 3-sigma confidence level was obtained.

The four-point-probe technique has received much wider application as a method

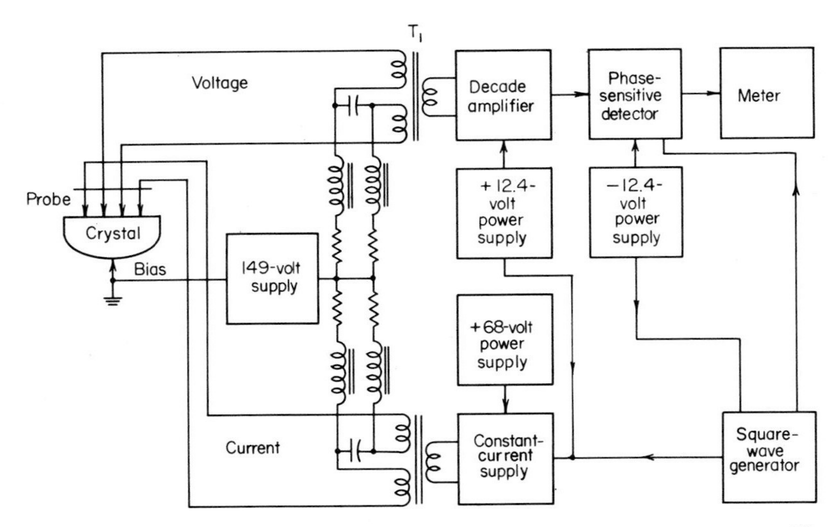

Fig. 4-18. Block diagram of ac system for four-point-probe resistivity measurements. *(Courtesy of Texas Instruments Incorporated.)*

for resistivity on slices, and ASTM has issued a proposed method.[63] In this proposed method the precision has been determined on 10 to 20 ohm-cm silicon as ±2 percent (relative percentage error at the 3-sigma level). Earleywine et al.[68] report a precision of ±3 percent, 95 percent of the time, on slices with resistivities between 0.01 and 100 ohm-cm. They feel that the two-point-probe technique should be used on material with resistivities higher than 300 ohm-cm.

One of the distinct advantages of the four-point probe, particularly with slices, is the ability to determine the resistivity profile radially across a slice. Because of probe spacing, it is difficult to detect changes in resistivities over distances smaller than 0.5 mm. To determine resistivity variations on a very small scale, the one-point probe is used.

One-point Probe. Mazur and Dickey[69,70] refined the one-point-probe or spreading-resistance technique to the point where a spatial resolution of 1 μ was obtained on silicon. By using a probe with an osmium tip to probe the face of the slice and employing a large-area ultrasonically soldered contact on the backface, it was possible to determine the resistivity in a sampled volume of 10^{-10} cm^3. Mazur reports a probable error of ±15 percent for sample resistivities in the range 0.001 to 500 ohm-cm. He was able to show that significant resistivity variations did exist in silicon slices.

4-17. MOBILITY AND CARRIER CONCENTRATION

The Hall mobility is determined by measuring the Hall coefficient and resistivity on a semiconductor sample. Section 2-12 describes the theory and the equations associated with the determination of the Hall mobility. From these same measurements the majority-carrier concentration is also obtained. ASTM[63] has issued a tentative procedure, Method F76-67T, for the determination of Hall mobility in extrinsic semiconductors on both shaped and lamellar specimens. This ASTM procedure is a system which has evolved from the discovery of the Hall effect,[71] by E. H. Hall in 1879, and the subsequent circuits developed[72–75] to measure this effect in semiconductors. The ASTM method is a dc system, direct current in a dc magnetic field. Other systems include alternating current in a dc magnetic field and alternating current in an ac magnetic field. In the Texas Instruments Incorporated laboratories the first two systems, either direct or alternating current in a dc magnetic field, are used.

The dc-dc system is the simplest and the one most often used in the industry. Generally this system is used when the resistance across the sample is 10^3 to 10^9 ohms. For lower-resistance samples the ac-dc system is used. The ac system is more sensitive than the dc system, and smaller voltage drops across the sample can be measured. The ac system also minimizes sample heating, which is a problem in low-resistivity samples.

DC-DC Hall System. The dc standard Hall system used by Texas Instruments Incorporated is like the six-lead-configuration ASTM system. The data-collection sheet with the outline for subsequent calculations is shown in Fig. 4-19. The measurement procedure is straightforward. A standard resistor R_s, approximately ten times the total resistance of the sample, is connected in series with the sample. A

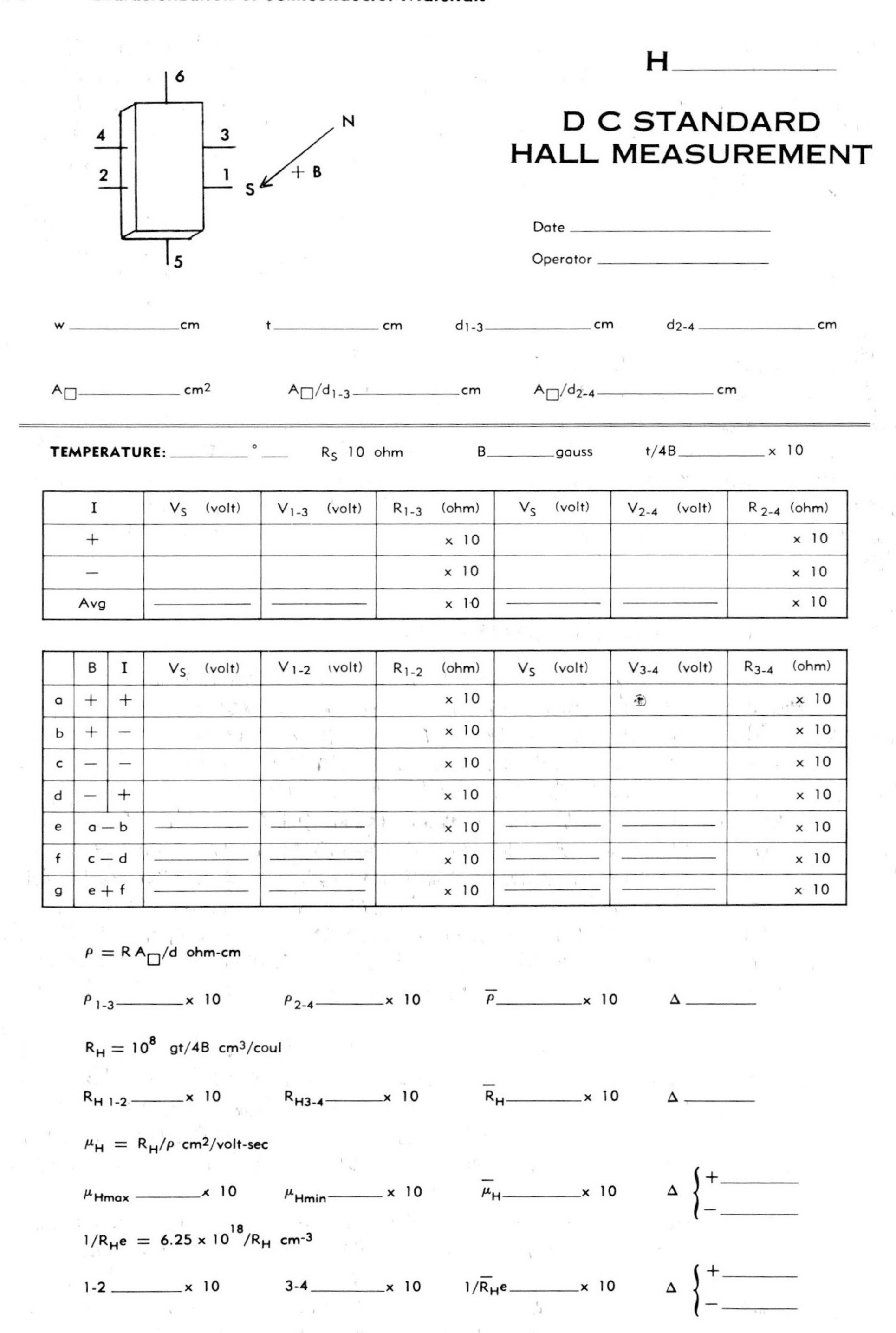

H ____________

D C STANDARD HALL MEASUREMENT

Date ____________

Operator ____________

w ________ cm   t ________ cm $d_{1\text{-}3}$ ________ cm $d_{2\text{-}4}$ ________ cm

$A_\square$ ________ cm^2 $A_\square/d_{1\text{-}3}$ ________ cm $A_\square/d_{2\text{-}4}$ ________ cm

TEMPERATURE: ________ ° ____ R_S 10 ohm   B ________ gauss   t/4B ________ × 10

I	V_S (volt)	$V_{1\text{-}3}$ (volt)	$R_{1\text{-}3}$ (ohm)	V_S (volt)	$V_{2\text{-}4}$ (volt)	$R_{2\text{-}4}$ (ohm)
+			× 10			× 10
−			× 10			× 10
Avg	————	————	× 10	————	————	× 10

	B	I	V_S (volt)	$V_{1\text{-}2}$ (volt)	$R_{1\text{-}2}$ (ohm)	V_S (volt)	$V_{3\text{-}4}$ (volt)	$R_{3\text{-}4}$ (ohm)
a	+	+			× 10			× 10
b	+	−			× 10			× 10
c	−	−			× 10			× 10
d	−	+			× 10			× 10
e	a − b		————	————	× 10	————	————	× 10
f	c − d		————	————	× 10	————	————	× 10
g	e + f		————	————	× 10	————	————	× 10

$\rho = R A_\square / d$ ohm-cm

$\rho_{1\text{-}3}$ ________ × 10 $\rho_{2\text{-}4}$ ________ × 10 $\bar{\rho}$ ________ × 10   Δ ________

$R_H = 10^8$ gt/4B cm^3/coul

$R_{H\,1\text{-}2}$ ________ × 10 $R_{H3\text{-}4}$ ________ × 10 $\bar{R}_H$ ________ × 10   Δ ________

$\mu_H = R_H/\rho$ cm^2/volt-sec

μ_{Hmax} ________ × 10 μ_{Hmin} ________ × 10 $\bar{\mu}_H$ ________ × 10   Δ { + ________ − ________

$1/R_He = 6.25 \times 10^{18}/R_H$ cm^{-3}

1-2 ________ × 10   3-4 ________ × 10 $1/\bar{R}_He$ ________ × 10   Δ { + ________ − ________

Fig. 4-19. Data sheet for dc standard Hall measurement. *(Continued on facing page.)*

TEMPERATURE: ________ ° ____ R_S 10 ohm B ________ gauss t/4B ________ x 10

I	V_S (volt)	$V_{1\text{-}3}$ (volt)	$R_{1\text{-}3}$ (ohm)	V_S (volt)	$V_{2\text{-}4}$ (volt)	$R_{2\text{-}4}$ (ohm)
+			x 10			x 10
−			x 10			x 10
Avg	————	————	x 10	————	————	x 10

	B	I	V_S (volt)	$V_{1\text{-}2}$ (volt)	$R_{1\text{-}2}$ (ohm)	V_S (volt)	$V_{3\text{-}4}$ (volt)	$R_{3\text{-}4}$ (ohm)
a	+	+			x 10			x 10
b	+	−			x 10			x 10
c	−	−			x 10			x 10
d	−	+			x 10			x 10
e	a − b		————	————	x 10	————	————	x 10
f	c − d		————	————	x 10	————	————	x 10
g	e + f		————	————	x 10	————	————	x 10

$\rho = RA_{\square}/d$ ohm-cm

$\rho_{1\text{-}3}$ ________ x 10 $\rho_{2\text{-}4}$ ________ x 10 $\bar{\rho}$ ________ x 10 Δ ________

$R_H = 10^8$ gt/4B cm³/coul

$R_{H\,1\text{-}2}$ ________ x 10 $R_{H3\text{-}4}$ ________ x 10 $\bar{R}_H$ ________ x 10 Δ ________

$\mu_H = R_H/\rho$ cm²/volt-sec

μ_{Hmax} ________ x 10 μ_{Hmin} ________ x 10 $\bar{\mu}_H$ ________ x 10 Δ { + ________ , − ________ }

$1/R_He = 6.25 \times 10^{18}/R_H$ cm⁻³

1-2 ________ x 10 3-4 ________ x 10 $1/\bar{R}_He$ ________ x 10 Δ { + ________ , − ________ }

TI—5588

Remarks: ____

____

____

Material: ________________ Originator's No.: ________________ Crystal No.: ________________

Fig. 4-19 *(Continued).* **Data sheet for dc standard Hall measurement.**

constant current I is applied through the sample and standard resistor so that the voltage drop V_s across R_s is 1 volt. Then the resistivity of the sample is determined by accurately measuring the voltages $V_{1\text{-}3}$ and $V_{2\text{-}4}$ in the forward and reverse directions.

A magnetic field is applied and the voltages $V_{1\text{-}2}$ and $V_{3\text{-}4}$ are measured in the

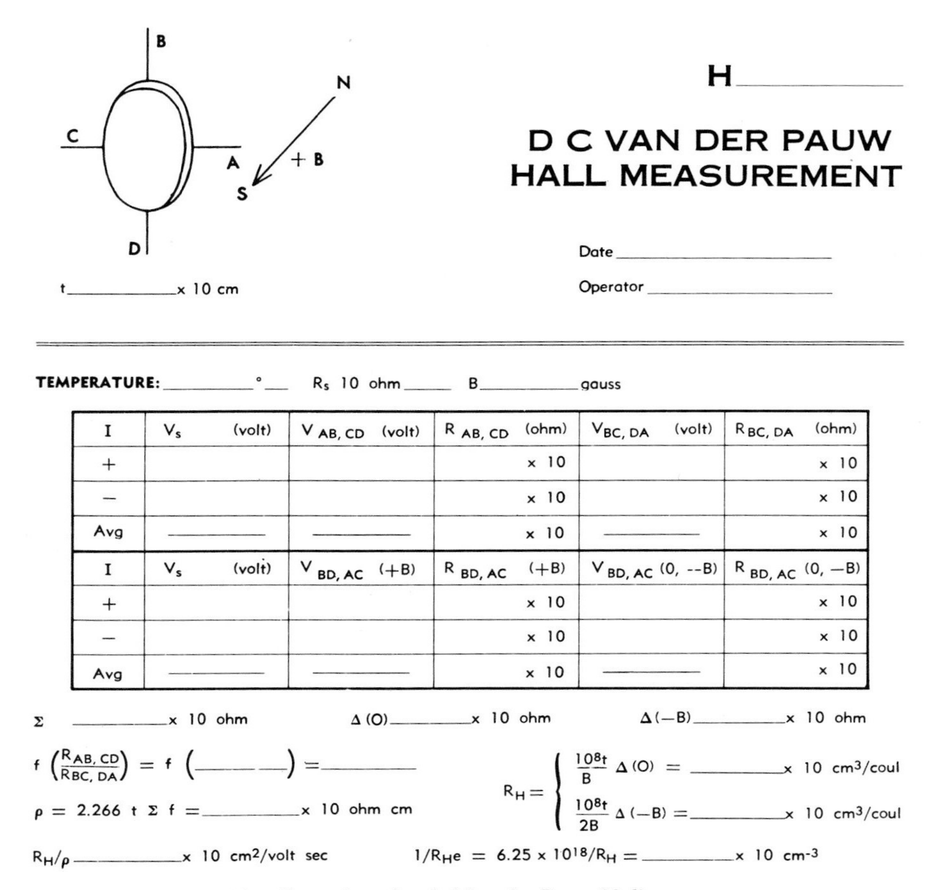

H ______

D C VAN DER PAUW
HALL MEASUREMENT

Date ______

Operator ______

TEMPERATURE: ______ ° ___ R_s 10 ohm ___ B ______ gauss

I	V_s (volt)	$V_{AB, CD}$ (volt)	$R_{AB, CD}$ (ohm)	$V_{BC, DA}$ (volt)	$R_{BC, DA}$ (ohm)
+			x 10		x 10
−			x 10		x 10
Avg	______	______	x 10	______	x 10
I	V_s (volt)	$V_{BD, AC}$ (+B)	$R_{BD, AC}$ (+B)	$V_{BD, AC}$ (0, −B)	$R_{BD, AC}$ (0, −B)
+			x 10		x 10
−			x 10		x 10
Avg	______	______	x 10	______	x 10

Σ ______ x 10 ohm   Δ (O) ______ x 10 ohm   Δ (−B) ______ x 10 ohm

$f\left(\frac{R_{AB, CD}}{R_{BC, DA}}\right) = f$ (______) = ______

$\rho = 2.266\ t\ \Sigma\ f =$ ______ x 10 ohm cm

$R_H = \begin{cases} \frac{10^8 t}{B}\ \Delta(O) = \text{______ x 10 cm}^3/\text{coul} \\ \frac{10^8 t}{2B}\ \Delta(-B) = \text{______ x 10 cm}^3/\text{coul} \end{cases}$

R_H/ρ ______ x 10 cm²/volt sec $1/R_He = 6.25 \times 10^{18}/R_H =$ ______ x 10 cm⁻³

Fig. 4-20. Data sheet for dc Van der Pauw Hall measurement.

forward and reverse directions for both the current and the magnetic field. The Hall coefficient, Hall mobility, and carrier concentration ($1/R_He$) are then calculated as shown in Fig. 4-19.

The dc Van der Pauw Hall measurements are made on samples that cannot be cut into a shape suitable for a six-lead Hall configuration. The Van der Pauw method uses four contacts, as shown in Fig. 4-20. Once again a standard resistor R_s is connected in series, a constant current I applied, and the voltages measured in the forward and reverse directions. This yields the sample resistivity. The magnetic field is applied and the voltages measured in the forward and reverse directions for both current and magnetic field. The factor f is obtained graphically, as described by Van der Pauw.[72] The Hall mobility, Hall coefficient, and carrier concentration are then calculated as shown in Fig. 4-20.

AC-DC Hall System. The ac system is generally used in the Texas Instruments Incorporated laboratories when the total resistance of the sample is less than 1,000 ohms. This technique eliminates errors due to thermal emfs and the Ettinghausen

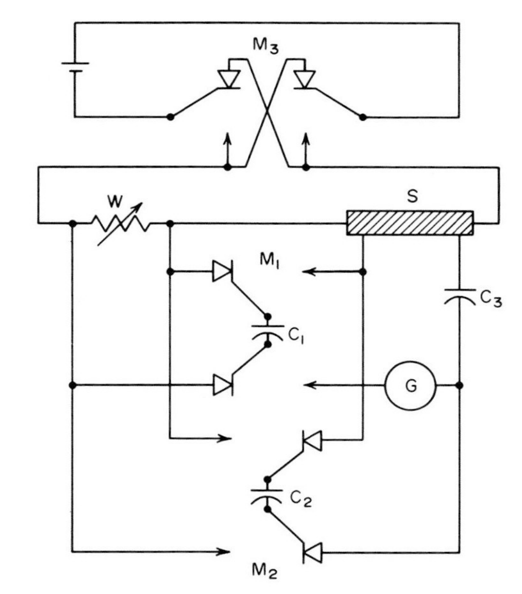

Fig. 4-21. Basic measuring circuit for the ac-dc Hall effect measurement. (*Adapted from Dauphinee and Mooser.*[74])

effect which tend to appear in low-resistivity samples. The basic measuring circuit is that developed by Dauphinee and Mooser[74] and is shown schematically in Fig. 4-21. Choppers M_1 and M_2 are driven synchronously with chopper M_3 so that the voltages (generated with a small constant current I) across the working resistor W and any two leads on the sample are compared and then balanced by using a galvanometer G. The resistance is read directly off the calibrated working resistor and recorded as shown in Fig. 4-22. The resistance is measured between leads 1-3 and 2-4 and then used to calculate the resistivity. The resistance is determined in the forward and reverse magnetic fields between leads 1-2 and 3-4 and used to calculate the Hall coefficient, Hall mobility, and carrier concentration.

The data sheet used for the ac Van der Pauw Hall measurements is shown in Fig. 4-23. The Van der Pauw ac measurements are made on the same Dauphinee circuit and calculated as shown on the data sheet.

4-18. LIFETIME

As was pointed out in Sec. 2-14, the minority-carrier lifetime of a semiconductor crystal can have a pronounced effect on the operation of a device, particularly a transistor. The lifetime is defined as the average time interval between the generation and recombination of minority carriers in the crystal. The lifetime is an indirect measure of the physical perfection and the presence of electrically neutral impurities.

ASTM[63] has issued a standard method, F28-66, for measuring the minority-carrier lifetime in bulk germanium and silicon. This is a photoconductive-decay method in which ohmic contact is made to a sample of the semiconductor material and a constant current passed through the sample. An intense flash of light of short duration is used to generate carriers, and an oscilloscope is used to measure the decay time.

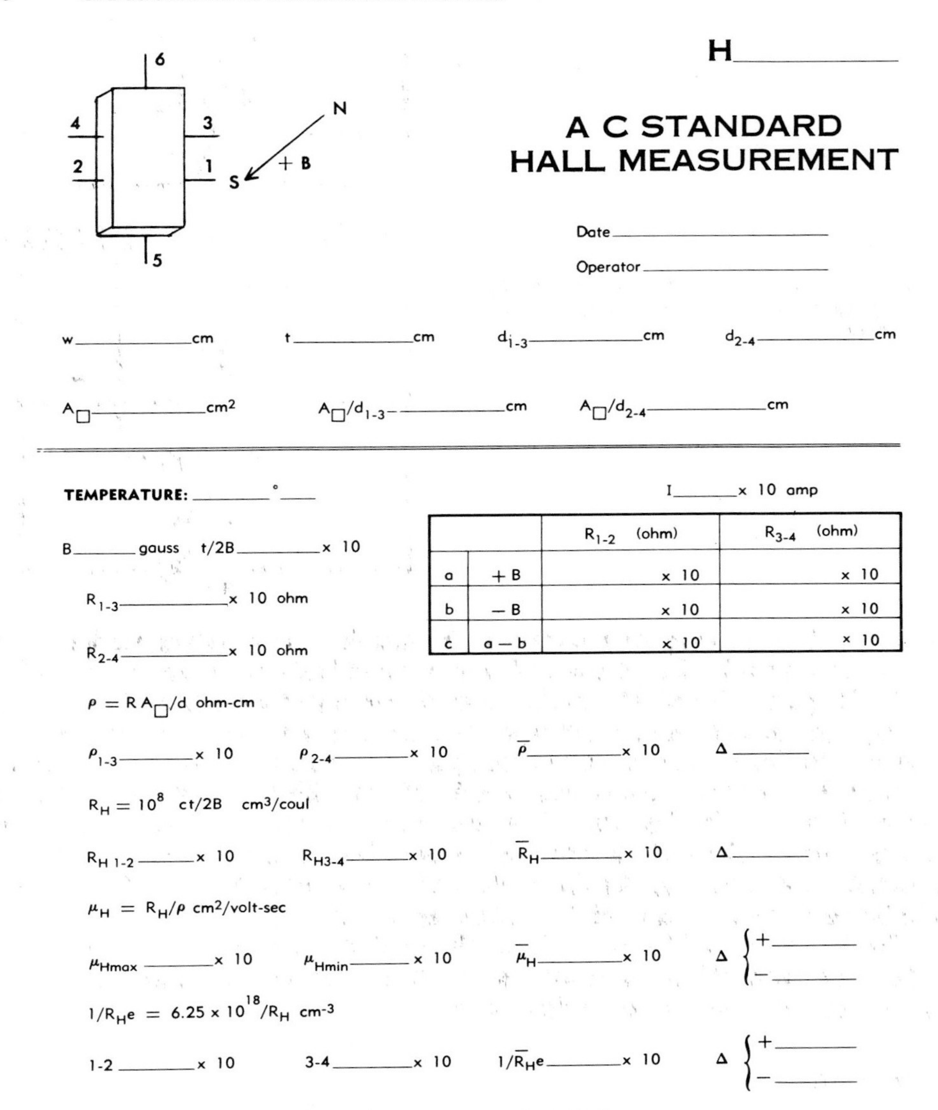

Fig. 4-22. Data sheet for ac standard Hall measurement.

For this measurement the minority-carrier lifetime is defined as the time required for the voltage pulse to decay to $1/e$ of its starting value.

Earleywine et al.[68] favor a contactless measurement method, as illustrated in Fig. 4-24. A high-frequency current is passed through the crystal by capacitive coupling. An intense light flash is used to generate carriers, which changes the conductivity of the crystal, which is then reflected in a voltage change across the crystal. This voltage change is monitored on an oscilloscope and the decay time recorded. This technique has the distinct advantage that it is unnecessary to apply ohmic contacts to the ends of the crystal. Further, the entire crystal can be tested rather than a smaller piece as required in the ASTM procedure.

Typical lifetime values for silicon and germanium range from 2 to 1,000 μsec.

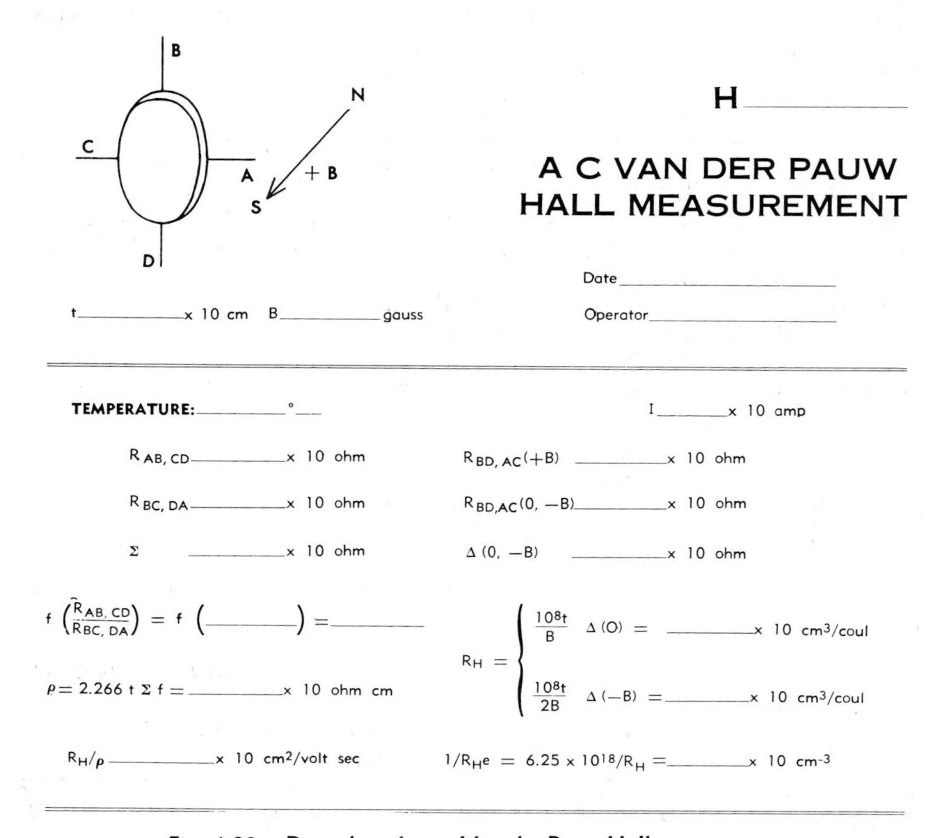
H ____

A C VAN DER PAUW
HALL MEASUREMENT

B, C, A, D, N, + B, S

Date ____

Operator ____

t ____ x 10 cm B ____ gauss

TEMPERATURE: ____ ° ____

I ____ x 10 amp

$R_{AB, CD}$ ____ x 10 ohm

$R_{BD, AC}(+B)$ ____ x 10 ohm

$R_{BC, DA}$ ____ x 10 ohm

$R_{BD,AC}(0, -B)$ ____ x 10 ohm

Σ ____ x 10 ohm

Δ (0, −B) ____ x 10 ohm

$f\left(\frac{R_{AB, CD}}{R_{BC, DA}}\right) = f$ (____) = ____

$R_H = \frac{10^8 t}{B}\ \Delta(O) =$ ____ x 10 cm³/coul

$R_H = \frac{10^8 t}{2B}\ \Delta(-B) =$ ____ x 10 cm³/coul

$\rho = 2.266\ t\ \Sigma\ f =$ ____ x 10 ohm cm

R_H/ρ ____ x 10 cm²/volt sec

$1/R_H e = 6.25 \times 10^{18}/R_H =$ ____ x 10 cm⁻³

Fig. 4-23. Data sheet for ac Van der Pauw Hall measurement.

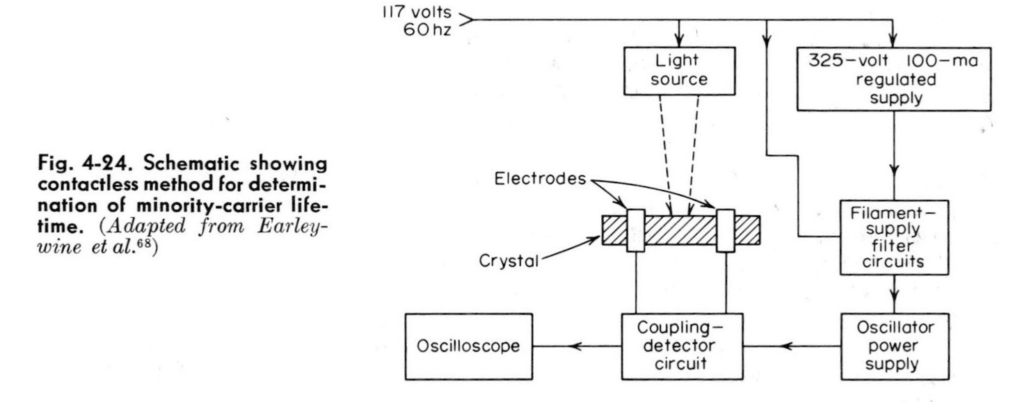

Fig. 4-24. Schematic showing contactless method for determination of minority-carrier lifetime. (*Adapted from Earleywine et al.*[68])

REFERENCES

1. Czochralski, J.: *Z. Phys. Chem.*, **92**:219 (1917).
2. Teal, G. K., and J. B. Little: *Phys. Rev.*, **78**:298 (1954).
3. Runyan, W. R.: "Silicon Semiconductor Technology," McGraw-Hill Book Company, New York, 1965.
4. Teal, G. K., M. Sparks, and E. Buehler: *Phys. Rev.*, **81**:637 (1951).
5. Pfann, W. G.: *Trans. AIME*, **194**:747 (1952).
6. Willardson, R. K., and H. L. Goering (eds.): "Compound Semiconductors," vol. 1, "Preparation of III-V Compounds," Reinhold Publishing Corporation, New York, 1962.
7. Pfann, W. G.: "Zone Melting," John Wiley & Sons, Inc., New York, 1958.
8. Richards, J. L.: in R. K. Willardson and H. L. Goering (eds.), "Compound Semiconductors," vol. 1, "Preparation of III-V Compounds," p. 279, Reinhold Publishing Corporation, New York, 1962.
9. Bridgman, P. W.: *Proc. Amer. Acad. Sci.*, **60**:305 (1925).
10. Woodall, J. M.: *Electronics*, **40**(23):110 (1967).
11. Whelan, J. M., J. D. Struthers, and J. A. Ditzenberger: in "Proceedings of the International Conference on Semiconductor Physics, Prague, 1960," p. 943, Academic Press, Inc., New York, 1960.
12. Kern, W.: *J. Electrochem. Soc.*, **109**:700 (1962).
13. Gansauge, P., and W. Hoffmeister: *Solid-State Electron.*, **9**:89 (1966).
14. Heinen, K. G., and G. B. Larrabee (Texas Instruments Incorporated): Unpublished work, 1965.
15. Ekstrom, L., and L. R. Weisberg: in M. S. Brooks and J. K. Kennedy (eds.), "Ultrapurification of Semiconductor Materials," p. 568, The Macmillan Company, New York, 1962.
16. Wolfstirn, K. B.: *Solid-State Electron.*, **6**:453 (1963).
17. Scace, R. I., and G. A. Slack: "Silicon Carbide, A High Temperature Semiconductor," Pergamon Press, New York, 1960.
18. Larrabee, G. B. (Texas Instruments Incorporated): Unpublished work, 1961.
19. Newman, R. C., and J. B. Willis: *J. Phys. Chem. Solids*, **26**:373 (1965).
20. Ducret, L., and C. Cornet: in M. S. Brooks and J. K. Kennedy (eds.), "Ultrapurification of Semiconductor Materials," p. 461, The Macmillan Company, New York, 1962.
21. Schink, N.: *Solid-State Electron.*, **8**:767 (1965).
22. Boltaks, B. I.: "Diffusion in Semiconductors," Academic Press, Inc., New York, 1963.
23. Fuller, C. S., J. D. Struthers, J. A. Ditzenberger, and K. B. Wolfstirn: *Phys. Rev.*, **93**:1182 (1954).
24. McCaldin, J. O.: *J. Appl. Phys.*, **34**:1748 (1963).
25. Struthers, J. D.: *J. Appl. Phys.*, **27**:1560 (1956).
26. Hurle, D. T. J.: *Solid-State Electron.*, **3**:37 (1961).
27. Bardsley, W., J. M. Callan, H. A. Chedzey, and D. T. J. Hurle: *Solid-State Electron.*, **3**:142 (1961).
28. Larrabee, G. B., G. R. Cronin, O. W. Wilson, and K. G. Heinen (Texas Instruments Incorporated): Unpublished work, 1966.
29. Schildknecht, H.: "Zone Melting," Academic Press, Inc., New York, 1966.
30. Harman, T. C.: *J. Electrochem. Soc.*, **103**:128 (1956).
31. Mullin, J. B.: *J. Electron. Controls*, **4**:358 (1958).
32. Strauss, A. J.: *J. Appl. Phys.*, **30**:559 (1959).
33. Burton, J. A., R. C. Prem, and W. P. Slichter: *J. Chem. Phys.*, **21**:1987 (1953).
34. Burton, J. A., E. D. Kolb, W. P. Slichter, and J. D. Struthers: *J. Chem. Phys.*, **21**:1991 (1953).

35. Slichter, W. P., and J. A. Burton: in H. E. Bridgers, J. H. Scaff, and J. N. Shive (eds.), "Transistor Technology," p. 107, D. Van Nostrand Company, Inc., Princeton, N.J., 1958.
36. Weisberg, L. R.: *J. Appl. Phys.*, **33**:1817 (1962).
37. Dikhoff, J. A. M.: *Philips Tech. Rev.*, **25**:195 (1963/64).
38. Camp, P. R.: *J. Appl. Phys.*, **25**:459 (1954).
39. Morizane, K., A. F. Witt, and H. C. Gatos: *J. Electrochem. Soc.*, **113**:51 (1966).
40. Witt, A. F., and H. C. Gatos: *J. Electrochem. Soc.*, **113**:808 (1966).
41. Witt, A. F., and H. C. Gatos: *J. Electrochem. Soc.*, **113**:413 (1966).
42. Morizane, K., A. F. Witt, and H. C. Gatos: *J. Electrochem. Soc.*, **114**:738 (1967).
43. Witt, A. F.: *J. Electrochem. Soc.*, **114**:298 (1967).
44. Witt, A. F., and H. C. Gatos: *J. Electrochem. Soc.*, **115**:70 (1968).
45. Cronin, G. R., G. B. Larrabee, and J. F. Osborne: *J. Electrochem. Soc.*, **113**:292 (1966).
46. Heinen, K. G. (Texas Instruments Incorporated): Unpublished work, 1967.
47. Ueda, H.: *J. Phys. Soc. Japan*, **16**:61 (1961).
48. Dikhoff, J. A. M.: *Solid-State Electron.*, **1**:202 (1960).
49. Benson, K. E.: *Electrochem. Tech.*, **3**:332 (1965).
50. Mullin, J. B., and K. F. Hulme: *J. Phys. Chem. Solids*, **17**:1 (1960).
51. Allred, W. P., and R. T. Bate: *J. Electrochem. Soc.*, **108**:258 (1961).
52. Banus, M. D., and H. G. Gatos: *J. Electrochem. Soc.*, **109**:829 (1962).
53. LeMay, C. Z.: *J. Appl. Phys.*, **34**:439 (1963).
54. Burton, J. A., R. C. Prem, and W. P. Slichter: *J. Chem. Phys.*, **21**:1987 (1953).
55. Plaskett, T. S., and A. H. Parsons: *J. Electrochem. Soc.*, **112**:965 (1965).
56. Meieran, E. S.: *J. Appl. Phys.*, **36**:2544 (1965).
57. Drougard, M. E., and J. B. Gunn: *J. Electrochem. Soc.*, **111**:155C (1964).
58. Massengale, J., and H. M. Klein, 3d National Meeting of the Society for Applied Spectroscopy, Cleveland, Ohio, 1964.
59. Cronin, G. R.: *J. Electrochem. Soc.*, **108**:178C (1961).
60. Allerton, G. L., and J. R. Seifert: *IRE Trans. Instrum.*, **I–9**:175 (1960).
61. Bryant, C. A., and J. B. Gunn: *Rev. Sci. Instrum.*, **36**:1614 (1965).
62. Lindmayer, J., and M. Kutsko: *Solid-State Electron.*, **6**:377 (1963).
63. "1968 Book of ASTM Standards," ASTM, Philadelphia, Pa., 1968.
64. Valdes, L. B.: *Proc. IRE*, **42**:420 (1954).
65. Uhlir, A., Jr.: *Bell System Tech. J.*, **34**:105 (1955).
66. Smits, F. M.: *Bell System Tech. J.*, **37**:711 (1958).
67. Logan, M. A.: *Bell System Tech. J.*, **40**:885 (1961).
68. Earleywine, E., L. P. Hilton, and D. Townley: *Semicond. Prod. Solid State Tech.*, **8** (10): 17 (1965).
69. Mazur, R. G., and D. H. Dickey: *J. Electrochem. Soc.*, **113**:255 (1966).
70. Mazur, R. G.: *J. Electrochem. Soc.*, **114**:255 (1967).
71. Hall, E. H.: *Amer. J. Math.*, **2**:287 (1879).
72. Van der Pauw, L. J.: *Philips Res. Rep.*, **13**:1 (1958).
73. Van der Pauw, L. J.: *Philips Tech. Rev.*, **20**(8):220 (1958/59).
74. Dauphinee, T. M., and E. Mooser: *Rev. Sci. Instrum.*, **26**:660 (1955).
75. Lindberg, O.: *Proc. IRE*, **40**:1414 (1952).

5

Analysis of Single Crystals for Chemical Imperfections

5-1. INTRODUCTION

In Chap. 2, the significance of foreign atoms in the lattice was discussed and the overriding importance of dopants to the properties of the material explained. In the elemental semiconductors, germanium and silicon, the elements of groups IIIA and VA are usually employed as dopants. Boron, gallium, and indium have been added as p-type dopants, and phosphorus, arsenic, antimony, and bismuth as n-type. For the III-V compounds, the situation is somewhat more complex; stoichiometry becomes important. An excess of gallium, for example, in a gallium arsenide crystal may imply gallium at arsenic sites, or a p-type doping. Group IVA elements are amphoteric, their action depending on which particular lattice sites they occupy. A silicon atom on a gallium site could dope n type, on an arsenic site p type, although these particular elements usually dope n type. The dopants normally added are, for p type, group IIA elements such as zinc, cadmium, or mercury and, for n-type material, group VIA elements such as sulfur, selenium, or tellurium.

Dopants have energy levels quite close to the conduction band in the case of donors, or to the valence band in the case of acceptors (Sec. 2-8). The difference in energy is about 0.01 to 0.05 ev, and these types of impurity are called shallow donors or acceptors. Another type of impurity has an energy level which is considerably further into the forbidden gap, say 0.3 to 0.4 ev. Such levels are due to so-called "deep donors" or "traps." Figure 5-1 illustrates the position for an n-type material. Since the deep-donor level is below the Fermi level, it follows that there is a high probability of this level's being filled by electrons; or, put another way, electrons do not leave atoms which form such levels. The electrons are said to be trapped. It can be shown statistically that there is a much higher probability of a hole's meeting such an electron than of its meeting an electron which is also free. In other words, the minority-carrier lifetime is drastically reduced by the presence of such traps. In fact, it is reduced by orders of magnitude.

The case for a p-type material is shown in Fig. 5-2. Here the deep-acceptor level

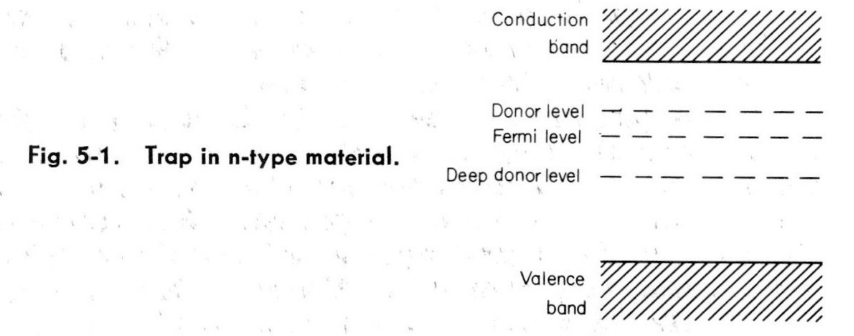

Fig. 5-1. Trap in n-type material.

is above the Fermi level, so there is a high probability of its being unfilled. The atoms forming the level are really positive ions or, put another way, stationary or trapped holes. Again, the probability of an electron's meeting a trapped hole is much higher than that of its meeting a free hole, so again the minority-carrier lifetime is reduced. The importance of the minority carrier in p-n junctions was explained in Sec. 2-15, and it follows that this degradation of lifetime is a serious problem. In germanium, copper, gold, nickel, manganese, and iron have all been shown to form traps.

In discussing mobility in Sec. 2-7, it was pointed out that this property was governed by the number of collisions occurring between the electrons and the lattice ions. This phenomenon is termed *scattering*, and it is affected not only by the temperature, which modifies the lattice vibration, but by imperfections in the lattice. Foreign atoms in the lattice will not only give rise to an electrostatic interaction but will also tend to distort the lattice, giving rise to strain fields around this point defect (see Sec. 6-2). The mobility can therefore be influenced by atoms substituting in the lattice; but, in addition, interstitials can also have an effect since these too will tend to distort the lattice. Generally, these effects are not important at room temperature or above, but are significant at low temperatures, e.g., liquid nitrogen or below. Such temperatures are employed for radiation detectors, e.g., copper-doped germanium or lithium-drifted silicon. For material for these purposes, dissolved elements such as oxygen or the halogens may be important. Moreover, many elements (iron is an example) may enter the lattice both substitutionally and interstitially. Only the substitutional atoms are electrically active, contributing carriers to the conduction band. The carrier concentration calculated electrically will not correspond with the actual concentration level as shown by

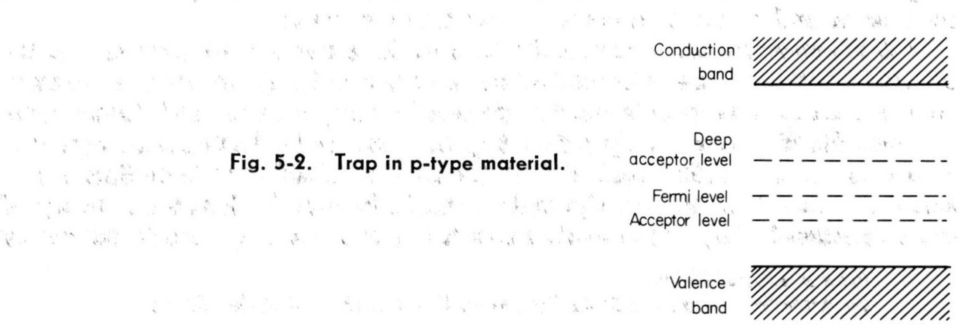

Fig. 5-2. Trap in p-type material.

analytical determinations. Not only is the carrier concentration lower than predicted from concentration, but the lower mobility brought about by impurity scattering will lead to increased discrepancy.

All these considerations emphasize the importance of reliable and sensitive methods for a wide range of elements. Almost any element in the periodic table may have an effect on the performance of a semiconductor. Only in a few cases are the dopant levels as high as the ppm range. In all other cases, the dopants and the important substitutional impurities will be in the low-ppb range. A few interstitial impurities such as oxygen or chlorine may be in the ppm range.

In the following sections, the current techniques for determining these impurities will be reviewed. However, it will become apparent that in some cases the present state of the art is inadequate. For germanium and silicon, methods are available for most of the metallic elements, and the survey methods (emission and mass spectrography) can give a good idea of the overall quality of the material. However, for the determination of boron, carbon, nitrogen, oxygen, and phosphorus, better methods are urgently required. For the III-V compounds, the situation is much the same, with the added difficulty that the extremely low-level activation analyses applicable to the elemental semiconductors cannot be used.

5-2. WORKING WITH SEMICONDUCTORS†

For most analysts, a pure material is one with impurities in the ppm range, and their techniques are geared to this level. In undertaking the analysis of semiconductor materials, however, the impurities an analyst is seeking are in the nanogram or even picogram range, and he must refine his laboratory methods accordingly. He must cultivate habits that will guard against adventitious contamination on a vanishingly small scale.

Ideally, the working area should be a dust-free room. However, a normally air-conditioned laboratory is acceptable with certain precautions. Bench tops must be kept scrupulously clean and reagent racks over the working surface avoided. Preferably, a laminar-flow bench should be used in which a flow of filtered air is recirculated over the surface. Fume hoods are a frequent source of contamination due to corrosion products and deposits falling from the upper surfaces; hoods with makeup air are particularly prone to this since there is a draft down toward the bench top. Fiber-glass hoods are preferable to metal, and they should be cleaned frequently. In laboratories which are not air-conditioned, the problems become more acute and it may be necessary to work in a glove box.

Just as important as a scrupulously clean working area is meticulous care of the analyst's working tools. Contamination of solutions by glassware is a common source of error. Soft glass is readily attacked by many reagents and should never be used. Hard glass is usually satisfactory but must be leached by aqua regia and rinsed thoroughly before use. PTFE (Teflon) is suitable for hydrofluoric acid solutions, and either this or polyethylene should be used for long-term storage of other solutions.[2] Very dilute solutions, such as standards, may tend to deplete by

†Adapted from Kane.[1]‡

‡Superscript numbers indicate References listed at the end of the chapter.

adsorption on the walls of the container and must, therefore, be freshly prepared. Conversely, containers that have been used for stronger solutions may desorb ions into weaker ones. Very dilute solutions are best prepared in new, freshly leached containers.

Reagents present a constant problem; of these, water is the most important, as might be expected. A very-high-purity deionized or distilled water is essential. A resistivity of 14 megohms, measured on a boiled-out sample, is a good criterion, although care must be taken to ensure that it is also free from suspended matter such as resin. Other reagents should be the purest available, and many manufacturers are now supplying a special grade for this industry which we shall refer to as semiconductor grade. However, even this is not always good enough, and where large volumes are required, for example of hydrofluoric acid for treating silicon, it may be necessary to redistill or otherwise purify them.

In every case, it can be assumed that the samples submitted for analysis have surface contamination, and this must be removed by a preliminary etch. All subsequent operations must include precautions to avoid recontaminating the sample, including the use of plastic-tipped tweezers for handling.

5-3. EMISSION SPECTROGRAPHY

Emission spectrography is probably the most widely used tool for assessing the quality of pure materials and has been extensively applied in the metals industry. However, when it is applied to semiconductors, its sensitivity with the usual direct techniques is found to be inadequate. Extension of this method into a usable range for these high-purity materials has followed two routes: a preconcentration step or refinement of the source conditions.

The preconcentration method has been used by several workers. Karabash et al.[3] dissolved a germanium sample in aqua regia and distilled off the chloride. The residue was evaporated on germanium oxide and arced to determine 23 metal impurities at sensitivities varying from 0.01 to 1 ppm. This procedure was simplified by Vasilevskaya et al.[4] for routine analysis, chiefly by the omission of the oxide carrier. Dvorak and Dobremyslova[5] applied the same principle to germanium oxide, adding an internal standard to the hydrochloric acid used for evaporation; their sensitivities were about 0.1 ppm. For detecting elements with volatile chlorides, this technique is obviously unsuitable. Veleker[6] dissolved the germanium in a peroxide-oxalate solution and extracted arsenic and bismuth into chloroform as the diethyldithiocarbamates. The organic phase was evaporated on graphite, mixed with a buffer, and arced by using a boiler cap. Sensitivities of 60 ppb for arsenic and 5 ppb for bismuth were obtained. Malkova et al.[7] used mannitol to retain boron during volatilization of the germanium chloride from aqua regia. The solution was evaporated on to carbon, which was arced. A sensitivity of 1 ppm was obtained on 10-mg samples.

Preconcentration methods for silicon have been largely employed by the Russians. Peizulaev et al.[8] determined 18 elements by volatilizing as the tetrachloride and evaporating the residue on strontium sulfate for arcing. Martynov et al.[9] volatilized silica samples with hydrofluoric acid, evaporating on to carbon in a nitrogen atmosphere prior to examination in a dc arc. Several metals were determined in the 10-ppb range by Zil'bershtein et al.[10] by treating the sample with a mixture of hydrofluoric and nitric acid vapors in a specially designed hollow cathode which was then subjected to a discharge. Morachevskii et al.[11,12] used the same concentration step but a more conventional dc arc source to obtain a similar sensitivity. Both liquid- and vapor-phase volatilizations were used by Rudnevskii et al.[13] as a preconcentration step. Keck et al.[14] distilled the impurities from a silicon rod by heating the tip by RF induction in a quartz tube in vacuum. The dopant collecting on the cool part of the tube was arced. This proved to be a semiquantitative method, sensitivity about 10 ppb, for aluminum, calcium, copper, magnesium, zinc, and titanium. A similar method was employed by Borovskii et al.[15] for bismuth, lead, zinc, and cadmium although they sublimed in air and condensed directly on an electrode. This same procedure was used[16] for several elements in silicon carbide. A specific method for tantalum was described by Tarasevich and Zheleznova[17] for silica in which a solution in hydrofluoric acid was reacted with Rhodamine 6G and the complex extracted into benzene or dichloroethane. The organic phase was evaporated on carbon with silver as the internal standard and arced to give a sensitivity of 0.2 ppm.

Boron, which is of considerable interest in silicon because it is a p-type dopant, presents some problems, since the halides are volatile and its boiling point, 2550°C, is even higher than that of silicon, 2355°C. Morrison and Rupp[18] applied an electrolytic preconcentration step. The sample was dissolved in sodium hydroxide solution and transferred to the anolyte compartment of a polyethylene cell; this was separated from a more dilute sodium hydroxide solution in the catholyte compartment by an anion-permeable membrane. After a 5-hr electrolysis, the anolyte was evaporated to dryness and the powdered residue mixed with an indium internal standard and arced in argon. The sensitivity was 1 ppb. Vasilevskaya et al.[19] dissolved the sample in a hydrofluoric acid–hydrogen peroxide mixture and added mannitol to retain the boron prior to volatilizing the silicon as the tetrafluoride (this is the method adapted by Malkova et al.[7] to germanium). A sensitivity of 1 ppm was obtained on arcing the residue. The same procedure was used by Semov,[20] who increased the sensitivity to 20 ppb by omitting the carbon powder used as a collector.

Two preconcentration methods have been described for gallium arsenide. Oldfield and Mack[21] removed the arsenic by dissolving the sample in hydrochloric acid and adding carbon tetrachloride and bromine. The bromine dissolves in the carbon tetrachloride layer and moderates the oxidation reaction. The solution was evaporated to small volume to volatilize the arsenic and then extracted with diisopropyl ether to remove the gallium. The remaining solution was evaporated and the residue arced in an argon-oxygen atmosphere. Several metals were determined with sensitivities ranging from 5 ppb upward. Kataev and Otmakhova[22] dissolved the sample in aqua regia, evaporated, and redissolved in hydrochloric acid. The gallium was extracted with isobutyl acetate and the solution passed through a cation-exchange resin to separate the impurities. These were eluted in 3 N hydrochloric acid, the eluate evaporated, and the residue arced. A sensitivity of 1 ppb was reported for six metallic impurities.

The use of reagents in many of these preconcentration methods, added to the fact that many impurities can be lost during treatment, renders them open to criticism. Direct methods are to be preferred if the sensitivity can be achieved. For silicon and germanium, the spectrum is obscured to a great extent by oxide bands. Babadag[23] used alternate controlled arc discharges in argon and air to determine arsenic at 1 ppb and phosphorus and selenium at the low-ppm range in germanium dioxide. Several workers[24-28] analyzed silicon carbide by arcing or sparking in air, and there is one reference[29] to the use of an argon atmosphere, but the application was to refractory material and the sensitivities were of ppm level or higher. Shvangiradze and Mozgovaya[30] arced silicon in air to determine six metals at a sensitivity around 1 ppm, and Vecsernyes[31,32] used an argon atmosphere for 18 elements with about the same sensitivity. For the special case of boron in silicon, an atmosphere of nitrogen has been used[30,33,34] and a sensitivity of 1 ppm obtained. Karpel and Shaparova[35] mixed gallium arsenide with graphite and arced to determine eight elements down to about 0.1 ppm.

The comparatively poor sensitivity makes all these direct methods of doubtful value for semiconductor-grade materials. However, Morrison et al.,[36] in analyzing silicon carbide, introduced a variation from the total-burn technique, used by the above workers. They pointed out that the impurities are selectively volatilized into the arc and, by moving the plate during the burn, the background could be reduced with respect to any one impurity. With an argon atmosphere, many elements were determined with sensitivities between 10 and 30 ppb. This same principle was applied by Massengale et al.[37] to the analysis of gallium arsenide. The principle is illustrated in Fig. 5-3. If we take the case of magnesium, we can obtain about 80 percent of the signal in the first half of the burn but, in the same period, only half the background, an improvement in the signal-to-noise ratio of 1.6. Iron, on the other hand, radiates better than 90 percent of its energy in the last half of the burn so that we can obtain an improvement of 2. In practice the burn is split into three periods, usually over about a 2-min burn. The method is referred to as the *split-burn technique* and is described in detail by Kane.[1]

For the III-V compounds gallium arsenide, indium antimonide, and indium

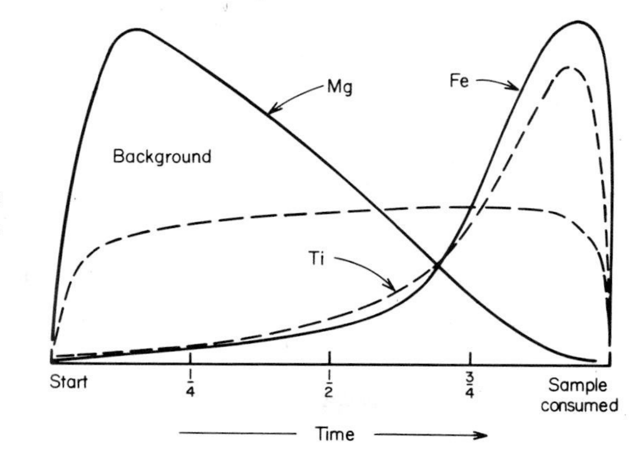

Fig. 5-3. Emission as a function of time during a spectrographic burn. (*From Burkhalter.*[38])

Table 5-1. Sensitivity Levels for GaAs, in ppm Atomic, Using the Emission Spectrographic Split-burn Technique†

Period	IA	IIA	IIIB	IVB	VB	VIB	VIIB	VIII	VIII	VIII	IB	IIB	IIIA	IVA	VA	VIA	VIIA	
1	H																	He
2	Li	Be											B	C	N	O	F	Ne
	0.1	0.08											335					
3	Na	Mg											Al	Si	P	S	Cl	A
	0.15	0.0015											0.05	0.025	1.2			
4	K	Ca	Sc	Ti	V	Cr	Mn	Fe	Co	Ni	Cu	Zn	Ga	Ge	As	Se	Br	Kr
	2.0	0.9	2.0	0.03	70	0.02	0.015	0.007	0.03	0.12	0.0006	5.6		0.1		90		
5	Rb	Sr	Y	Zr	Nb	Mo	Tc	Ru	Rh	Pd	Ag	Cd	In	Sn	Sb	Te	I	Xe
	0.8	0.4		40	40	0.08		0.7	0.6	0.7	0.0004	20	0.06	0.3	0.3	6.0		
6	Cs	Ba	La	Hf	Ta	W	Re	Os	Ir	Pt	Au	Hg	Tl	Pb	Bi	Po	At	Rn
	270	0.5	0.25		40	80		40	185	2	0.002	0.2	18	0.007	0.2			
7	Fr	Ra	Ac															

Period														
6	Ce	Pr	Nd	Pm	Sm	Eu	Gd	Tb	Dy	Ho	Er	Tm	Yb	Lu
	100	225	250		480	2.5	230	225	135	220	2.0		0.2	
7	Th	Pa	U											

†Adapted from Kane.[39]

arsenide, the basic procedure employs a 200-mg sample loaded into an undercut electrode. The sample is arced for 40 sec in the first period. The middle period is judged by the residual size and is terminated when the sample reaches about 1 mm in diameter, just sufficient to give a final 20- to 30-sec period. The first exposure is used to determine aluminum, beryllium, bismuth, boron, cadmium, lead, magnesium, and silicon and the last exposure to determine chromium, cobalt, copper, gold, iron, manganese, nickel, silver, tin, and titanium. The middle period acts as a check against the other two; significant line intensities can be measured and added to either. Alkali metals will also appear in the first group, but since, on a large spectrograph, a second exposure is usually necessary to detect them, a separate procedure is described using only an initial 20-sec period. The very volatile elements mercury, phosphorus, tellurium, and zinc are also best dealt with separately by using a boiler cap on the electrode to enhance the sensitivity. In this case, only the first 30 sec of burn is utilized.

This method is very useful for gallium arsenide, and the sensitivities obtainable are given in Table 5-1. These values, due to Klein and quoted by Kane,[39] range down to 1 ppb atomic for some elements of interest.

A method for silicon and germanium is also given in detail by Kane.[1] It is a modification of the method of Morrison et al.[36] The first period of 60 sec is carried out in an atmosphere of argon, and the middle and last periods of 10 to 30 sec in air. The use of argon throughout, as recommended by Morrison et al., leads to exposures running into several minutes. The use of two atmospheres is a compromise between sensitivity and speed; as a consequence, this method is considerably less successful than that for gallium arsenide. The problems are increased by the lower densities of the materials. Samples of only 20 mg of silicon or 40 mg of germanium can be used, a loss of one order of magnitude immediately. For the less volatile elements, the heavier background from band structure lowers the signal-to-noise ratio. By and large, the sensitivities are ten to one hundred times poorer than for gallium arsenide, making this method for semiconductors of doubtful value. It may well be that use of an alternative atmosphere, e.g., nitrogen or helium, would be advantageous.

Two other applications of emission spectroscopy have been described in which a gaseous discharge is the exciting medium. Babko and Get'man[40] passed oxygen over germanium heated to 950 to 1000°C. Hydrogen was combined to form water, which was frozen out in a special tube at liquid-air temperature. After 1 hr, the combustion was stopped and the water allowed to vaporize into an electrodeless gas discharge tube at low pressure. The intensity of a hydrogen line was measured from a photographic plate and the hydrogen content of the germanium calculated. Andrychuk and Jones[41] devised an excitation source for several elements in gallium arsenide in which the sample is contained in a hollow anode and a discharge is initiated between this and a heated cathode in a pressure of 150 μ of helium. Oxygen, nitrogen, hydrogen, phosphorus, sulfur, and halogens could be detected by directing the emission to a spectrograph; but the sensitivities were poor, about 0.05 percent for most of them. Generally, the important halogens and sulfur cannot be determined by emission spectrography.

5-4. MASS SPECTROSCOPY

As was seen in Sec. 5-3, emission spectrography is a very useful technique for determining many elements in the III-V compounds, although for silicon and germanium it is of decidedly less value. Even for the III-Vs it is restricted essentially to the metallic elements. Of broader application and, in general, of higher sensitivity is mass spectroscopy.

The first attempt to apply this technique to analysis of semiconductors appears to be due to Honig,[42] who heated germanium stepwise from 500 to 1200°C inside a conventional 180° mass spectrometer. The impurities were vaporized and ionized by 45-volt electrons. The total ion current for each impurity was related to concentration. Levels in general were high, above 10 ppm. A few months later, Hannay and Ahearn[43] published a paper describing the application of a double-focusing mass spectrograph of the Mattauch type to the analysis of germanium, silicon, and antimony, and all subsequent work on bulk material has used the same technique.

The instrument used by Hannay and Ahearn was originally designed by Shaw and Rall[44] and consisted of a high-voltage, high-frequency vacuum spark source, an electrostatic sector to provide a monoenergetic beam of ions to the magnetic sector, and a photographic plate detector. This instrument was manufactured subsequently by Associated Electrical Industries, Ltd. (AEI), as their Model MS7 and became available commercially around 1958. Consolidated Electrodynamics Corporation (CEC) followed with their version, Model 21-110, and this led to a widespread application of the technique to semiconductor materials. Craig et al.[45] have described the AEI instrument, and a schematic is given in Fig. 5-4. The CEC instrument is basically the same and has been described by Robinson et al.[46] It has narrower slits, so that it has somewhat better resolution but requires longer exposure times. The paper by Craig et al.[45] in which the instrument was described also included a method for solid samples which is now generally used. The method was applied to silicon by Craig et al.,[45,47] Duke,[48] and of course,

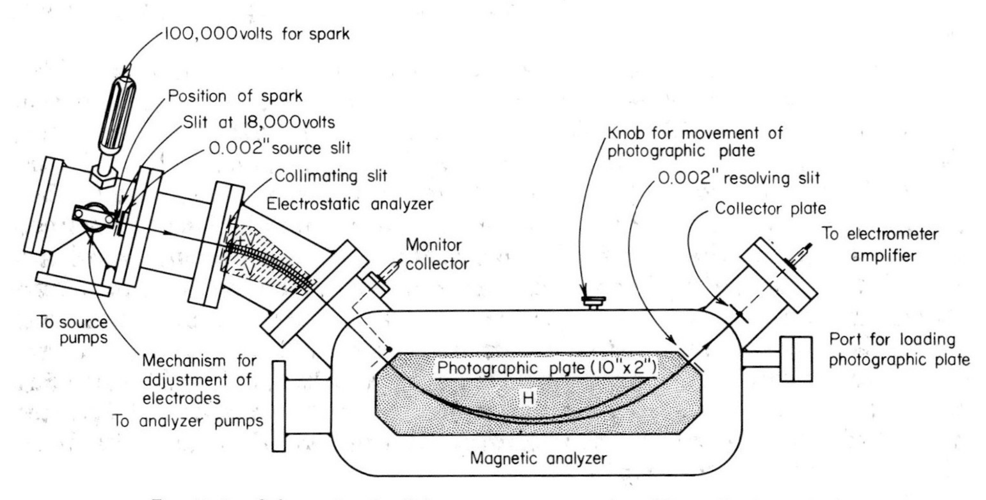

Fig. 5-4. Schematic of solids mass spectrograph. (*From Craig et al.*[45])

Ahearn.[43,49] Germanium has been examined by Ahearn[43,49] and several III-V compounds by Craig et al.,[47,50,51] Willardson,[52] and Brice et al.[53]

The procedure is given in detail by Kane.[1] The sample is in the form of bars, 1 mm² cross section and 2 cm long, held in adjustable holders in the source section of the instrument. After pumping down to a good vacuum, a spark is generated between the two self-electrodes held about 1 mm apart. The spark is from a radiofrequency oscillator of 500 khz, and the pulse duration and frequency, as well as the voltage, can be adjusted. The pulse duration may be varied from 25 to 200 μsec, and the frequency from 10 to 10,000 hz. The voltage can be varied up to 100 kv on a percentage scale. The parameters are chosen on an empirical basis to give the best impurity response in the particular matrix under investigation. It is judged by the response of the ion integrator, i.e., by the achievement of a satisfactory level of ion generation.

When the source parameters have been determined and the spark established, a series of exposures is made, based on the ion-integrator readings. These exposures usually vary from 0.0003 to 1,000 ncoul in a 3:10 series. The pulse repetition rate is increased for the longer exposures in order to maintain a reasonable elapsed time. The spectrum is recorded on a photographic plate. A typical series of spectra for silicon is shown in Fig. 5-5.

By using the known isotopic masses for the matrix element, it is possible to construct a scale of mass-to-charge ratios in order to identify the lines of the impurity elements. With these identified, a semiquantitative estimate of the amount can be made visually. A calibration factor, or so-called "plate sensitivity factor," S_p is first calculated. S_p is defined as the least amount, in ppm atomic, of any isotope which is just detectable at the longest exposure on the plate (usually 1,000 ncoul). It is defined regardless of the element, and this points up one of the basic assumptions in this treatment: that all elements have an equal chance of reaching the plate. As we shall see later, this may not be completely true. However, S_p is determined by finding a line due to a minor isotope of the matrix element which is just detectable in one of the exposures, say E_s. Then

$$S_p = \frac{E_s}{E_{\max}} \times \frac{I_s}{100} \times 10^6$$

where $E_{\max}$ = maximum exposure
I_s = isotopic abundance

This is the case for an elemental semiconductor; for compound semiconductors

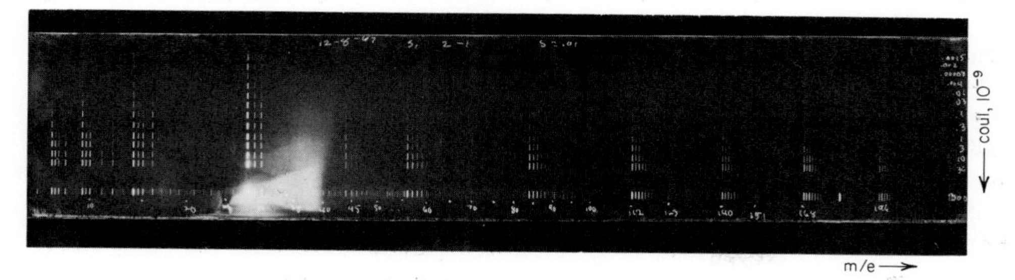

Fig. 5-5. Graded series of mass spectrographic exposures for silicon sample.

such as the III-V compounds the value obviously must be further corrected for the atomic percentage of the selected matrix element. To determine the content of an impurity, the same relationship is applied to a just detectable line from an impurity isotope by substituting the appropriate isotopic abundance, i.e.,

$$\text{Concentration (ppm atomic)} = S_p \times \frac{E_{\max}}{E_i} \times \frac{100}{I_i}$$

where E_i = just detectable exposure
I_i = isotopic abundance

The use of atomic ppm follows from the treatment and the basic assumption of independence of element. It is related to ppm by weight by the expression

$$\text{ppm by weight} = \text{ppm atomic} \times \frac{M_i}{M_b}$$

where M_i = atomic weight of impurity
M_b = atomic (or equivalent) weight of matrix

If we take the example of 2 ppb atomic boron in silicon,

$$\text{ppm by weight} = 0.002 \times \frac{10.8}{28} = 0.00075$$
$$(= 0.75 \text{ ppb})$$

This, as was shown in Sec. 3-8, is 10^{14} atoms/cm^3. More directly,

$$\text{Atoms/cm}^3 = \text{ppb atomic} \times \frac{A \times d}{M_b \times 10^9}$$

where A = Avogadro's number = 6×10^{23}
d = density of bulk material

Again, for our example,

$$\text{Atoms/cm}^3 = 2 \times \frac{6 \times 10^{23} \times 2.4}{28 \times 10^9} = 10^{14}$$

An alternative procedure uses a densitometer, and this photometric procedure is claimed to be more reproducible. However, it is less sensitive and, in view of the many uncertainties in the ion generation, it hardly seems to warrant the extra effort.

Sensitivity data have been given by several workers. The most comprehensive are due to the AEI personnel and are given in a series of technical bulletins available in the United States from Picker-Nuclear.† Among the semiconductors investigated were silicon,[54] gallium arsenide,[55] gallium phosphide,[56] and indium antimonide.[57] Woolston and Honig[58] gave sensitivity figures for gallium arsenide which were referenced by Honig[59] in a later publication. Sensitivities of gallium arsenide have also been given by Brice et al.[53] and by Klein, quoted by Kane.[39] These latter figures are given in Table 5-2. With one or two exceptions, they are not significantly different from the other sources given nor markedly different from those for other semiconductors. About 3 ppb atomic is probably the commonest sensitivity level,

†Picker-Nuclear, 1275 Mamaroneck Avenue, White Plains, N.Y.

although several exceptions will be found. Compared with Table 5-1, they are generally of the order of 100 times more sensitive than emission spectrographic values and, of course, cover a wider range of elements. However, it should be pointed out that this is not entirely the case; several electrically important elements show equal or even better sensitivity with emission spectrography. Copper is an important case in point.

The quantitative treatment outlined above includes a plate calibration, which is a threshold ion sensitivity, and an internal standard treatment in which an isotope of a matrix element provides the reference line. The basic assumptions are (1) that all ions affect the photographic plate equally and (2) that the ion-source parameters are such that all the elements present are sampled equally. Hannay and Ahearn[43] believed these to be generally valid, and their results for boron in silicon and germanium in antimony appeared to confirm this within a factor of 3. Owens and Giardino[60] investigated several sources of error, working with III-V compounds and a stainless steel, and confirmed the equal response of the photographic plate to ions of different elements, at least for the Ilford Q2 plates which are usually used in this work. They ascribed major discrepancies which had been encountered to variations in the ion source. Woolston and Honig[61] studied the energy distribution in the RF spark for several different matrices, and distribution curves for four are given in Fig. 5-6. It is evident that the elements do not respond equally, and since the bandpass of the electric sector is, for this case, 20 kv ± 300 volts different fractions of the total ions are sampled for the magnetic sector. Better than 90 percent of the germanium is passed, but only about one-third of the silicon. If these same values were valid for silicon in a germanium sample, calculated values would be too high by a factor of 3. While this is still within the generally accepted range of accuracy, the information available is extremely scanty, and it may well be that other elements are very much further off.

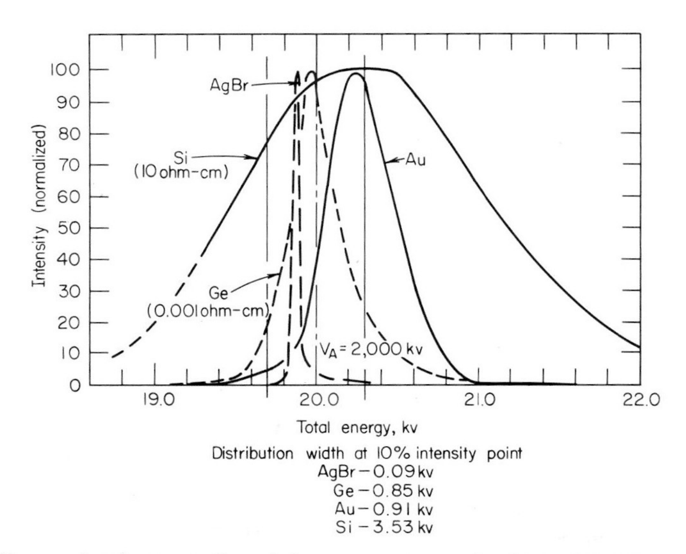

Fig. 5-6. Energy distribution in the solids mass spectrograph. (*After Woolston and Honig.*[61])

Table 5-2. Sensitivity Levels for GaAs, in ppm Atomic, Using the Solids Mass Spectrograph†

Period	IA	IIA	IIIB	IVB	VB	VIB	VIIB	VIII	VIII	VIII	IB	IIB	IIIA	IVA	VA	VIA	VIIA	
1	H																	He
2	Li 0.003	Be 0.002											B 0.003	C 5.0	N 0.2	O 1.0	F 0.006	Ne
3	Na	Mg 10											Al 0.002	Si 0.1	P 0.002	S 0.05	Cl 0.01	A
4	K 0.002	Ca 0.004	Sc 0.01	Ti 0.03	V 0.002	Cr 0.02	Mn 0.005	Fe 0.03	Co 0.005	Ni 0.01	Cu 0.1	Zn 0.1	Ga	Ge 1.0	As	Se 0.5	Br 0.5	Kr
5	Rb 0.4	Sr 0.3	Y 0.2	Zr 0.5	Nb 0.2	Mo 0.06	Tc	Ru 0.06	Rh 0.1	Pd 0.007	Ag 0.004	Cd 0.01	In 0.01	Sn 0.006	Sb 0.004	Te 0.01	I 0.002	Xe
6	Cs 0.002	Ba 0.04	La 0.02	Hf 0.01	Ta 0.2	W 0.007	Re 0.003	Os 0.005	Ir 0.003	Pt 0.006	Au 0.002	Hg 0.005	Tl 0.003	Pb 0.004	Bi 0.002	Po	At	Rn
7	Fr	Ra	Ac															

Period														
6	Ce 1.0	Pr 0.06	Nd 0.02	Pm	Sm 0.008	Eu 0.004	Gd 0.02	Tb 0.002	Dy 0.02	Ho 0.002	Er 0.006	Tm 0.002	Yb 0.006	Lu 0.002
7	Th 0.002	Pa 0.002	U											

†Adapted from Kane.[39]

More direct estimates of the accuracy have been made. Short and Keene[62] used a number of standard metal samples to determine the "ionization efficiency factor" for different elements in the same matrix and for the same element in different matrices. They found these factors, which were simply ratios of found to known impurity contents, to vary from 0.7 to about 5, with one or two exceptional values as high as 10. A series of three indium antimonide samples containing 3 and 10 ppm Zn, determined by radioactivation and Hall measurement, was also examined by three different laboratories. In this case, values of 1.1, 1.4, and 2.1 were obtained, really quite good agreement. Ahearn et al.[63] compared mass spectrographic values for zinc, silicon, germanium, tin, selenium, tellurium, and sulfur in gallium phosphide with those obtained spectrophotometrically and obtained "relative sensitivity coefficients" (identical to the "ionization efficiency factors" above) between 0.5 and 1.5. Brice et al.[53] obtained relatively good agreement for a number of dopants in gallium arsenide as compared with electrical evaluation.

In general, there is good reason to believe that for III-V compounds the accuracy of the method is within the generally accepted factor of 3. This is borne out by results obtained by Klein and Larrabee[64] for a restricted number of gallium arsenide samples. In these, the impurities were added as radiotracers during crystal growth, and their concentrations determined by counting. They were then used to check the electrical evaluation and mass and emission spectrographic values. The results are given in Table 5-3. The agreement, generally, is acceptable. Similar experiments were carried out on silicon, with the exception that the impurities were added in the inactive form, determined by activation analysis, and compared with the electrical evaluation and mass spectrographic analyses. The results are given in Table 5-4. While the agreement between the activation analyses and electrical evaluations is good, the mass spectrographic values are high, in one case by a factor of 100. There is no information on the accuracy of determinations in germanium. Values for both these elemental semiconductors must be accepted with caution although, since correction factors generally are high, they can usually be assumed to represent upper limits of impurity.

There is an undoubted need for reliable standards in this technique. An empirical calibration, similar to that used for emission spectrographic calibration curves, would considerably enhance the confidence in the results.

Table 5-3. Results of Analyses for Dopants in GaAs in ppm Atomic

Crystals	Dopant	Radiotracer	Electrical	Emission	Mass
OW 138/4	Sn	0.81	0.74	1.5	1.5
OW 149/9	Sn	6.7	6.0	9.0	12
OW 164/5	Sn	53	47	97	95
OW 168/9	Te	47	48	34	43
GC 237/24	Zn	12	7.0	8.4	11
3-51/12	Fe	0.57		2.7	2.4
555-217/10	Cr	0.51		0.6	1.6

Table 5-4. Results of Analysis for Dopants in Silicon in ppm Atomic

Crystals	Dopant	Radiotracer	Electrical	Mass
08602	P	1.4	1.2	2.2
0.3–0.4	Ga	1.5	1.6	41
0.6–1.0	Ga	0.4	0.54	17
0.6–0.7	In	0.58	0.50	54
S-3477	As	100	57	350
RSB 4451	Sb	2.5	2.3	21
NSB 02356	Sb	26	12	71

5-5. ACTIVATION ANALYSIS

While the emission and mass spectrographic methods are probably the most generally useful inasmuch as they provide a survey of the material in question, they are really borderline in their sensitivity. A content of 2 ppb atomic corresponds to 10^{14} atoms/cm^3 in silicon, and this is a not unusual doping range. Dopants, possibly adventitious, of a tenth this level could be significant, and traps of very much less can be detrimental. For specific elements, values much below this are of interest, and only activation analysis has the necessary sensitivity.

Activation is accomplished by bombarding the sample either with fast or thermal neutrons or with high-energy particles such as protons, tritons, or doubly ionized helium nuclei. The ultimate sensitivity is dependent on the amount of radioactivity induced in the element being determined, and this is given by the following equation:

$$N_{\text{dps}} = \phi\sigma N(1 - e^{-\lambda t})$$

where N_{dps} = induced activity, decompositions/sec
ϕ = flux, neutrons/(cm^2)(sec)
σ = cross section, barns
N = number of atoms of target nuclide
λ = decay constant
= 0.693/half-life
t = time of irradiation

Since σ, N, and λ are properties of the material, the sensitivity is dependent on the flux and the time of irradiation. For work with semiconductors, only the high-thermal-neutron fluxes in the large nuclear reactors will give sensitivities of interest; and, in general, fluxes from other sources can be neglected.

There are a considerable number of applications of this technique to semiconductors mentioned in the literature, and a comprehensive review and summary has been given by Cali.[65] The earliest appears to be due to Smales and Brown,[66] who, in 1950, described a method for arsenic in germanium dioxide and specifically mentioned its use in semiconductor work. This method was amplified in a later publication by Smales and Pate.[67] After irradiation, the bulk of the germanium is distilled off from a hydrochloric acid–chlorine mixture (cf. Secs. 3-5 and 3-6) ,the residue reduced with hydrobromic acid, and the arsenic then distilled over. The distillate is counted through a 300 mg/cm^2 beta absorber by a scintillation β-counter. The method suffered one of the interferences commonly encountered in this technique. The active isotope of arsenic is formed by the reaction

$$^{75}As \xrightarrow{n,\gamma} {}^{76}As$$

However, one of the germanium isotopes undergoes the reaction

$$^{76}Ge \xrightarrow{n,\gamma} {}^{77}Ge \xrightarrow{\beta-} {}^{77}As$$

and another

$$^{74}Ge \xrightarrow{n,\gamma} {}^{75}Ge \xrightarrow{\beta-} {}^{75}As \xrightarrow{n,\gamma} {}^{76}As$$

^{76}As can be distinguished from ^{77}As by the difference in energy; beta counting through an absorber can eliminate much of this interference. ^{76}As cannot, of course, be identified as from any one source. Fortunately, this latter reaction, since it is second order, has a very low yield, and interference in Smales's method was less than 1 ppb. The same procedure was used by Jaskolska and Wodkiewicz.[68] Leliaert[69] determined arsenic and phosphorus in diffusion studies by passing the fluoride solution over Dowex-1, an anion exchanger. The germanium complex is retained, and arsenic and phosphorus, in the pentavalent form, are eluted. Arsenic is precipitated as the trisulfide and phosphorus as the phosphomolybdate for counting. The arsenic determination in germanium has been refined more recently by De Soete et al.,[70] who preferred to separate the arsenic by homogeneous precipitation as the sulfide, using thioacetamide. They used the ^{77}As as an internal standard and employed γ-ray spectroscopy for levels above 50 ppb. Below this, and down to one ppb, they retained the β-counting technique.

Of other specific analyses, Szekely[71] determined copper in germanium by evaporating an aqua regia solution of the irradiated sample to dryness to drive off germanium and then reducing any arsenic to the trichloride to volatilize it. The residues were reduced with sulfur dioxide and copper precipitated as cuprous thiocyanate for β-counting. This particular determination is essentially free from interfering reactions. Gottfried and Yakovlev[72] used the same general procedure with some additional steps to remove other possible interfering elements. Extraction of the neocuproine complex of copper from aqueous fluoride solution into chloroform was applied by Leliaert.[73] Lloyd[74] devised a method for tellurium in which the ^{131}I produced by

$$^{130}Te \xrightarrow{n,\gamma} {}^{131}Te \xrightarrow{\beta-} {}^{131}I$$

was separated. This has a half-life of 8 days, so that the shorter-lived germanium activities can be allowed to decay before handling. The iodine was separated by oxidation and extraction into carbon tetrachloride prior to β-counting. On a 1-g sample, a sensitivity of 4 ppb was obtained. Rommel[75] activated boron in germanium to ^{11}C using protons and deuterons to give a sensitivity of 1 ppb; nitrogen interferes. However, this determination requires irradiation in a cyclotron; moreover, the half-life of ^{11}C is only 20 min, so that the combustion separation of the product must be carried out on the site.

Somewhat simpler chemical procedures have been devised by Ruzicka and his coworkers. Ruzicka and Stary[76] introduced the principle of substoichiometric separations to activation analysis. In this, for an element to be determined, the induced activity A is given by

$$A = a\frac{x}{m}$$

where a = activity of recovered fraction of weight m
x = amount of carrier added

Similarly, for a standard sample simultaneously irradiated,

$$A_s = a_s\frac{x_s}{m_s}$$

If y and y_s are the respective amounts of the elements to be determined, then

$$\frac{y}{y_s} = \frac{A}{A_s}$$

and if $$x = x_s$$

and $$m = m_s$$

then $$y = y_s\frac{a}{a_s}$$

If the amounts of carrier added are equal and the amounts of element separated are equal, then the ratio of the two activities will give the unknown concentration y. The first requirement is easily met since this is the amount of inactive element added after irradiation. The second is met by adding an amount of reagent, e.g., precipitant, insufficient for the amount of carrier added, i.e., a substoichiometric amount. This procedure has two advantages: (1) the reagent is more selective than when added in excess, and (2) no chemical yield need be determined since the conditions are the same for both sample and standard. Ruzicka et al. subsequently applied this to the determination of zinc and copper[77] in germanium dioxide by extracting with dithizone in carbon tetrachloride from suitably complexed solutions, of indium[78] by first separating in dithizone in carbon tetrachloride, then extracting this with aqueous EDTA, and of molybdenum[79] by extracting with 8-hydroxyquinoline in chloroform. In each case, the extracting reagent was in substoichiometric amount. Sensitivities better than 1 ppm were obtained by using a flux of 5×10^{12} neutrons/(cm^2)(sec) without any great effort toward ultimate levels.

A comprehensive scheme was devised by Morrison and Cosgrove[80] for the analysis of germanium. After distilling off germanium chloride in the usual way, the residue was reduced and arsenic trichloride distilled over. Total arsenic activity was measured with a single-channel recording γ-ray spectrometer at 0.55 and 1.22 Mev. The interference from ^{77}As was not considered, but the small interference due to the second-order reaction on ^{74}Ge producing ^{76}As was calculated and corrected for. The residue in the flask was evaporated to dryness and submitted to γ-ray spectroscopy.

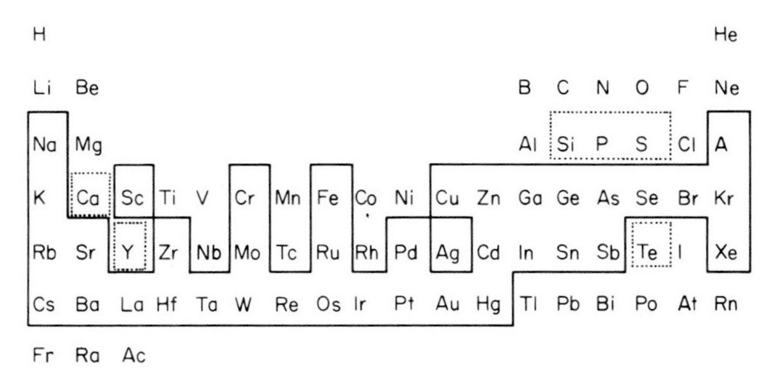

Fig. 5-7. Elements producing measurable activity after irradiation. *(After Morrison and Cosgrove.*[80]*)*

Almost all the heavy metals could be determined by this procedure, as shown in Fig. 5-7. The only other interference encountered was from the reaction

$$^{70}\mathrm{Ge} \xrightarrow{n,\gamma} {}^{71}\mathrm{Ge} \xrightarrow{\mathrm{K}} {}^{71}\mathrm{Ga} \xrightarrow{n,\gamma} {}^{72}\mathrm{Ga}$$

This also is a second-order reaction, and its contribution to the gallium-72 activity was calculated and a correction made. The sensitivity was between 1 ppb and 1 ppm for the majority of elements, but the flux used was only 3.4×10^{12} neutrons/(cm^2)(sec). Apparently identical procedures were used by Yakovlev et al.[81] and by Rytchkov and Glukhareva.[82] Robertson[83] used the separation shown in Fig. 5-8 prior to β-counting. However, it is difficult to achieve radiochemical purity, and γ-ray spectroscopy is a valuable addition if the equipment is available.

The direct γ-ray spectroscopy of germanium presents some problems inasmuch as the residual activity is due to relatively long-lived isotopes. Germanium-76 has an isotopic abundance of 7.76% and produces ^{77}Ge with a half-life of 11 hr. In addition, it decays to another unstable species, ^{77}As with a 36-hr half-life. The nuclear reactions are shown as

$$^{76}\mathrm{Ge} \xrightarrow{n,\gamma} {}^{77}\mathrm{Ge} \xrightarrow{\beta-} {}^{77}\mathrm{As} \xrightarrow{\beta-} {}^{77}\mathrm{Se}$$

De Neve et al.[84] have shown the significant interferences to be due to two fast-neutron reactions:

$$^{72}\mathrm{Ge} \xrightarrow{n,p} {}^{72}\mathrm{Ga}$$

and

$$^{72}\mathrm{Ge} \xrightarrow{n,\alpha} {}^{69m}\mathrm{Zn}$$

^{72}Ge has an isotopic abundance of 27.37%, ^{72}Ga has a half-life of 14.2 hr, and ^{69m}Zn a half-life of 13.8 hr. The contribution of both these reactions was not considered in the method of Morrison and Cosgrove.[80]

If we assume a 24-hr period for return of samples from the reactor, the original high level of activity due to the matrix has decayed to only one-fourth of its original value, based on ^{72}Ge, and by considerably less in total activity. A preliminary

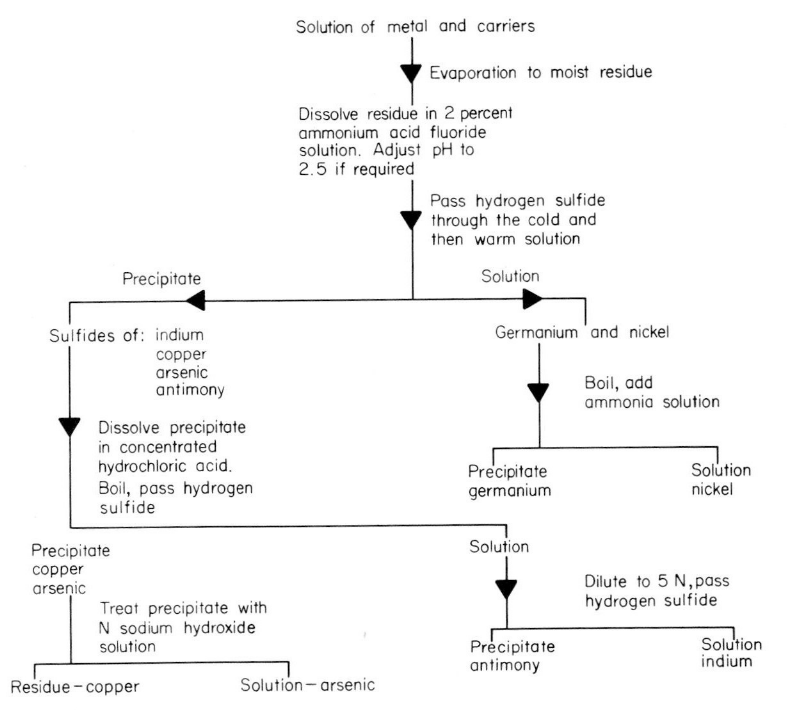

Fig. 5-8. Radiochemical separation of impurities in germanium. (*From Robertson.*[83])

chemical separation is almost mandatory even if γ-ray spectrometry is used. On the other hand, the sili con activity is due to

$$^{30}\text{Si} \xrightarrow{n,\gamma} {}^{31}\text{Si} \xrightarrow{\beta-} {}^{31}\text{P}$$

^{30}Si has only a 3% isotopic abundance and ^{31}Si only a 2.6-hr half-life. In 24 hr, the activity has decayed to 2^{-9}, that is, to 0.2 percent of its original value, which was lower to start with. This makes a chemical separation often unnecessary if γ-ray spectroscopy is used. The attractiveness of this procedure is reflected in the considerably larger volume of literature dealing with this matrix.

The earliest applications to semiconductor silicon appeared almost simultaneously in 1955. James and Richards[85] applied Smales and Pate's method for arsenic in germanium virtually unchanged to silicon. They reported a sensitivity better than 1 ppb. Morrison and Cosgrove[86] published a comprehensive scheme for γ-ray spectrometry in which no chemical separation was necessary. No interference was encountered with the γ-emitters, and the elements detectable were again those shown in Fig. 5-7. Sensitivities, using a flux of 3.4×10^{12} neutrons/(cm^2)(sec), were between 1 ppb and 1 ppm. Figure 5-7 also shows that several important elements form β-emitters on irradiation, and these cannot, of course, be determined by γ-ray spectroscopy. They were determined by β-counting by using calibrated aluminum absorbers in a Feather analysis. Kant et al.[87] carried out a separation into five groups, from which the individual elements were separated and β-counted. The phosphorus content was corrected for the ^{31}P formed from ^{30}Si. James and

Richards[88] gave values for 12 elements in two samples of silicon and described their method as a radiochemical separation which was different from that of Kant et al. but neglected to give details. A detailed scheme for 29 elements was given by Thompson et al.[89] and is essentially a sulfide separation. The group separations take into account the necessity of dealing with the short-lived isotopes first.

The direct application of γ-ray spectroscopy is naturally very attractive and has been applied by a number of workers.[81,82,90-92] However, for the highest sensitivity, a separation procedure is to be recommended. Yakovlev et al.[81] used a hydroxide procedure, but perhaps the most useful of the comprehensive schemes

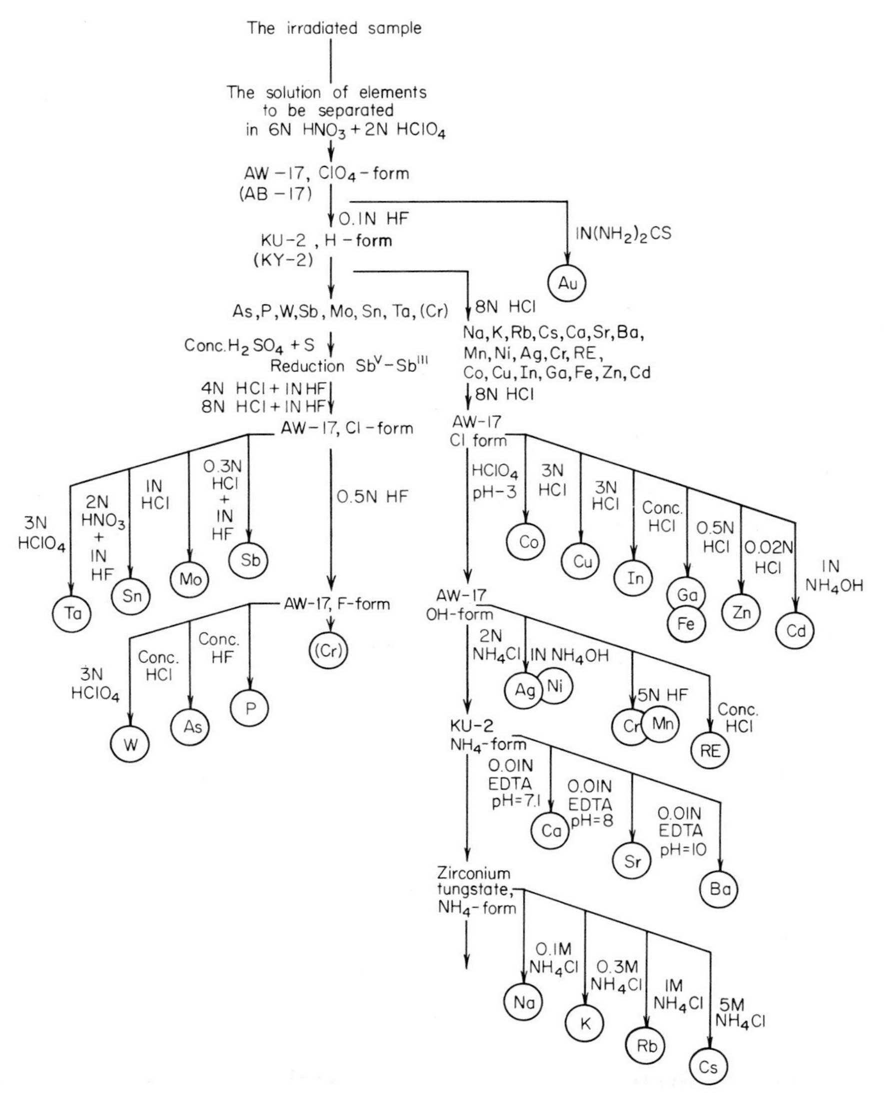

Fig. 5-9. Chromatographic separation in the activation analysis of silicon. (*After Moiseev et al.*[96])

is that due to Gebauhr et al.[93] They divided their determinations into three groups based on the half-lives of the isotopes formed. The short-lived isotopes are those with $t_{1/2}$ from 0.3 to 15 hr and the separation scheme is shown in Table 5-5. Medium-lived isotopes are those with $t_{1/2}$ 12 to 90 hr, and long-lived isotopes those above 90 hr. The separation schemes for these groups are given in Tables 5-6 and 5-7, respectively. Sensitivities as high as 0.001 ppb were obtained on 30 elements.

An alternative separation procedure was first applied to silicon by Nakai et al.,[94] namely, separation of the halogen complexes of a number of elements by anion exchange. The method was developed by Kalinin et al.,[95] using the fluorocomplexes, and is given in detail by Moiseev et al.[96] The separation scheme is shown in Fig. 5-9. A nitric-perchloric acid solution of the sample is passed first over an anion

Table 5-5. Scheme for the Separation of Short-lived Radioisotopes†

Radioisotopes	^{139}Ba	$^{115+117m}Cd$	^{64}Cu	^{116m}In	^{56}Mn	^{65}Ni	$^{197+199}Pt$	^{131}Te	^{69}Zn
Half-life, hr	1.4	543.0	12.8	0.9	2.58	2.6	190.5	0.42	0.98
Carrier, mg	10	5	5	5	5	2	5	5	5

Digestion (NaOH/KOH + carrier)
Dissolve in distilled water
Add Na_2CO_3, $N_2H_4.HCl$, $Na_2S_2O_4$
Boil $\xrightarrow{\text{Soln}}$ Add Na_2S $\xrightarrow[\text{Reppt}]{\text{Ppt}}$ Zn
↓ Ppt
Dissolve in aqua regia
Add 50 mg Si carrier
Twice to dryness with HCl $\xrightarrow{\text{Ppt}}$ (SiO_2)
↓ Soln
Add 2 ml satd alcoholic rubeanic acid $\xrightarrow{\text{Ppt}}$ Pt
↓ Soln
Add 2 N H_2SO_4 $\xrightarrow{\text{Ppt}}$ Ba
↓ Soln
Add 2 N HCl, SO_2, $N_2H_4.HCl$ $\xrightarrow{\text{Ppt}}$ Te
↓ Soln
Add NH_4OH (cold, pH 9) $\xrightarrow{\text{Ppt}}$ In
↓ Soln
Add 3 N H_2SO_4, boil, add H_2S $\xrightarrow{\text{Ppt}}$ Cu
↓ Soln
Add H_2S at pH 1 $\xrightarrow{\text{Ppt}}$ Cd
↓ Soln
Add NH_4OH, H_2S $\xrightarrow{\text{Ppt}}$ MnS + NiS
↓
Add HNO_3, $KClO_3$ $\xrightarrow{\text{Ppt}}$ Mn
↓ Soln
Add NH_4OH, dimethylglyoxime $\xrightarrow{\text{Ppt}}$ Ni

†Translated from Gebauhr et al.[93]

Table 5-6. Scheme for the Separation of Radioisotopes of Medium Half-life (12–90 hr)†

The elements in parentheses were determined only occasionally

RADIOISOTOPES	^{76}As	^{198}Au	(^{82}Br)	(^{115}Cd)	^{64}Cu	^{77}Ge	^{203}Hg	^{99}Mo	^{24}Na	^{32}P	^{122}Sb	(^{187}W)	^{69}Zn
Half-life, days	1.1	2.7	1.5	3.4	0.53	0.51	47	2.8	0.62	14	2.8	1.0	0.58
Carrier	5 mg each. (^{131}Ba), ^{60}Co, ^{51}Cr, ^{59}Fe, and rare earths (total) were also included for information.												

Volatilization of SiF_4 from $HF/HNO_3/H_2SO_4$ mixture (100°C–315°C) $\xrightarrow{\text{Dist}}$ Si (Hg) (Br)

↓ Residue

Add HCl, Br_2 (140°C) $\xrightarrow{\text{Dist}}$ $GeCl_4 + H_2S \xrightarrow{\text{Ppt}}$ Ge

↓ Residue

Add hydroxylamine, HCl (230°C) ⟶ $AsCl_3 + SbCl_3$

Add HCl (9 N), $H_2S \xrightarrow{\text{Ppt}}$ As

↓ Solution

Add HCl (2 N), $H_2S \xrightarrow{\text{Ppt}}$ Sb

↓ Residue (from hydroxylamine step)

Filter $\xrightarrow{\text{Ppt}}$ Au (Ba) (W)

↓ Filtrate

Add 3 N H_2SO_4, $H_2S \xrightarrow{\text{Ppt}}$ Cu

↓ Solution

Add HNO_3, boil off H_2S, add $ZrO(NO_3)_2 \xrightarrow{\text{Ppt}}$ P

(Rare earths) $\xleftarrow{\text{Ppt}}$ Add NH_4F(pH 2)

↓ Solution

(Mo) $\xleftarrow{\text{Ppt}}$ Add H_2SO_4 (5%), benzoinoxime

↓ Solution

(Cd) $\xleftarrow{\text{Ppt}}$ Add H_2S

Add alcohol (boil), $NH_4OH \xrightarrow{\text{Ppt}}$ Fe, Cr

↓ Solution

Add NH_4OH (pH 9) + $H_2S \xrightarrow{\text{Ppt}}$ Co, Zn

↓ Solution

Boil off H_2S, add HCl gas (ice cold) $\xrightarrow{\text{Ppt}}$ Na

†Translated from Gebauhr et al.[93]

Table 5-7. Separation Scheme of Long-lived Radioisotopes†

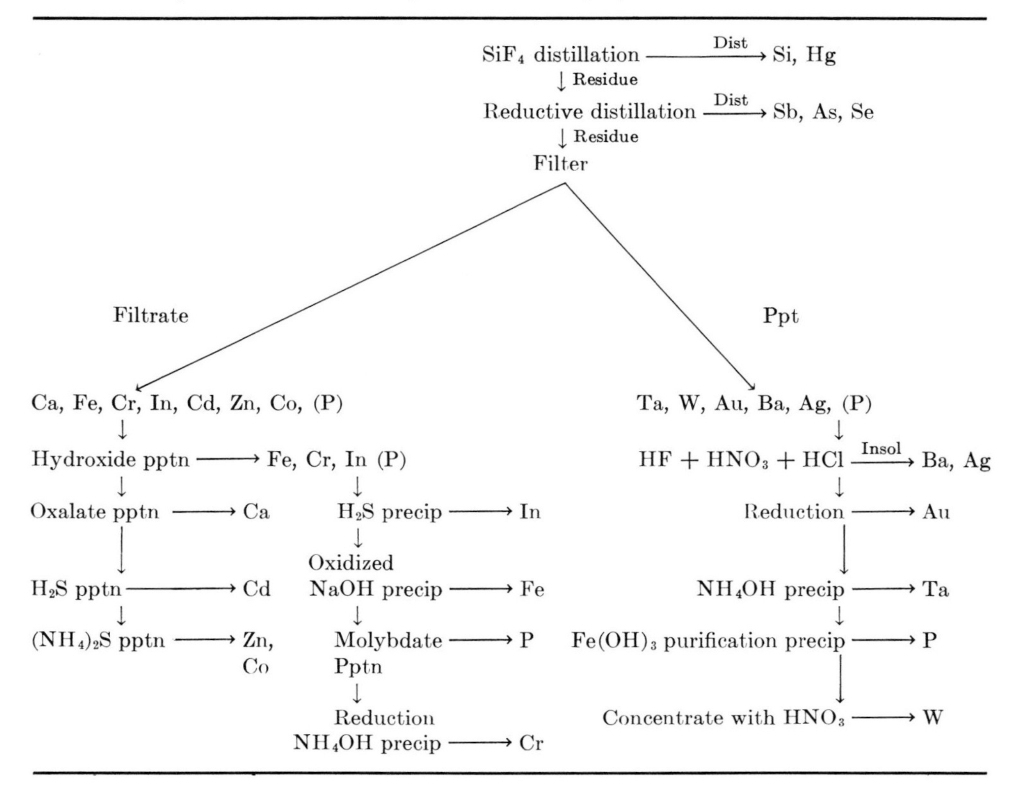

†Translated from Gebauhr et al.[93]

exchanger on which gold is adsorbed. The eluate is obtained in a hydrofluoric-nitric acid mixture and then passed through a cation exchanger. This separates the elements into two major groups: (1) those forming fluorocomplexes, e.g., tungsten and antimony, or anions, e.g., phosphate and arsenate; and (2) those remaining as cations, e.g., alkalis and alkaline earths. The two groups are further separated on anion exchangers with the alkaline earths being separated in a final fraction on a cation exchanger and the alkali metals on zirconium tungstate, an inorganic anion exchanger. The complete separation takes about 3.5 to 4 hr.

Generally, it is not necessary to carry out a complete separation since many of the elements either are very unlikely contaminants or are electrically inactive. Moiseev et al.[96] have shortened their method to a 2-hr operation by separating only into groups. Heinen and Larrabee have devised a shortened chemical separation which was given in detail by Kane.[1] It followed the previously published comprehensive schemes in separating the short-lived isotopes first. It covers eight of the more commonly encountered impurities or dopants in silicon. Subsequently, Heinen and Larrabee[97] modified this procedure somewhat, and this modified scheme is given in Fig. 5-10. They compared it with a γ-ray spectroscopic technique in which the output was treated by a computer program. The flowsheet for this

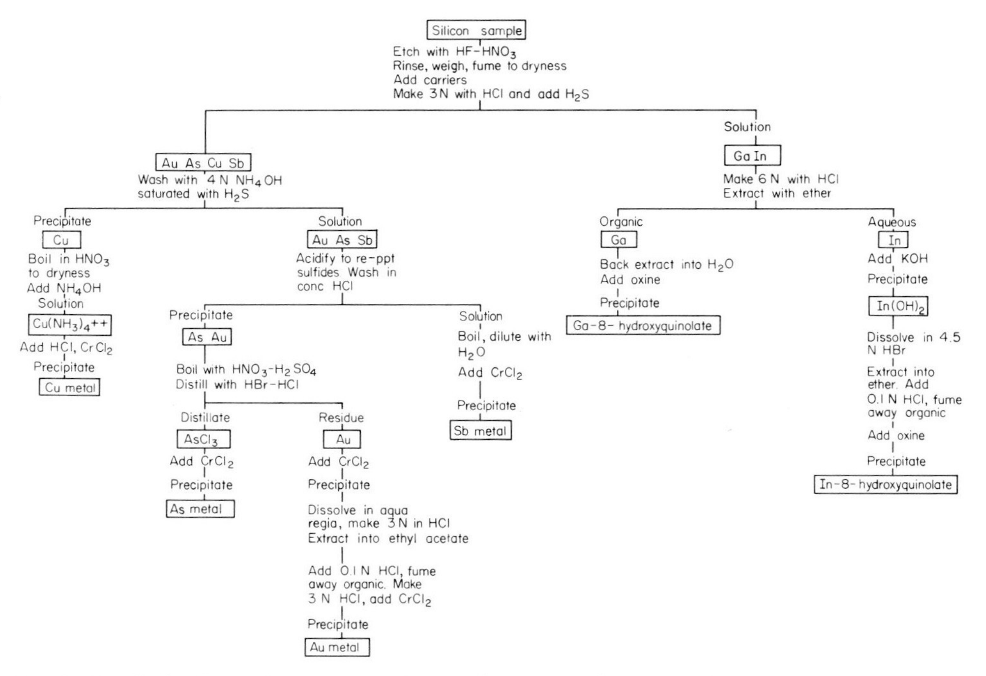

Fig. 5-10. Radiochemical separation of trace elements in silicon. (*Adapted from Kane,*[1] *modified by Heinen and Larrabee.*[97])

treatment, which was based on a linear least-squares fitting program developed by Helmer et al.,[98] was reported by Kane[39] and is given in Fig. 5-11. The program gain-shifts the spectrum in relation to the exact position of a well-defined photopeak in the standard supplied in the input. The gain-shift subroutine finds this peak in the sample spectrum and, using a three-point parabolic fit, shifts the spectrum to match that of the standard. This corrects for shifts within the analyzer and photomultiplier shifts due to differences in count rate. The ratio of sample to standard intensity is found by a linear least-squares fit. The intensity ratio is converted to concentration by using the weights of sample and monitor, count times of each, and the decay time elapsed between irradiation and analysis of monitor and sample. The γ-ray spectroscopic results were biased about 5 percent higher than the radiochemical values. One exception was noted with arsenic, and it was found that dissolution of silicon in hydrofluoric acid can lead to losses of arsenic trifluoride. In general, the spectroscopic method was preferred since it is more rapid. Its sensitivity is better than 0.1 ppb, adequate for much electronic material.

In many cases, only one element is of interest, usually a dopant. One of the more important is arsenic, and a method by James and Richards[85] has already been mentioned. They dissolved the silicon in sodium hydroxide containing hydrogen peroxide with arsenic trioxide as carrier. After acidification with hydrochloric acid and reduction to small volume, hydrobromic acid was added and arsenic distilled as the trichloride, after which it was reduced to the metal and β-counted.

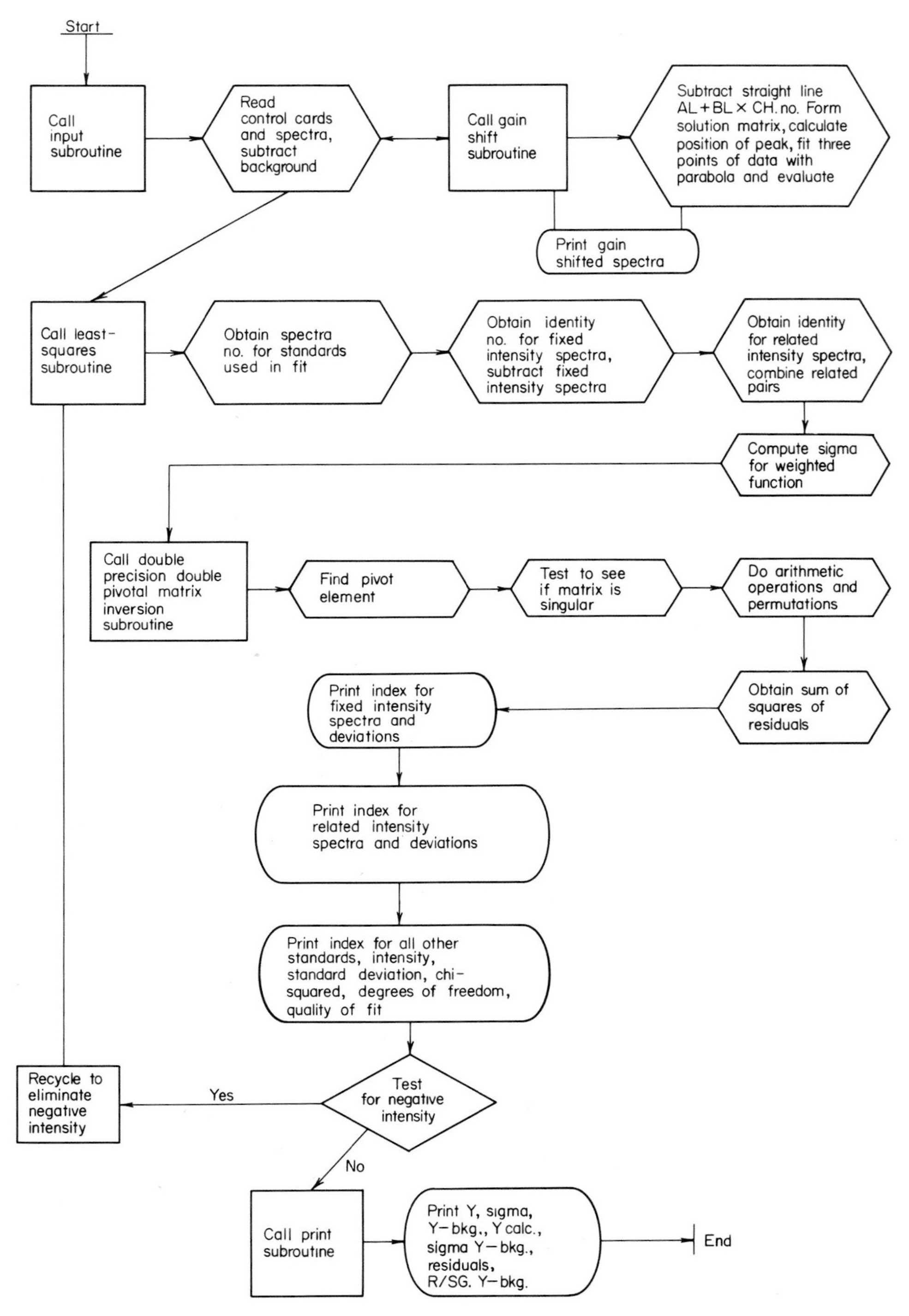

Fig. 5-11. Linear least-squares fitting program. (*From Kane.*[39])

Harvey and Smith[99] used the same solution procedure but employed a trap for any arsine evolved, which proved to be as high as 50 percent; they also determined antimony on the same sample. Smales et al.[100] criticized James and Richards' method on the basis of this arsine loss and recommended a fusion with sodium hydroxide and potassium nitrate. Their method included determinations for antimony and copper also. After fusion, the aqueous alkaline solution was treated with hydrosulfite; metallic copper and antimony were precipitated. The arsenic was distilled as the trichloride and precipitated as metal for counting. The copper and antimony were redissolved in hydrochloric acid from which copper was precipitated by alkali sulfide and counted as the thiocyanate. Antimony was recovered as the trisulfide following homogeneous precipitation by thiocyanate. β-counting gave sensitivities of 1 ppb or better. γ-ray spectroscopy gave comparable results for all three elements without preliminary chemical separation but with reduced sensitivity. However, Heinen and Larrabee[97] feel that there is some possibility of arsenic losses during any alkaline treatment, and, in view of the current sensitivity and speed, γ-ray spectroscopy is to be preferred for this determination.

James and Richards[101] determined phosphorus by dissolving the sample after irradiation in a hydrofluoric-nitric acid mixture and subsequently precipitating magnesium ammonium phosphate for β-counting. Berthel et al.[102] used the same procedure and found good agreement between their values and those obtained by calculation from Hall measurements assuming a bias due to chlorine. Harvey and Smith[99] dissolved in the same way but then separated the phosphate from cations by passage through a cation-exchange resin; iron and copper were subsequently eluted and determined. In all cases, corrections must be made for the reaction

$$^{30}\mathrm{Si} \xrightarrow{n,\gamma} {}^{31}\mathrm{Si} \xrightarrow{\beta-} {}^{31}\mathrm{P} \xrightarrow{n,\gamma} {}^{32}\mathrm{P}$$

and the effect of this has been calculated by Heinen and Larrabee,[97] based on a treatment due to Cali.[65] The theoretical sensitivity and interference from the secondary reaction are given in Table 5-8 for a flux of 10^{13} neutrons/(cm^2)(sec). The optimum irradiation time at this flux is 6 to 24 hr. For a flux of 10^{12} neutrons/(cm^2)(sec) it is 3 days; for 10^{14}, 2.4 hr. Both Cali[65] and Berthel et al.[102] pointed out the possible interferences from sulfur and chlorine which may be present by the reactions

$$^{32}\mathrm{S} \xrightarrow{n,p} {}^{32}\mathrm{P}$$

and

$$^{35}\mathrm{Cl} \xrightarrow{n,\alpha} {}^{32}\mathrm{P}$$

These are brought about by high-energy neutrons in the flux and are consequently less important than the secondary silicon reaction. Heinen and Larrabee[97] experimentally determined that, in the particular flux of 10^{13} neutrons/(cm^2)(sec) which they employed, the contribution of 1 ppb sulfur in a 1-g sample would be 0.185 ppb and of chlorine 0.035 ppb to the phosphorus content. Sulfur would not be expected to be high in high-purity silicon, although chlorine might be a significant factor in some samples since the usual manufacturing process is from the silicon chlorides.

Nozaki et al.[103] dissolved the irradiated sample by fusion in potassium hydroxide and added potassium iodide as a carrier in the determination of iodine. After

Table 5-8. Phosphorus Detection Limits and Interference during Irradiation at a Flux of 1 × 10^{13} neutrons/(sec)(cm^2)†

Irradiation time, days	Detection limit—primary reaction, ppb	Amount of interference from secondary reaction, ppb
0.1	1.52	0.2
0.5	0.31	1.0
1.0	0.16	1.9
3.0	0.05	6.0
7.0	0.03	14.5
14.0	0.02	30.6
21.0	0.01	48.1

† From Heinen and Larrabee.[97]

acidification with sulfuric acid, sodium nitrite was added and the liberated iodine distilled into sodium sulfite from which it was extracted into xylene, back-extracted into sulfite, and finally precipitated as silver iodide for β-counting. A sensitivity of 5 ppb was obtained using a 6×10^{11} neutrons/(cm^2)(sec) flux. Essentially the same procedure was later applied[104] in a separation of chlorine, bromine, and iodine by adding chromic oxide to the acidified sample solution prior to distillation. Bromine distills first, the receiver is changed, and, on continuation of the distillation, chlorine is evolved. Oxalic acid is added, after which iodine distills over. Each is collected in sulfite and treated by much the same procedure as that described for iodine. About 1 ppm was reported on their samples for the halogen from which the raw material was made.

Lobanov et al.[105] determined manganese both by γ-ray spectroscopy and by radiochemical separation as manganese dioxide followed by β-counting. A series of substoichiometric determinations has been published for heavy metals. Bismuth was determined by Ruzicka et al.[106] by extracting into dithizone in chloroform, using an insufficient amount of reagent. Several interferences, including copper and gold, are possible in this case. Krivanek et al.[107] determined copper by extracting the diethyldithiocarbamate into chloroform from a suitably masked solution. The substoichiometric yields were submitted to γ-ray spectroscopy. Beardsley et al.[108] extracted gold substoichiometrically with copper diethyldithiocarbamate in chloroform. They claimed high radiochemical purity and a sensitivity of better than 1 ppb. Zeman et al.[109] used a substoichiometric amount of 8-hydroxyquinoline in chloroform to determine gallium. Harvey and Smith[110] have applied the classical perchlorate separation to the determination of potassium and sodium.

Boron is of considerable interest in silicon, as it is in germanium, since it is a p-type dopant. Unfortunately, the only suitable reactions, as was pointed out above, are

$$^{11}\text{B} \xrightarrow{p,n} {}^{11}\text{C}$$

and

$$^{10}\text{B} \xrightarrow{d,n} {}^{11}\text{C}$$

and a cyclotron is required for generation of these high-energy particles. Rommel[75] applied the same method to silicon as he used for germanium with about the same 1-ppb sensitivity. Gill,[111] some time earlier, had used the same proton reaction to obtain 3-ppb sensitivity. Oxygen also falls into this same class of determination. Saito et al.[112] used the reactions

$$^{16}O \xrightarrow{\alpha,pn} {}^{18}F$$

and

$$^{16}O \xrightarrow{\alpha,d} {}^{18}F$$

to determine the oxygen content of silicon.

The problems of matrix activity encountered with germanium are multiplied when III-V compounds are considered. The irradiations produce the nuclear reactions given in Table 5-9. The long-lived active species induced in the matrix make safe handling a problem. For some of the longer-lived impurity species, the ^{72}Ga half-life is such that it can be allowed to decay for, say, 2 or 3 weeks prior to separation, and this was the approach taken in some of the methods for the metal described in Sec. 3-26. However, the addition of arsenic in gallium arsenide makes this virtually unworkable since the half-life of ^{76}As is twice that of ^{72}Ga. In 2 weeks, the gallium activity has dropped by a factor of 2^{24} or to 6×10^{-6} percent of its original level (ignoring ^{69}Ga), whereas the arsenic activity is down only by a factor of 2^{13} or to 0.01 percent of its original activity. As Lloyd[74] pointed out, 1 g of arsenic irradiated for 1 week at a flux of only 10^{12} neutrons/(cm^2)(sec) will produce 1 curie of radioactivity, so that after 2 weeks the activity is still 0.1 mC. Moreover, the gamma radiation is of high energy, ranging up to 2 Mev, making this even more hazardous. Very few impurity isotopes have half-lives that will allow their determination at a realistic level after this length of time.

The half-lives for the active indium and antimony species are even longer. Moreover, the indium isotopes have very high capture cross sections which make it virtually impossible to activate the sample beyond a few microns of the surface.

As a consequence of this high matrix activity, very little work has been attempted with III-V compounds. Green et al.[113] determined silicon, zinc, and magnesium by using their short-lived radionuclides. The active species were formed by

$$^{26}Mg \xrightarrow{n,\gamma} {}^{27}Mg \qquad t_{1/2} = 9.5 \text{ min}$$

$$^{30}Si \xrightarrow{n,\gamma} {}^{31}Si \qquad t_{1/2} = 2.6 \text{ hr}$$

$$^{68}Zn \xrightarrow{n,\gamma} {}^{69}Zn \qquad t_{1/2} = 52 \text{ min}$$

Table 5-9. Nuclear Reactions Which Produce Long-lived Radiation in Neutron Activation of III-V Semiconductor Compounds

Reaction	*Half-life*
$^{71}Ga(n,\gamma)^{72}Ga$	14.2 hr
$^{75}As(n,\gamma)^{76}As$	26.6 hr
$^{113}In(n,\gamma)^{114}In$	49 days
$^{121}Sb(n,\gamma)^{122}Sb$	2.8 days
$^{123}Sb(n,\gamma)^{124}Sb$	60 days

In one half-life, usable activities can be induced, while in the same period the activity of the matrix can be held at a safe level. However, the impurity elements must be separated and counted within one half-life also. A rapid transfer system was used and fast separation schemes devised. One-hundred-milligram samples were used for each determination; they were dissolved electrolytically in nitric acid in 3 to 4 min. Silica was precipitated by ammonium carbonate, redissolved, twice precipitated with perchloric acid, precipitated as molybdenum silicate, and finally precipitated again as silica from perchloric acid. This takes about 2 1/2 hr. Magnesium was precipitated as the hydroxide with sodium hydroxide, redissolved in hydrochloric acid, and passed over an anion exchanger to remove gallium. Magnesium hydroxide was precipitated from the eluate, redissolved, and finally precipitated as ammonium magnesium phosphate. This takes about 20 min. Zinc was precipitated as the carbonate, redissolved in hydrochloric acid, and passed over an anion exchanger to remove gallium. The zinc removed in the eluate was precipitated again as the carbonate, redissolved, precipitated as the mercurithiocyanate, redissolved again, and finally precipitated as the quinaldate. This takes about 1 hr. With these decay times and an irradiation time corresponding to the half-life of the particular species, the sensitivity for silicon is 10 ppb, for magnesium and zinc about 1 ppm. The sensitivity for zinc is reduced by the fast-neutron reaction

$$^{69}\mathrm{Ga} \xrightarrow{n,p} \underset{14\ \mathrm{hr}}{^{69m}\mathrm{Zn}} \xrightarrow{\mathrm{IT}} \underset{52\ \mathrm{min}}{^{69}\mathrm{Zn}}$$

Lloyd[74] was able to use the same procedure for tellurium in gallium arsenide which he had applied for germanium and which was described earlier. ^{131}I is formed as a daughter of ^{131}Te, formed by an n,γ reaction from ^{130}Te. ^{131}I has an 8-day half-life, so that it is possible to allow the arsenic to decay without appreciably reducing the sensitivity, which proved to be 4 ppb on a 100-mg sample.

A determination of oxygen in gallium arsenide has been described by Bailey and Ross[114] which depends on the reaction

$$^{16}\mathrm{O} \xrightarrow{\mathrm{T},n} {^{18}\mathrm{F}}$$

The tritons are formed in a thermal neutron flux by wrapping the sample in lithium foil; they are generated by the reaction

$$^{6}\mathrm{Li} \xrightarrow{n,\mathrm{T}} {^{4}\mathrm{He}}$$

The fluorine was counted after the gallium had been removed from the nitric-hydrochloric acid solution of the sample by ether extraction and the chloride and arsenate precipitated with silver. Alternatively, it could be precipitated as the lanthanum salt after the ether extraction. γ-ray spectroscopy was used for the determination, and a sensitivity of about 20 ppm was obtained.

5-6. SPECTROPHOTOMETRIC ANALYSIS

There is a considerable volume of literature on spectrophotometric analysis for specific elements in semiconductors, and a selection has been reviewed by Parker and

Rees.[115] They are of limited value since generally the sensitivities lie somewhere in the 0.1- to 1-ppm range. However, they can be used for dopants in some cases, and for this reason they are reviewed here. It should be borne in mind that in most determinations the reagents will have to be specially purified from the particular element being sought since the usual analytical reagents contain levels exceeding that in the semiconductor. High-quality water, specially distilled solvents, and reagents with very low blanks are essential and must be assumed in all the procedures outlined.

The most important elements in germanium and silicon, as has already been pointed out, are the group III and V dopants and the lifetime killers such as copper and gold. It is not surprising, therefore, that one of the earliest applications of these spectrophotometric methods was to the determination of arsenic in germanium dioxide intended for crystal rectifiers. Payne[116] dissolved the oxide in ammonium oxalate solution and extracted the arsenic from this into a chloroform solution of diethyldithiocarbamate. Under these conditions, germanium is not extracted. The organic solution was decomposed with perchloric acid and the arsenic determined by a Gutzeit test. Luke and Campbell[117] applied this procedure to germanium, using an oxalic acid–hydrogen peroxide dissolution as suggested by Payne. They replaced the Gutzeit step by a molybdenum-blue finish. Goto and Kakita[118] used the same procedure, except that solution of the germanium was made in hydrogen peroxide. Fowler[119] modified the Payne method by replacing the Gutzeit test by an absorption of arsine in silver diethyldithiocarbamate solution in pyridine, the optical density of which was measured. This eliminated interferences from silica, derived from glassware, which he encountered with Luke and Campbell's method. A procedure similar to Payne's was carried out by Tumanov et al.,[120] except that the arsenic was separated from germanium by coprecipitation with manganese dioxide. Rezac and Ditz[121] simplified Fowler's method by removing the germanium by distilling it from a hydrochloric acid solution in a stream of chlorine followed by the evolution of arsine into silver diethyldithiocarbamate in pyridine. Luke and Campbell avoided this distillation because they found results to be about 5 percent too low, and the results of Rezac and Ditz, although they noted no loss, appear to substantiate this. However, at the levels encountered, this hardly seems significant, and the method is the simplest.

Several of the procedures for arsenic also included methods for antimony. Luke and Campbell[117] dissolved the sample in nitric-hydrochloric acid containing perchloric acid and boiled to expel germanium chloride, keeping the hot-plate temperature below 200°C. After evaporating to fumes with sulfuric acid, the antimony was reduced with sulfur and extracted as the cupferrate into chloroform, the organic matter destroyed with perchloric acid, and the solution extracted with cupferron in chloroform, which leaves pentavalent antimony essentially isolated except for gold, which is removed by coprecipitation with selenium. The final solution is oxidized with ceric sulfate and the complex with Rhodamine B extracted into benzene for determination. Goto and Kakita[118] dissolved the sample in sodium hydroxide, acidified with sulfuric acid, and coprecipitated the antimony with manganese dioxide; after redissolving in hydrochloric acid and oxidizing with ceric sulfate, the complex with methyl violet was extracted into amyl acetate and its optical density measured. Rezac and Ditz[121] distilled off the germanium chloride as they did for the arsenic determination, reduced, and extracted the complex with crystal violet into toluene for extinction measurements. The volatilization of germanium chloride as used by Luke and Campbell was also employed by Roberts et al.,[122] but they found it necessary to reduce the hot-plate temperature to 130°C to avoid losses of antimony. The resultant solution in hydrochloric acid was treated first with formic acid (to reduce any Sb^{IV} to Sb^{III}) and then oxidized with ceric sulfate (to Sb^{V}). The chloro compound was extracted into diisopropyl ether and the organic solution shaken with an aqueous Rhodamine B solution; after separation, the extinction of the ether layer was measured. This procedure is probably the easiest.

The determination of phosphorus was effected by Luke and Campbell[117] after removing the germanium by volatilization as before, taking care to keep the phosphate oxidized and not to heat beyond the fumes of perchloric acid in the final evaporation. Hydrobromic acid was added, and arsenic, antimony, and selenium were boiled off as the bromides. A little lead was added to plate out copper and gold, and fluoborate was added to complex zirconium before finishing with the molybdenum-blue procedure. Ishihara and Taguchi[123] used much the same procedure for germanium oxide, except that they introduced an extraction with 8-hydroxyquinoline in chloroform to remove vanadium and carried out the molybdenum-blue reduction after extracting the phosphomolybdate into a butanol-chloroform mixture. The same method was arrived at by Roberts et al.,[122] apparently independently of Ishihara and Taguchi.

Boron in germanium was determined by Luke[124] by dissolving the sample in sodium hydroxide solution containing hydrogen peroxide, precipitating the germanate by adding methanol, distilling off methyl borate which was trapped, and determining the boron with curcumin. Gallium was determined by Luke and Campbell[125] by removing the germanium by volatilization, as in their phosphorus method. Gallium chloride was extracted into ether and then back into water, sodium cyanide added to complex interferences such as iron, and the 8-hydroxyquinolinate extracted into chloroform for determination. Indium was also determined by these same authors.[125] After removal of the germanium as before, the solution was masked with citrate and extracted with chloroformic dithizone; this removes bismuth. The solution was neutralized, cyanide added, and the dithizone extraction repeated. The chloroform solution was evaporated and treated with perchloric acid to remove organic matter, after which the 8-hydroxyquinolinate was formed and extracted into chloroform for an extinction measurement.

A method for copper was employed by Luke and Campbell[117] in which, after removal of germanium as the chloride, copper was reduced to Cu^{I} with hydroxylamine hydrochloride and reacted with neocuproine in a citrate buffer, and the complex was extracted into chloroform for absorptiometry. Baba[126] confirmed the superiority of neocuproine over dithizone in determining copper in germanium dioxide. He used much the same method as Luke and Campbell, extracting into pentanol instead of chloroform. Titanium has been determined by Nazarenko and

Biryuk,[127] using the reagent disulfophenylfluorone, 9-(2′,4′-disulfophenyl)-2,3,7-trihydroxy-6-fluorone,

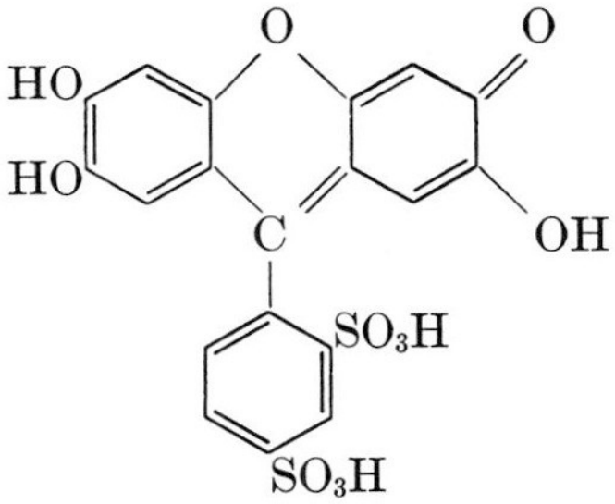

After removal of the germanium in the usual way, thioglycolic acid and EDTA are added as masking agents, and addition of the reagent in the presence of pyridine gives a color with an absorption maximum at 570 mμ. Luke[128] dissolved in hydrochloric-nitric acid, boiled to expel germanium, then reduced any sulfate to sulfide. After distilling into ammonia, sulfide was determined as colloidal lead sulfide. An interesting method for iodine in germanium was described by Tumanov et al.,[129] in which the reaction between ceric ion and arsenite is used as a measurement. After solution of the sample in potassium hydroxide and hydrogen peroxide, the two reagents are added to the sample and to a blank. The extinctions of the two solutions are measured after 60 min; iodine catalyzes the reaction, and the difference can be related to its concentration. Unfortunately, several other possible impurities also affect the rate. Ducret and Cornet[130] determined carbon in germanium by heating with sulfur in an evacuated tube to 1100°C. Germanium formed the sulfide while carbon formed carbon disulfide. After cooling, the tube was opened under benzene, diethylamine added to form diethyldithiocarbamate, and this reacted with copper to give a color. Babko et al.[131] have determined oxygen by a somewhat similar approach. The sample was fused with sulfur at 700°C when oxygen formed sulfur dioxide, which was determined colorimetrically with fuchsine-formaldehyde reagent.

With a few obvious exceptions, all the above methods can be adapted to germanium oxide or to germanium halides with minor modifications in the sample solution.

For the determination of arsenic in silicon, Luke and Campbell[117] adapted the molybdenum-blue method they had devised for germanium, dissolving the sample in sodium hydroxide with hydrogen peroxide and removing the silica after dehydration with perchloric acid. Nazarenko et al.,[132] after dissolving the sample in sodium hydroxide, acidified and distilled off the arsenic as arsine, trapping it in mercuric chloride solution. Molybdenum blue was formed by a molybdate-hydrazine reaction and extracted into isoamyl alcohol for color determination. Tumanov et al.[120] used a Gutzeit test after dissolving the silicon in sodium hydroxide, and a variation given by Rigin and Melnichenko[133] uses an electrolytic reduction to arsine. Phosphorus was determined by Pohl and Bonsels[134] by volatilizing the silicon as the tetrafluoride and reacting the residue to form molybdenum blue.

Boron was determined by Luke[124] by the same procedure that he used for germanium, adding a second precipitation from methanol for the sodium silicate and finishing, as before, with curcumin. Ducret and Seguin[135] described a rather involved procedure in which silicon was decomposed with an ammonium fluoride solution and the fluoborate extracted with tetraphenylarsonium chloride in chloroform. The extract was evaporated with sodium hydroxide and the residue treated with curcumin in ethanol containing trichloracetic acid. After recrystallization by drying, the complex was redissolved in methanol and its color measured. Ducret[136] simplified this procedure somewhat by extracting the fluoborate as the methylene-blue complex with 1,2-dichloroethane and determining its optical density directly. Luke and Flaschen[137] increased the sensitivity of Luke's earlier procedure by introducing a hydrothermal treatment of the silicon sample. As much as 1 g powdered sample was reacted with 14 ml 0.5% sodium hydroxide solution in a platinum-lined autoclave for 5 hr at 350°C and 5,000 psi. The silicon formed quartz, but the boron oxide remained in the mother liquor from which it was distilled as methyl borate. The distillate was reacted to form the colorimetric curcumin complex. This modification increased the sensitivity to 20 ppb. Barcanescu and Minasian[138] also separated the boron as methyl borate but used the blue color with carmine for determination. Pohl et al.[139] used a technique in which a sample, up to 20 g, was reacted with bromine at 750°C to form silicon tetrabromide, which was subsequently volatilized from the relatively nonvolatile boron tribromide. This residue was extracted as methyl borate into isopropyl ether, where it was reacted with curcumin. Marczenko and Kasiura[140] dissolved the silicon in nitric-hydrofluoric acid containing mannitol and evaporated to dryness. Aluminum sulfate solution was added and heated to dissolve the residue, after which it was neutralized with sodium carbonate and ignited. This residue was dissolved in sulfuric acid and methyl borate distilled over into aqueous sodium hydroxide containing glycerol. The distillate solution was evaporated, ignited, dissolved in concentrated sulfuric acid, and reacted in this medium with carmine. Boron was determined by Berthel et al.[102] by the method of Pohl et al., and they obtained good agreement with values calculated from Hall measurements. Roberts et al.[122] used Luke's method and were able to improve the sensitivity somewhat by precipitating the boron-curcumin complex from the final solution and redissolving in a smaller volume.

An interesting method for copper in silica was described by Dolmanova and Peshkova,[141] in which the catalytic effect of this element on the oxidation of hydroquinone by hydrogen peroxide was used as the determining factor. The reaction was followed by measuring the optical density of the solution. A similar approach had been mentioned earlier by Burkhalter,[38] quoting work by Baird, as being applicable to semiconductors. He used the reduction of iron by thiosulfate, following the reaction by the color of the ferric salicylate complex. These kinetic methods are extremely sensitive. Dolmanova and Peshkova obtained a sensitivity of 5 ppb and found it to be surprisingly free of interference. This is not always the case in these procedures, as witness the method of Tumanov et al.[129] for iodine. This procedure, which was described above for germanium, was also used for silicon, but chlorine in relatively large amounts can interfere, and other possible impurities such as mercury, silver, lead, and tellurium can inhibit the reaction. Nevertheless, this approach is a promising one for determinations at levels of interest in semiconductors and probably merits more attention.

Lebedeva and Nazarenko[142] determined tin in silicon, using a phenylfluorone reagent,

HO O O

HO C OH

NO_2

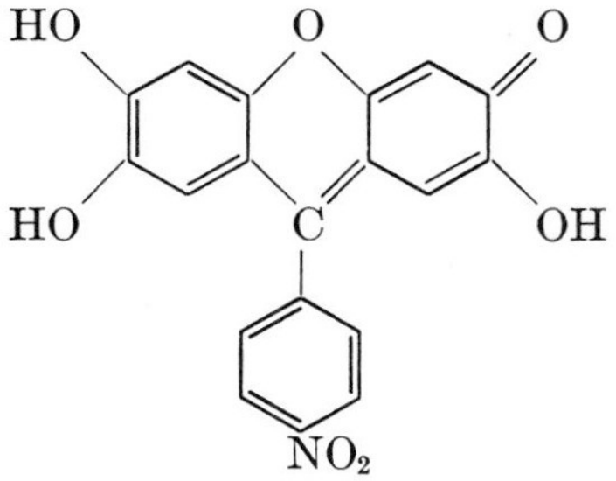

To avoid several interferences, the tin is separated first by an extraction with diethyldithiocarbamate in chloroform. Nazarenko and Biryuk[127] used their disulfophenylfluorone reagent for the determination of titanium in silicon, removing the matrix as silicon tetrafluoride and then proceeding as for germanium. A method for iodine was devised by Nazarenko and Shustova[143] in which, after solution of the sample in sodium hydroxide and acidification, the iodide was oxidized to iodine with nitrate and extracted into benzene. By oxidizing this to iodate and reacting with potassium iodide, a sixfold increase in the iodine was obtained; it was measured absorptiometrically in benzene. Carbon was also determined in silicon by Ducret and Cornet[130] by the procedure given above for germanium; silicon forms the disulfide, and the subsequent treatment is the same.

In examining the III-V compounds, the more important dopants are the group II and VI elements, the latter group, the n-type dopants, being of more concern. In addition, group IV elements such as silicon may also act as dopants. Sulfur was determined by Adler and Paff[144] in gallium arsenide by the methylene-blue method. After dissolving the sample in nitric-hydrochloric acid, arsenic was volatilized as the bromide. The sulfate present in the residue after evaporation was reduced by hypophosphorous acid in hydriodic acid in a stream of nitrogen, and the sulfide was trapped and reacted with *N*,*N*-dimethyl-*p*-phenylenediamine in the presence of ferric perchlorate to form methylene blue. A similar procedure was used by Goryushina and Biryukova,[145] except that the evolved sulfide was reacted to form colloidal lead sulfide. Selenium was determined by Bush and Cornish[146] by coprecipitating with tellurium as carrier and, after redissolving, reacting with asymmetric diphenylhydrazine. Roberts et al.[122] determined tellurium in gallium arsenide by first isolating it as the diethyldithiocarbamate complex in chloroform, then forming the iodotellurite yellow color for measurement. They also applied the method used for silicon in arsenic (Sec. 3-37) to gallium arsenide. The arsenic was removed by hydrochloric acid and bromine dissolved in carbon tetrachloride, and gallium in the residue was removed by extraction of the chloride into ether. The residual hydrochloric acid solution was evaporated to dryness; the residue was dissolved in potassium hydroxide, acidified, and reacted with molybdate; and the silicomolybdate was extracted into *n*-pentanol and reduced with stannous chloride to molybdenum blue. A similar procedure was used by Soldatova and Kristaleva[147] for phosphorus, the differences being an extraction into ether instead of *n*-pentanol and preliminary removal of the arsenic by a hydrochloric-hydrobromic acid evaporation. However, according to Goryushina and Esenina,[148] this gives erratic results, and a hydrochloric acid–bromine mixture is preferable.

Knizek and Galik[149] determined iron in gallium arsenide by dissolving the sample in hydrochloric-nitric acid, reducing the iron to ferrous with hydroxylamine, and extracting the complex with bathophenanthroline into chloroform for measurement (cf. Secs. 3-26 and 3-27). Copper was determined in gallium arsenide by Knizek and Pecenkova,[150] using neocuproine (cf. Secs. 3-26 and 3-27); after reduction with hydroxylamine to Cu^{I}, the reagent was added and the complex extracted into chloroform for determination of the optical density.

5-7. FLUORIMETRIC ANALYSIS

Although fluorimetry generally gives better sensitivity than straight absorptiometry, the number of elements that give suitable complexes is restricted. Consequently, the number of applications to semiconductors is small, although this might be a fruitful field for additional investigation.

A rapid method for gallium in germanium was devised by Shigematsu.[151] After dissolution of the sample in sodium hydroxide containing hydrogen peroxide, the solution was acidified and buffered to pH 3.9, and 8-hydroxy-2-methylquinoline was added. The complex was extracted into chloroform and its fluorescence measured.

Parker and Barnes[152] utilized the fluorescence of the borate-benzoin compound for the determination of boron in silicon. The silicon was submitted to the hydrothermal method of Luke and Flaschen,[137] in which the hydrolysis to silica forms an insoluble quartz, leaving borate in the sodium hydroxide mother liquor. It was separated as ethyl borate by using a high-vacuum distillation and reacted with benzoin to form the fluorescent end product. As ensitivity of 0.03 ppm on a 1.5-g sample was obtained.

Gallium was determined in silicon by Nazarenko et al.[153] after removal of the matrix as the fluoride by forming the fluorescent compound with sulfonaphtholazoresorcinol [1-(2,4-dihydroxyphenylazo)-2-naphthol-4-sulfonic acid]. A sensitivity of 10 ppb was obtained on a 1-g sample. Alimarin et al.[154] removed silica by evaporating with hydrofluoric acid and adding benzene and Rhodamine 6G solution. The fluorescent benzene solution was measured for intensity. The method was calibrated down to 25 ppb of tantalum as Ta_2O_5 (cf. Sec. 3-17).

5-8. POLAROGRAPHIC ANALYSIS

Conventional polarographic methods are generally insufficiently sensitive for application to semiconductor materials. The few applications that have been made with the dropping-mercury electrode have been in conjunction with one of the more sophisticated polarographs. Gokhshtein et al.[155] used an oscillographic polarograph to determine several impurities in germanium. The sample was distilled from hydrochloric acid to remove germanium and the residue taken up in a sodium thiocyanate–base solution. Copper, lead, zinc, and iron were determined with a sensitivity of 10 ppb and nickel with a sensitivity of 100 ppb. On a separate sample, the residue was dissolved in a sodium thiosulfate base and silver determined to the 10-ppb level. Pohl and Bonsels[156] also used an oscillographic polarograph for the analysis of silicon. After removing the matrix as the fluoride, the residue was oxidized with hydrogen peroxide and extracted with isopropyl ether; iron and thallium pass to the organic phase. After removal of organic matter, the aqueous phase was taken up in an ammonia base and polarographed for copper, cadmium, nickel, and zinc. In another run, the aqueous phase was taken up in a tartrate base to determine bismuth, lead, indium, and zinc. The organic phase was evaporated, organic matter removed, and the residue, after reduction, dissolved in a tartrate-base electrolyte and the iron and thallium determined. The sensitivity was only about 1 ppm. The same instrument was used by Bush and Cornish[146] to determine selenium in gallium arsenide, using arsenate derived from the sample as the base electrolyte. The sample was dissolved in aqua regia, nitric acid evaporated off by boiling with hydrochloric acid, and the gallium extracted as the chloride into diisopropyl ether. The aqueous phase was repeatedly evaporated with nitric acid to ensure oxidation to the pentavalent state, and the resulting aqueous solution polarographed. A sensitivity of 2.5 ppb could be obtained.

A square-wave polarograph was applied by Jennings[157] to the determination of copper and lead in indium arsenide. The sample, after solution in a nitric-hydrochloric acid mixture, was evaporated to dryness and the residue dissolved in phosphoric acid for polarography. Copper was determined down to 0.1 ppm and lead to 0.2 ppm. The same technique was applied[158] to gallium arsenide. After dissolution and evaporation as before, the residue was dissolved in hydrochloric acid containing sufficient potassium bromate to oxidize any remaining As^{III} to As^{V}. Polarography in this electrolyte gave sensitivities of 0.1 ppm for copper, indium, and cadmium and 1 ppm for bismuth.

Bush[159] states that he has applied this last method of Jennings with the addition of an extraction of the gallium with isopropyl ether. However, his subsequent description is of an extension of his and Cornish's earlier method for selenium. He gave half-wave potentials in a 25% arsenate base for copper, bismuth, antimony (III), lead, selenium (IV), cadmium, indium, and tellurium (IV). He does not mention the instrument, but in view of this earlier paper it is more probably an oscillographic instrument than a square-wave.

An attractive technique is one which we have referred to in Chap. 3 as stripping polarography, without further elaboration. This technique has also been described as amalgam polarography, stationary-drop polarography, hanging-mercury-drop-electrode (HMDE) polarography, and others, but is probably most accurately described as cathodic deposition and voltage-sweep stripping chronoamperometry. It has been reviewed by Kemula and Kublik.[160] The impurity being sought is concentrated from the base electrolyte by electrolysis into a mercury drop. The potential is then reversed and a rapid scan made to obtain an anodic wave. It has the advantage that sensitivities approximating those of the more exotic instruments can be obtained by using conventional polarographs. Kataev et al.[161] applied the method to gallium by dissolving in nitric-hydrochloric acid, evaporating to dryness, and dissolving the residue in potassium hydroxide solution, which was used as the base electrolyte. An HMDE was used as a cathode in an electrolysis at −1.0 volt for 30 min. The current was then reversed and scanned rapidly to +0.4 volt. Copper and lead were determined down to 0.1 ppm. Vinogradova and Kamenev[162] determined bismuth and antimony in germanium by dissolving in nitric-hydrochloric acid mixture, then distilling off both acid and germanium chloride. The residue was dissolved in dilute hydrochloric acid and electrolyzed at −0.3 volt for 30 min using an HMDE. An anodic sweep enabled as little as 1.3-ppb bismuth and 2-ppb antimony to be determined.

Procedures for a number of impurities have been devised by Burson and applied to germanium, silicon, gallium arsenide, indium arsenide, and indium antimonide; they have been given in detail by Kane.[1] The essential feature of a polarograph suitable for voltage-sweep stripping chronoamperometry is a fast sweep. Such instruments are available from Sargent,† their Model FS, or Metrohm,‡ their Polarecord E261-R. The Sargent instrument scans in 1 min, the Polarecord in 48 sec. It is a relatively simple job to convert any automatic recording polarograph to this rapid sweep. For example, the Sargent Model XV is adapted by substituting

†E. H. Sargent and Co., 4647 West Foster Ave., Chicago, Ill. 60630.

‡Metrohm A. G., Herisau, Switzerland. In the United States: Brinkmann Instruments, Cantiague Road, Westbury, N.Y. 11590.

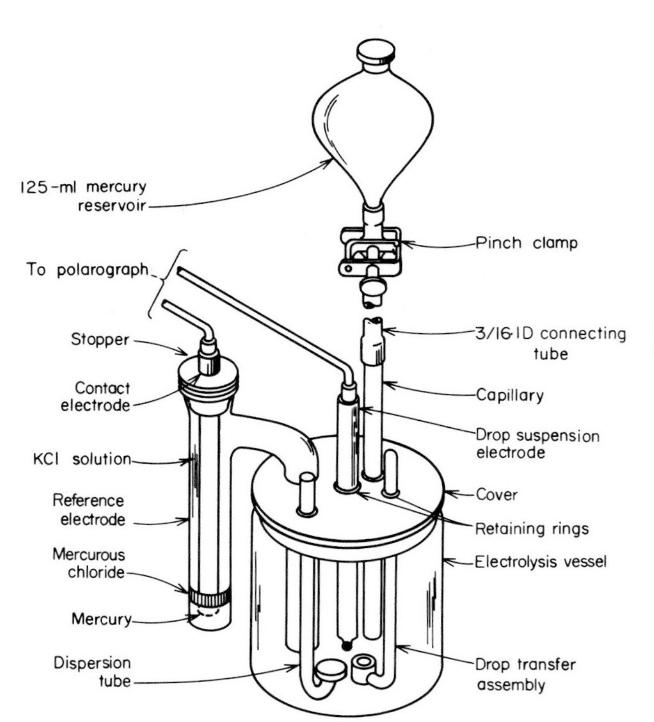

Fig. 5-12. The hanging-mercury-drop-electrode cell. *(Courtesy of E.H. Sargent and Co.)*

three motors: the synchronous sweep motor, converting the sweep time from 10 to 1 min; the chart speed motor, increasing the speed from 1 to 10 in./min; and the pen drive motor, increasing the response from 10 to 1 sec full scale.

The HMDE cell has taken several forms, but one of the more convenient designs, and one that is available commercially, is shown in Fig. 5-12. The HMDE itself is constructed from a piece of 26-gauge (0.404-mm) platinum wire mounted in 6-mm glass tubing. The wire is sealed and cut off, and the end is polished flat. It is then etched back to a depth of 0.5 to 1 mm by boiling in aqua regia. Before use, the platinum is mercury plated by electrolysis in a mercurous perchlorate solution. The mercury drop itself is formed after the test solution has been added to the cell by collecting a standard number, two or three, of drops from the dropping-mercury electrode in the drop transfer assembly and hanging the globule from the electrode.

Germanium is solubilized by reacting with a sulfuric-hydrofluoric acid mixture, bringing to the boil, adding nitric acid to dissolve, and evaporating to fumes. Silicon is treated with nitric-hydrofluoric acid and evaporated to dryness. The residue in either case is dissolved in 6 *M* ammonia and the solution transferred to the cell, deaerated, and electrolyzed for 30 min with standard stirring at −1.5 volts. An anodic scan is then made from −1.5 to −0.1 volt at 0.033 volt/sec. If present, impurity peaks will be detected at half-wave potentials as follows:

	Volts
Zinc	−1.04
Indium	−0.84
Cadmium	−0.80
Tin	−0.76, 0.61
Lead	−0.55
Copper	−0.45, 0.22
Thallium	−0.37
Bismuth	−0.24

The concentrations are determined by the method of standard additions.

At −1.5 volts gallium will plate out. To avoid this, a modified method was devised for gallium arsenide. After evaporation to dryness with nitric-hydrochloric acid and dissolution in ammonia, the electrolysis is carried out at −1.0 volt and the subsequent anodic scan from −1.0 to −0.1 volt. Since this does not include zinc, a separate determination is made in *M* sodium hydroxide, electrolyzing at −1.5 volts and scanning from −1.5 to −0.1 volt. In this medium, gallium does not plate out and zinc has a half-wave potential of −1.20 volts.

Indium arsenide and antimonide present something of a problem since indium is electrolyzed at −0.8 volt. The sample is dissolved in nitric-hydrochloric acid, evaporated to dryness, then evaporated to dryness repeatedly with hydrobromic acid to remove arsenic or antimony. The residue is dissolved in hydrobromic acid, and indium bromide extracted with isopropyl ether. The aqueous phase is evaporated, treated with perchloric acid to remove organic matter, and 6 *M* ammonia added for determination as before. A short method for copper, lead, bismuth, and tin avoids much of the treatment by dissolving the nitric-hydrochloric acid residue in 0.1 *M* phosphoric acid and electrolyzing at −0.49 volt before scanning from −0.49 to +0.5 volt. In this medium, the half-wave potentials in volts are

Tin	−0.38
Lead	−0.33
Bismuth	+0.05
Copper	+0.08

With the conditions given, sensitivities of about 50 ppb are attainable, although a micro cell, such as that used by Vinogradova and Kamenev,[162] would undoubtedly improve on this. Longer electrolysis times could also be used, although the lower limit is set in practice by the reagent impurities. However, these reagents are fewer in number than those in a chemical concentration, and there is no transfer required. Consequently, for a specific impurity, this method is well worth considering.

5-9. OTHER CHEMICAL METHODS

A volumetric method has been described by Galik and Knizek,[163] in which a total impurity level was estimated in gallium arsenide. After solution in a nitric-hydrochloric acid mixture, the solution was evaporated and dissolved in an ammonia-tartrate solution; the tartrate masks the gallium. An extraction was made with successive 2-ml portions of 10^{-4} *M* dithizone solution in chloroform until colorless. The combined extracts were washed with 0.01 *M* ammonia and the aqueous wash combined with the previous aqueous phase. The excess was back-titrated with standard mercuric solution, plotting a titration curve by absorptiometric measurements at 620 mμ. This gives a quantitative value for nine metals, viz., mercury, copper, bismuth, cadmium, lead, cobalt, nickel, zinc, and silver. The sensitivity is 2.3×10^{-8} mole, or the equivalent of 1.5 ppm Zn on a 1-g sample.

A method due to Schink[164] determines carbon in silicon by a combustion procedure. After dissolution of the sample in sodium hydroxide solution, finely divided silica is added to adsorb any undissolved carbon. The mixture is centrifuged and the solid phase transferred to a tube with a lead chromate combustion mixture. The tube carries a calibrated capillary with a drop of water held in it. The tube is heated to 600°C, then cooled, and the volume of carbon dioxide formed deduced from the drop position. A range of 25- to 100-ppm carbon was determined.

Gases in semiconductors have been determined by vacuum-fusion analysis. Briefly, this technique melts the sample at an elevated temperature in a graphite crucible under high vacuum. Oxygen is converted to carbon monoxide; hydrogen and nitrogen evolve as such. They are pumped to a volumetric detection system, where the carbon monoxide is converted over copper oxide to carbon dioxide and frozen out in a liquid-nitrogen trap. Hydrogen is diffused through palladium, and nitrogen is determined by difference. Since silicon attacks graphite, Beach and Guldner[165] used an iron bath to dilute it, and a similar procedure was used by Donovan et al.[166] By careful design of the apparatus, these latter workers were able to determine 1 ppm oxygen in silicon. This procedure has also been described by

Kane,[1] using a temperature of 1700°C for the bath. Turovtseva and Kunin[167] recommended a platinum bath at 1800°C for this determination since, on addition of silicon, dissolved carbon is precipitated in the molten iron, forming a pasty mass from which gases are slow to evolve. There is no comparable problem with germanium, and Beach and Guldner[165] used a dry bath for determining gases in this material. A similar procedure, using a temperature of 1550°C, has been reported by Kane.[1] Wilson et al.[168] used a copper bath for determining gases in gallium arsenide and indium antimonide. However, there must be some doubt about the retention of the probably volatile oxides of some of these elements in the bath long enough for reduction to carbon monoxide, and additional information is needed on this point.

5-10. INFRARED ABSORPTIOMETRY

In Sec. 2-4, it was pointed out that at absolute temperature the valence band was full and the conduction band empty. If, however, energy were supplied to such a system, electrons would move from the valence band to the conduction band and intrinsic conduction would take place. The energy necessary to bring about this transfer can be determined by infrared absorption. Figure 5-13 is a representation of the spectrum of a perfect crystal at absolute zero and is characterized by the absorption edge, the wavelength at which sufficient energy is supplied for electrons to cross the forbidden gap. At room temperature, the absorption edge is 1.8 μ for germanium, 1.1 μ for silicon, and 0.9 μ for gallium arsenide.

As well as intrinsic conduction, extrinsic conduction is possible because of the presence of n- or p-type dopants. These levels, as discussed in Sec. 5-1, can be at various points within the forbidden gap, that is, at energies less than that of the forbidden gap. It follows that, in a real crystal, other absorptions will occur at wavelengths longer than the absorption edge, corresponding to different impurity levels. It should be possible, therefore, to correlate absorption peaks with various impurities in the crystal. In practice, this is extremely difficult since the spectra are complicated by a number of other factors. At anything other than absolute zero, lattice vibrations occur and, in anything but a perfect lattice, free-charge-

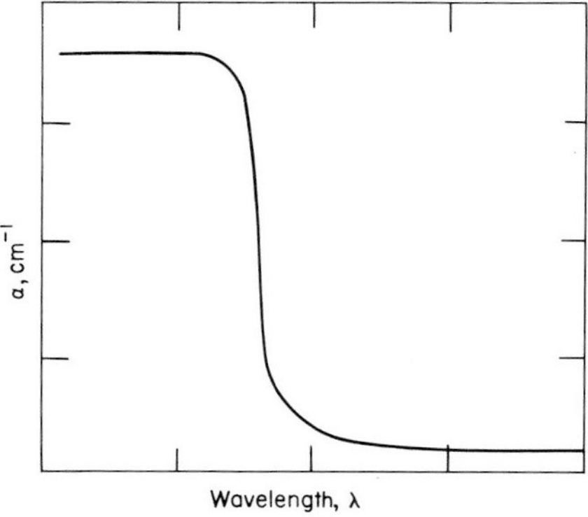

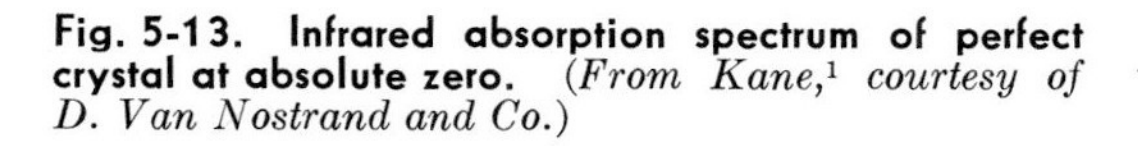
Fig. 5-13. Infrared absorption spectrum of perfect crystal at absolute zero. *(From Kane,[1] courtesy of D. Van Nostrand and Co.)*

carrier absorption occurs. Other interactions are also possible, and the result is an exceedingly complex field which has received considerable attention over many years. For a more detailed review of this subject, reference must be made elsewhere.[169-171] Suffice it to say that, although it is theoretically possible to identify dopants by this procedure, it cannot be recommended as a technique for their quantitative determination.

The case of oxygen in single-crystal germanium or silicon is significantly different. Kaiser et al.[172] were able to relate an absorption peak at 9 μ with the presence of oxygen in samples of single-crystal silicon and further correlated the absorption with the oxygen content as determined by vacuum-fusion analysis. They suggested that the oxygen was held interstitially and that it was bonded between two silicon atoms thus:

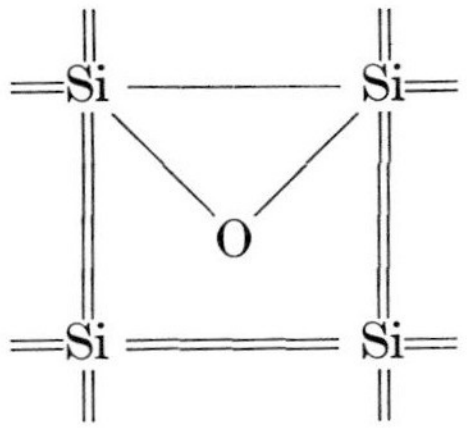

```
    ||          ||
  =Si ——————— Si=
    ||\        /||
    ||  \    /  ||
    ||    O     ||
    ||          ||
  =Si ═══════ Si=
    ||          ||
```

The absorption is due to the SiO stretching vibration. Absorption curves for two samples at room temperature are given in Fig. 5-14; *A* contained oxygen dissolved from a quartz crucible, whereas *B* was essentially free. A similar peak was found at 11.7 μ for oxygen in germanium. In both cases, high-resistivity material had to be used in order to reduce the free-charge-carrier absorption to a minimum. The detection limit for silicon was 10^{16} atoms oxygen per cubic centimeter or 0.1 ppm by weight. Kaiser and Keck[173] later calibrated this method against vacuum-fusion analyses and obtained a linear relationship from zero to 1.8×10^{18} atoms/cm^3 (22 ppm). Above this, silicon dioxide tends to precipitate from solid solution, and a correlation no longer exists. The relationship can be expressed as

$$\text{Concentration, ppm} = 3.2\ (\alpha - 0.8)$$

where α = absorption coefficient, cm^{-1}, and 0.8 is the coefficient due to lattice vibrations at 9.0 μ. This method forms the basis of ASTM Method F45-64T.[174] The calibration for germanium was made similarly by Kaiser and Thurmond[175] for the absorption maximum at 11.7 μ. In this case the linear relationship was

$$\text{Concentration, ppm} = 0.25\ (\alpha - 0.1)$$

The maximum solubility of oxygen in germanium is 2.2×10^{18} atoms/cm^3 (10 ppm). In practice, the method is relatively simple and has been described by Kane.[1] The sample is cut to a thickness of about 5 mm (a somewhat thicker specimen is possible for germanium) and optically polished on both sides. After a thickness measurement, the infrared spectrum is obtained and the absorption coefficient determined at the appropriate wavelength.

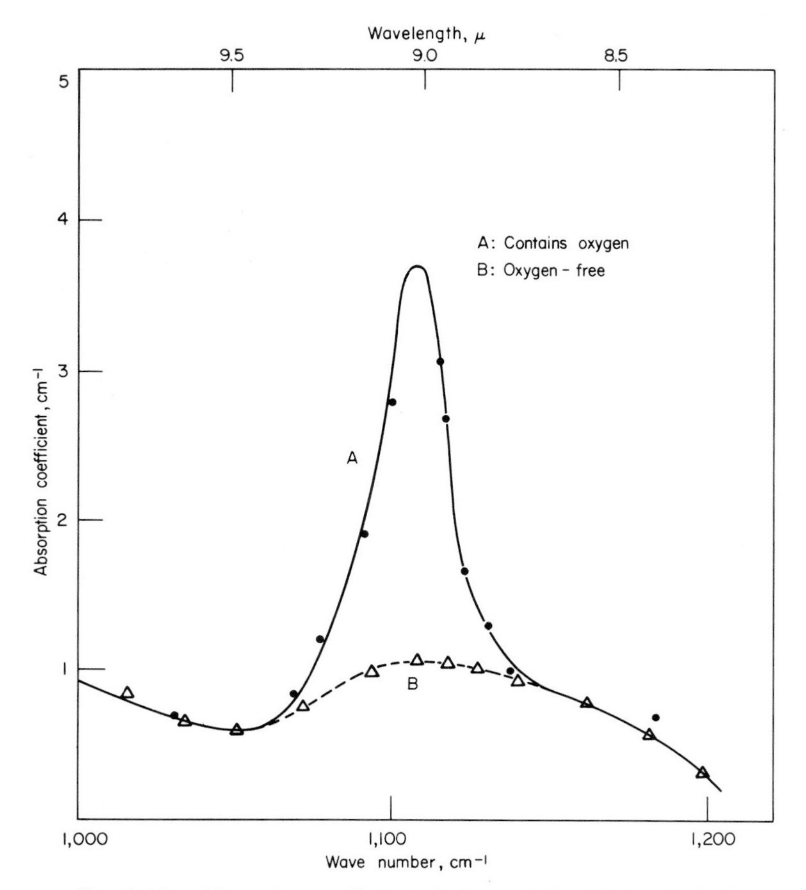

Fig. 5-14. Absorption coefficient of silicon. (*After Kaiser et al.*[172])

5-11. STOICHIOMETRY OF III-V COMPOUNDS

In the case of III-V compounds, a complicating factor arises inasmuch as an excess of one component over the other can be expected to lead to defects in the lattice. If there is a gross excess of material, and this may be only a few ppm, a second phase separates. This can be detected as segregates usually by conventional metallographic techniques or, for microsegregation in which the particular phase must be identified, by microprobe analysis or electron diffraction.

For nonstoichiometry above 0.1 percent, volumetric methods can be employed. Bachelder and Sparrow[176] fused indium antimonide with sodium carbonate and sulfur and dissolved in dilute hydrochloric acid. After oxidation of any unreacted sulfur with potassium chlorate, the antimony was titrated iodimetrically. Chernikhov and Cherkashina[177] dissolved indium antimonide, indium arsenide, or gallium arsenide in a sulfuric acid–ammonium sulfate mixture, diluted, and titrated the arsenic or antimony with potassium bromate. An amperometric titration was used by Gallai et al.[178] They dissolved the sample by Chernikhov and Cherkashina's method, diluted, and neutralized. The gallium was titrated amperometrically with ethanolic *N*-benzoylphenylhydroxylamine, with a graphite electrode at +1.1 volts. In Texas Instruments laboratories, EDTA titrations of gallium and indium are used with the copper-PAN indicator.

A somewhat better precision has been obtained by Kelly et al.[179] using differential spectrophotometry. An indirect method was used. A sample of gallium arsenide was dissolved in nitric-hydrochloric acid and taken to fumes with sulfuric acid. An aqueous solution of the residue was saturated with sulfur dioxide to reduce the arsenic to As^{III}. An aliquot of this solution was added to a standard copper-EDTA solution and compared by differential spectrophotometry with a series of standards, prepared by taking aliquots from a standard of gallium and arsenic in equivalent amounts. The position of the sample absorption against a reference (the highest standard) on the calibration curve obtained by measuring the standards against the reference was used to calculate the gallium content. A similar procedure, using potassium chromate as the reagent, was applied to the determination of arsenic. A precision of 0.01 percent could be obtained by using weight aliquots.

The above methods are useful in determining variations in stoichiometry in obviously unsuitable preparations. For example, losses of volatile arsenic may lead to a gallium-rich arsenide. Its appearance will indicate that it is nonstoichiometric, but correction will require, for the next run, a knowledge of the arsenic deficiency. A much more difficult problem arises when electrical properties are indicating a high carrier concentration but there is no evidence of dopants. If there is a small, ppm-order deviation from stoichiometry, the lattice may be able to adjust without precipitating a second phase. Vacancies may be created on the deficient component lattice sites, interstitial atoms of the excessive component may be accommodated, or there may even be substitution of the atoms in excess on lattice sites of the other type. All these possibilities will lead to strain in the lattice and generate carriers.

The problem of finding a few ppm of, say, arsenic in gallium arsenide is obviously a formidable one. What is being sought is really atoms of the same element but with a different energy environment. Chemical methods can take advantage of this. Tumanov et al.[120] extracted free arsenic with ethanol from gallium arsenide and determined it by a Gutzeit test. Their sensitivity was about 0.1 ppm, but since the

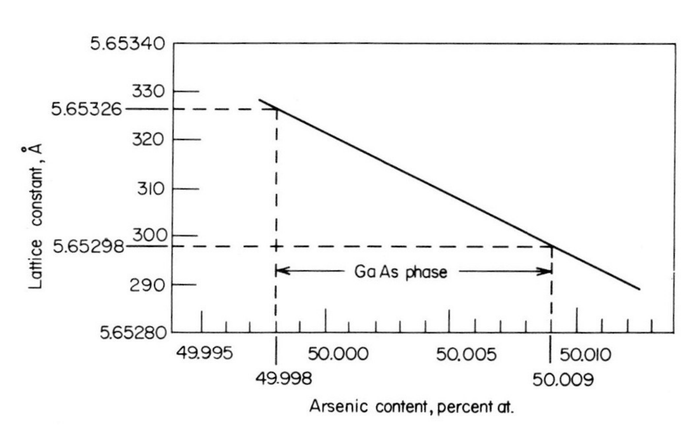

Fig. 5-15. The GaAs single phase and its relation to the lattice parameter. *(After Straumanis and Kim.*[180]*)*

few trials were done with "prepared" but otherwise unspecified standards, there must be some doubt as to the effectiveness of this method. Even if it does extract arsenic from gallium arsenide, the question arises as to exactly how this arsenic originated.

Physical methods would be expected to show more promise. According to Straumanis and Kim,[180] the gallium arsenide single phase extends from a lattice parameter of 5.65326 Å at the gallium-rich side to 5.65298 Å on the arsenic-rich side (see Fig. 5-15), with corresponding arsenic contents of 49.998 and 50.009%. Since the precision of their measurements was 0.00003 Å, it should be at least theoretically possible to obtain a precision in concentration of 20 ppm. However, this level of precision is very difficult to obtain, and the method is not recommended, at least not on today's instrumentation.

As is evident from the foregoing, there is no satisfactory method for stoichiometry in the single-phase region of III-V compounds. The x-ray diffraction method comes closest, but is extremely demanding and, with a precision of 20 ppm, somewhat borderline anyway. There is a pressing need for some method that will give this property to a precision of 1 ppm or better.

REFERENCES

1. Kane, P. F.: in F. J. Welcher (ed.), "Standard Methods of Chemical Analysis," vol. IIIB, p. 1764, D. Van Nostrand Company, Inc., Princeton, N.J., 1966.
2. Vasilevskaya, L. S., V. P. Muravenko, and A. I. Kondrashina: *Zh. Anal. Khim.*, **20**:540 (1965).
3. Karabash, A. G., Sh. I. Peizulaev, G. G. Morozova, and I. I. Smirenkina: *Tr. Kom. Anal. Khim., Akad. Nauk SSSR*, **12**:25 (1960).
4. Vasilevskaya, L. S., M. A. Notkina, S. A. Sadof'eva, and A. I. Kondrashina: *Zavodsk. Lab.*, **28**:678 (1962).
5. Dvorak, J., and I. Dobremyslova: *Chem. Prumysl*, **13**:136 (1963).
6. Veleker, T. J.: *Anal. Chem.*, **34**:87 (1962).
7. Malkova, O. P., A. N. Tumanova, and N. K. Rudnevskii: *Zh. Anal. Khim.*, **20**:130 (1965).
8. Peizulaev, Sh. I., A. G. Karabash, L. S. Krauz, F. A. Kostareva, N. I. Smirnova-Averina, F. L. Babina, L. I. Kondrateva, E. F. Voronova, and V. M. Meshkova: *Zavodsk. Lab.*, **24**:723 (1958).
9. Martynov, Yu. M., I. I. Kornblit, N. P. Smirnova, and R. V. Dzhagatspanyan: *Zavodsk. Lab.*, **27**:839 (1961).
10. Zil'bershtein, Kh. I., N. I. Kaliteevskii, A. N. Razumovskii, and Yu. F. Fedorov: *Zavodsk. Lab.*, **28**:43 (1962).
11. Morachevskii, Yu. V., Kh. I. Zil'bershtein, M. M. Piryutko, and O. N. Nikitina: *Zh. Anal. Khim.*, **17**:614 (1962).
12. Morachevskii, Yu. V., Kh. I. Zil'bershtein, M. M. Piryutko, and O. N. Nikitina: *Vestn. Leningr. Univ. Ser. Fiz. i Khim.*, **22**(4):140 (1962).
13. Rudnevskii, N. K., L. N. Sokolova, and S. G. Tsvetkov: *Tr. Khim. i Khim. Tekhnol. (Gor'kii)*, **1962**:341.
14. Keck, P. G., A. L. MacDonald, and J. W. Mellichamp: *Anal. Chem.*, **28**:995 (1956).
15. Borovskii, I. B., A. N. Shteinberg, and V. V. Bugulova: *Tr. Inst. Met., Akad. Nauk SSSR*, **1958**:283.

16. Kalinnikov, V. T., and A. N. Shteinberg: *Zavodsk. Lab.*, **30**:178 (1964).
17. Tarasevich, N. I., and A. A. Zheleznova: *Tr. Kom. Anal. Khim.*, **15**:121 (1965).
18. Morrison, G. H., and R. L. Rupp: *Anal. Chem.*, **29**:892 (1957).
19. Vasilevskaya, L. S., A. I. Kondrashina, and A. A. Shifrina: *Zavodsk. Lab.*, **28**:674 (1962).
20. Semov, M. P.: *Zavodsk. Lab.*, **29**:1450 (1963).
21. Oldfield, J. H., and D. L. Mack: *Analyst*, **87**:778 (1963).
22. Kataev, G. A., and Z. I. Otmakhova: *Zh. Anal. Khim.*, **18**:339 (1963).
23. Babadag, T.: *Z. Anal. Chem.*, **207**:328 (1965).
24. Hegemann, F., K. Giesen, and C. Von Sybel: *Ber. Deut. Keram. Ges.*, **32**:329 (1955).
25. Rost, F.: *Mikrochim. Acta*, **1956**:343.
26. Hegemann, F., K. Giesen, and C. Von Sybel: *Ber. Deut. Keram. Ges.*, **33**:378 (1956).
27. Ehrlich, A., R. Gerbatsch, and G. Freitag: *Chem. Anal. (Warsaw)*, **7**:435 (1962).
28. Pikhtin, A. N.: *Zavodsk. Lab.*, **31**:559 (1965).
29. Vasilevskaya, L. S., A. I. Kondrashina, G. A. Makarova, and N. A. Panarina: *Zavodsk. Lab.*, **31**:561 (1965).
30. Shvangiradze, R. R., and T. A. Mozgovaya: *Zh. Anal. Khim.*, **12**:708 (1957).
31. Vecsernyes, L.: *Magy. Kem. Folyoirat*, **66**:513 (1960).
32. Vecsernyes, L.: *Z. Anal. Chem.*, **182**:429 (1961).
33. Shteinberg, A. N.: *Tr. Inst. Met., Akad. Nauk SSSR*, **11**:229 (1962).
34. Karpel, N. G.: *Zavodsk. Lab.*, **30**:1078 (1964).
35. Karpel, N. G., and V. V. Shaparova: *Zavodsk. Lab.*, **30**:1459 (1964).
36. Morrison, G. H., R. L. Rupp, and G. L. Klecak: *Anal. Chem.*, **32**:933 (1960).
37. Massengale, J. F., D. Andrychuk, and C. E. Jones: Symposium on Impurities in III-V Elements and Compounds, Battelle Memorial Institute, June 2, 1960.
38. Burkhalter, T. S.: *Anal. Chem.*, **33**(5):21A (1961).
39. Kane, P. F.: *Anal. Chem.*, **38**(3):29A (1966).
40. Babko, A. K., and T. E. Get'man: *Tr. Kom. Anal. Khim., Akad. Nauk SSSR*, **12**:36 (1960).
41. Andrychuk, D., and C. E. Jones: in M. S. Brooks and J. K. Kennedy (eds.), "Proceedings of the Conference on Ultrapurification of Semiconductor Materials, Boston, Mass., 1961," p. 373, The Macmillan Company, New York, 1962.
42. Honig, R. E.: *Anal. Chem.*, **25**:1530 (1953).
43. Hannay, N. B., and A. J. Ahearn: *Anal. Chem.*, **26**:1056 (1954).
44. Shaw, A. E., and W. Rall: *Rev. Sci. Instrum.*, **18**:278 (1947).
45. Craig, R. D., G. A. Errock, and J. D. Waldron: in J. D. Waldron (ed.), "Advances in Mass Spectrometry," p. 136, Pergamon Press, New York, 1959.
46. Robinson, C. F., G. D. Perkins, and N. W. Bell: in "Instruments and Measurements," vol. 1, p. 260, Academic Press Inc., New York, 1961.
47. Brown, R., R. D. Craig, J. A. James, and C. M. Wilson: in M. S. Brooks and J. K. Kennedy (eds.), "Ultrapurification of Semiconductor Materials," p. 279, The Macmillan Company, New York, 1962.
48. Duke, J. F.: in M. S. Brooks and J. K. Kennedy (eds.), "Ultrapurification of Semiconductor Materials," p. 294, The Macmillan Company, New York, 1962.
49. Ahearn, A. J.: in "Eleventh Annual Conference on Mass Spectrometry and Allied Topics," p. 223, ASTM Committee E-14, 1963.
50. Brown, R., R. D. Craig, and J. D. Waldron: in R. K. Willardson and H. L. Goering (eds.), "Compound Semiconductors," vol. 1, "Preparation of III-V Compounds," p. 106, Reinhold Publishing Corporation, New York, 1962.
51. Brown, R., R. D. Craig, and R. M. Elliott: in R. M. Elliott (ed.), "Advances in Mass Spectrometry," vol. 2, p. 141, Pergamon Press, New York, 1963.

52. Willardson, R. K.: in M. S. Brooks and J. K. Kennedy (eds.), "Ultrapurification of Semiconductor Materials," p. 316, The Macmillan Company, New York, 1962.
53. Brice, J. C., J. A. Roberts, and G. Smith: *J. Mater. Sci.*, **2**:131 (1967).
54. *AEI Tech. Inform. Sheet* A16.
55. *AEI Tech. Inform. Sheet* A17.
56. *AEI Tech. Inform. Sheet* A43.
57. *AEI Tech. Inform. Sheet* A54.
58. Woolston, J. R., and R. E. Honig: Symposium on Impurities in III-V Elements and Compounds, Battelle Memorial Institute, June 2, 1960.
59. Honig, R. E.: in J. P. Cali (ed.), "Trace Analysis of Semiconductor Materials," p. 169, Pergamon Press, New York, 1964.
60. Owens, E. B., and N. A. Giardino: *Anal. Chem.*, **35**:1172 (1963).
61. Woolston, J. R., and R. E. Honig: *Rev. Sci. Instrum.*, **35**:69 (1964).
62. Short, H. G., and B. J. Keene: *Talanta*, **13**:297 (1966).
63. Ahearn, A. J., F. A. Trumbore, C. J. Frosch, C. L. Luke, and D. L. Malm: *Anal. Chem.*, **39**:350 (1967).
64. Klein, H. M., and G. B. Larrabee: 17th Annual Pittsburgh Conference on Analytical Chemistry and Applied Spectroscopy, 1966.
65. Cali, J. P.: in J. P. Cali (ed.), "Trace Analysis of Semiconductor Materials," p. 6, Pergamon Press, New York, 1964.
66. Smales, A. A., and L. O. Brown: *Chem. Ind.* (*London*), **1950**:441.
67. Smales, A. A., and B. D. Pate: *Anal. Chem.*, **24**:717 (1952).
68. Jaskolska, H., and L. Wodkiewicz: *Chem. Anal.* (*Warsaw*), **6**:161 (1961).
69. Leliaert, G.: *Pure Appl. Chem.*, **1**:121 (1960).
70. De Soete, D., R. De Neve, and J. Hoste: "Proceedings of the 1965 International Conference on Modern Trends in Activation Analysis," p. 31, Texas A&M University Press, 1966.
71. Szekely, G.: *Anal. Chem.*, **26**:1500 (1954).
72. Gottfried, J., and Y. V. Yakovlev: *Chem. Prumysl*, **9**:179 (1959).
73. Leliaert, G.: *Int. J. Appl. Radiat. Isotopes*, **12**:63 (1961).
74. Lloyd, K. W.: "Proceedings of the SAC Conference, Nottingham, 1965," p. 180, W. Heffer & Sons, Ltd., Cambridge, England, 1965.
75. Rommel, H.: *Anal. Chim. Acta*, **34**:427 (1966).
76. Ruzicka, J., and J. Stary: *Talanta*, **10**:287 (1963).
77. Zeman, A., J. Ruzicka, and J. Stary: *Talanta*, **10**:685 (1963).
78. Zeman, A., J. Stary, and J. Ruzicka: *Talanta*, **10**:981 (1963).
79. Stary, J., J. Ruzicka, and A. Zeman: *Anal. Chim. Acta*, **29**:103 (1963).
80. Morrison, G. H., and J. F. Cosgrove: *Anal. Chem.*, **28**:320 (1956).
81. Yakovlev, Y. V., A. E. Kulak, V. A. Ryabukhin, and R. S. Rytchkov: *Proc. U.N. Int. Conf. Peaceful Uses At. Energy, 2nd, Geneva*, **28**:496 (1958).
82. Rytchkov, R. S., and N. A. Glukhareva: *Zavodsk. Lab.*, **27**:1246 (1961).
83. Robertson, D.: *Nature*, **210**:1357 (1966).
84. De Neve, R., D. De Soete, and J. Hoste: *Radiochim. Acta*, **5**:188 (1966).
85. James, J. A., and D. H. Richards: *Nature*, **175**:769 (1955).
86. Morrison, G. H., and J. F. Cosgrove: *Anal. Chem.*, **27**:810 (1955).
87. Kant, A., J. P. Cali, and H. D. Thompson: *Anal. Chem.*, **28**:1867 (1956).
88. James, J. A., and D. H. Richards: *J. Electron. Contr.*, **3**:500 (1957).
89. Thompson, B. A., B. M. Strause, and M. B. Leboeuf: *Anal. Chem.*, **30**:1023 (1958).
90. Erokhina, K. I., I. Kh. Lemberg, I. E. Makasheva, I. A. Maslow, and A. P. Obukhov: *Zavodsk. Lab.*, **26**:821 (1960).

91. Chiba, M.: *Trans. Nat. Res. Inst. Met. (Tokyo)*, 4:143 (1962).
92. Upor-Juvancz, V., and M. Ordogh: *Kozlemeny*, **12**:365 (1964).
93. Gebauhr, W., J. Martin, and E. Haas: *Z. Anal. Chem.*, **200**:266 (1964).
94. Nakai, T., S. Yajima, I. Fujii, and M. Okada: *Japan Analyst*, **8**:367 (1959).
95. Kalinin, A. I., R. A. Kuznetsov, and V. V. Moiseev: *Radiokhimiya*, 4:575 (1962).
96. Moiseev, V. V., R. A. Kuznetsov, and A. I. Kalinin: "Proceedings of the 1965 International Conference on Modern Trends in Activation Analysis," p. 164, Texas A&M University Press, 1966.
97. Heinen, K. G., and G. B. Larrabee: *Anal. Chem.*, **38**:1853 (1966).
98. Helmer, R. G., R. L. Heath, D. D. Metcalf, and G. A. Cazier: *U.S. Govt. Rep.* IDO 17015, 1964.
99. Harvey, J. T., and A. J. Smith: *U.K. Govt. Rep.* AML-B/120(M), 1957.
100. Smales, A. A., D. Mapper, A. J. Wood, and L. Salmon: *UKAEA Rep.* C/R 2254, 1957.
101. James, J. A., and D. H. Richards: *Nature*, **176**:1026 (1955).
102. Berthel, K.-H., H.-G. Doge, G. Ehrlich, A. Kothe, and A. Schmidt: *Mikrochim. Ichnoanal. Acta*, **1963**:702.
103. Nozaki, T., H. Baba, and H. Araki: *Bull. Chem. Soc. Japan*, **33**:320 (1960).
104. Nozaki, T., T. Kawashima, H. Baba, and H. Araki: *Bull. Chem. Soc. Japan*, **33**:1438 (1960).
105. Lobanov, E. M., V. I. Zvyagin, A. A. Kist, B. P. Zverev, A. I. Sviridova, and G. A. Moskovtseva: *Zh. Anal. Khim.*, **18**:1349 (1963).
106. Ruzicka, J., A. Zeman, and I. Obrusnik: *Talanta*, **12**:401 (1965).
107. Krivanek, M., F. Kikula, and J. Slunecko: *Talanta*, **12**:721 (1965).
108. Beardsley, D. A., G. B. Briscoe, J. Ruzicka, and M. Williams: *Talanta*, **12**:829 (1965).
109. Zeman, A., J. Ruzicka, and V. Kuvik: *Talanta*, **13**:271 (1966).
110. Harvey, J. T., and A. J. Smith: *U.K. Govt. Rep.* AML-B/121(M), 1957.
111. Gill, R. A.: *UKAEA Rep.* C/R 2758, 1958.
112. Saito, K., T. Nozaki, S. Tanaka, M. Furukawa, and H. Cheng: *Int. J. Appl. Radiat. Isotopes*, **14**:357 (1963).
113. Green, D. E., J. A. B. Heslop, and J. E. Whitley: *Analyst*, **88**:522 (1963).
114. Bailey, R. F., and D. A. Ross: *Anal. Chem.*, **35**:791 (1963).
115. Parker, C. A., and W. T. Rees: in J. P. Cali (ed.), "Trace Analysis of Semiconductor Materials," p. 208, Pergamon Press, New York, 1964.
116. Payne, S. T.: *Analyst*, **77**:278 (1952).
117. Luke, C. L., and M. E. Campbell: *Anal. Chem.*, **25**:1588 (1953).
118. Goto, H., and Y. Kakita: *J. Chem. Soc. Japan, Pure Chem. Sect.*, **77**:739 (1956).
119. Fowler, E. W.: *Analyst*, **88**:380 (1963).
120. Tumanov, A. A., A. N. Sidorenko, and F. S. Taradenkova: *Zavodsk. Lab.*, **30**:652 (1964).
121. Rezac, R., and J. Ditz: *Zavodsk. Lab.*, **29**:1176 (1963).
122. Roberts, J. A., J. Winwood, and E. J. Millett: "Proceedings of SAC Conference, Nottingham, 1965," p. 528, W. Heffer & Sons, Ltd., Cambridge, England, 1965.
123. Ishihara, Y., and Y. Taguchi: *Japan Analyst*, **6**:724 (1957).
124. Luke, C. L.: *Anal. Chem.*, **27**:1150 (1955).
125. Luke, C. L., and M. E. Campbell: *Anal. Chem.*, **28**:1340 (1956).
126. Baba, H.: *Japan Analyst*, **5**:631 (1956).
127. Nazarenko, V. A., and E. A. Biryuk: *Zh. Anal. Khim.*, **15**:306 (1960).
128. Luke, C. L.: *Anal. Chem.*, **21**:1369 (1949).
129. Tumanov, A. A., A. N. Sidorenko, and Ya. I. Korenman: *Zavodsk. Lab.*, **30**:1058 (1964).
130. Ducret, L., and C. Cornet: *Anal. Chim. Acta*, **25**:542 (1961).

131. Babko, A. K., A. I. Volkova, and O. F. Drako: *Tr. Kom. Anal. Khim., Akad. Nauk SSSR*, **12**:53 (1960).
132. Nazarenko, V. A., G. V. Flyantikova, and N. V. Lebedeva: *Zavodsk. Lab.*, **23**:891 (1957).
133. Rigin, V. I., and N. N. Melnichenko: *Zavodsk. Lab.*, **32**:394 (1966).
134. Pohl, F. A., and W. Bonsels: *Mikrochim. Acta*, **1962**:97.
135. Ducret, L., and P. Seguin: *Anal. Chim. Acta*, **17**:207 (1957).
136. Ducret, L.: *Anal. Chim. Acta*, **17**:213 (1957).
137. Luke, C. L., and S. S. Flaschen: *Anal. Chem.*, **30**:1406 (1958).
138. Barcanescu, V., and H. Minasian: *Rev. Chim. (Bucharest)*, **9**:316 (1968).
139. Pohl, F. A., K. Kokes, and W. Bonsels: *Z. Anal. Chem.*, **174**:6 (1960).
140. Marczenko, Z., and K. Kasiura: *Chem. Anal. (Warsaw)*, **8**:185 (1963).
141. Dolmanova, I. F., and V. M. Peshkova: *Zh. Anal. Khim.*, **19**:297 (1964).
142. Lebedeva, N. V., and V. A. Nazarenko: *Tr. Kom. Anal. Khim., Akad. Nauk SSSR*, **11**:287 (1960).
143. Nazarenko, V. A., and M. B. Shustova: *Zavodsk. Lab.*, **27**:15 (1961).
144. Adler, S. J., and R. J. Paff: in R. K. Willardson and H. L. Goering (eds.), "Compound Semiconductors," vol. 1, "Preparation of III-V Compounds," p. 123, Reinhold Publishing Corporation, New York, 1962.
145. Goryushina, V. G., and E. Ya. Biryukova: *Zavodsk. Lab.*, **31**:1303 (1965).
146. Bush, E. L., and E. H. Cornish: in M. S. Brooks and J. K. Kennedy (eds.), "Ultrapurification of Semiconductor Materials," p. 454, The Macmillan Company, New York, 1962.
147. Soldatova, L. A., and L. B. Kristaleva: *Tr. Tomsk. Univ.*, **157**:279 (1963).
148. Goryushina, V. G., and N. V. Esenina: *Zh. Anal. Khim.*, **21**:239 (1966).
149. Knizek, M., and A. Galik: *Z. Anal. Chem.*, **213**:254 (1965).
150. Knizek, M., and V. Pecenkova: *Zh. Anal. Khim*, **21**:260 (1966).
151. Shigematsu, T.: *Japan Analyst*, **7**:787 (1958).
152. Parker, C. A., and W. J. Barnes: *Analyst*, **85**:828 (1960).
153. Nazarenko, V. A., S. Ya. Vinkovetskaya, and R. V. Ravitskaya: *Ukr. Khim. Zh.*, **28**:726 (1962).
154. Alimarin, I. P., A. P. Golovina, I. M. Gibalo, and Yu. A. Mittsel: *Zh. Anal. Khim.*, **20**:339 (1965).
155. Gokhshtein, Ya. P., M. P. Volynets, and V. D. Yukhtanova: *Tr. Kom. Anal. Khim., Akad. Nauk SSSR*, **12**:5 (1960).
156. Pohl, F. A., and W. Bonsels: *Mikrochim. Acta*, **1960**:641.
157. Jennings, V. J.: *Analyst*, **85**:62 (1960).
158. Jennings, V. J.: *Analyst*, **87**:548 (1962).
159. Bush, E. L.: *J. Polarogr. Soc.*, **11**:41 (1965).
160. Kemula, W., and Z. Kublik: in C. N. Reilley (ed.), "Advances in Analytical Chemistry and Instrumentation," vol. 2, p. 123, Interscience Publishers, Inc., New York, 1963.
161. Kataev, G. A., E. A. Zakharova, and L. I. Oleinik: *Tr. Tomsk. Univ.*, **157**:261 (1963).
162. Vinogradova, E. N., and A. I. Kamenev: *Zh. Anal. Khim.*, **20**:183 (1965).
163. Galik, A., and M. Knizek: *Talanta*, **13**:589 (1966).
164. Schink, N.: *Z. Anal. Chem.*, **216**:319 (1966).
165. Beach, A. L., and W. G. Guldner: in Symposium on Determination of Gases in Metals, *ASTM Spec. Tech. Publ.* 222, p. 15, Philadelphia, 1958.
166. Donovan, P. D., J. L. Evans, and G. H. Bush: *Analyst*, **88**:771 (1963).
167. Turovtseva, Z. M., and L. L. Kunin: "Analysis of Gases in Metals," English translation, p. 336, Consultants Bureau, New York, 1961.

168. Wilson, C. M., D. Hazelby, M. L. Aspinal, and J. A. James: *CVD Res. Rep. RP* 4/3, *Rep.* G1374, Associated Electrical Industries (Rugby) Ltd., 1962.
169. Hrostowski, H. J.: in N. B. Hannay (ed.), "Semiconductors," p. 437, Reinhold Publishing Corporation, New York, 1959.
170. Moss, T. S.: "Optical Properties of Semiconductors," Butterworth & Co. (Publishers), Ltd., London, 1959.
171. Hilsum, C., and A. C. Rose-Innes: "Semiconducting III-V Compounds," Pergamon Press, New York, 1961.
172. Kaiser, W., P. H. Keck, and C. F. Lange: *Phys. Rev.*, **101**:1264 (1956).
173. Kaiser, W., and P. H. Keck: *J. Appl. Phys.*, **28**:882 (1957).
174. "1968 Book of ASTM Standards," ASTM, Philadelphia, Pa., 1968.
175. Kaiser, W., and C. D. Thurmond: *J. Appl. Phys.*, **32**:115 (1961).
176. Bachelder, M. C., and P. M. Sparrow: *Anal. Chem.*, **29**:149 (1957).
177. Chernikhov, Yu. A., and T. V. Cherkashina: *Zavodsk. Lab.*, **24**:1047 (1958).
178. Gallai, Z. A., N. M. Sheina, and I. P. Alimarin: *Zh. Anal. Khim.*, **20**:1093 (1965).
179. Kelly, R. S., A. Eldridge, E. W. Lanning, and R. Bastian: *U.S. Govt. Rep.* AFCRL 65–50, 1965.
180. Straumanis, M. E., and C. D. Kim: *Acta Cryst.*, **19**:256 (1965).

6

Characterization of Single Crystals for Physical Imperfections

6-1. INTRODUCTION

In Sec. 5-1, the effect of foreign atoms in the lattice on the electrical properties of a semiconductor was discussed. The emphasis was mainly on the ability of these atoms to change the number of charge carriers and hence the resistivity. The presence of these foreign atoms can be determined by compositional methods, and these, of course, formed the basis of Chap. 5. There are, however, purely physical imperfections in the crystal which can also affect the characteristics of the material. The absence of an atom or the presence of unsatisfied bonds can also give rise to changes in charge-carrier concentration, mobility, or lifetime.

In dealing with semiconductor materials, we are fortunate in that the crystal lattice is one of the simpler cubic structures. It is generally referred to as the *diamond structure* and consists of two interpenetrating face-centered-cubic lattices. Silicon, germanium, and the III-V compounds all have this same structure. This cubic habit means that Miller indices are relatively easy to use; the directions are normal to the planes, and there is no confusion as to the angles generated between planes.

6-2. POINT DEFECTS

When a foreign atom enters a crystal lattice, it can do so either substitutionally or interstitially, as was pointed out in Sec. 5-1. Either case will set up strain within the crystal, but it is localized and affects essentially only one lattice site. Such a region of strain is termed a *point defect;* the lattice can accommodate this without any interruption in the crystal perfection. Other types of point defects are also known. If an atom is simply missing from a lattice site, the defect is termed a *vacancy* or a *Schottky defect.*[1,†] This vacancy, which is the absence of an atom from a

†Superscript numbers indicate References listed at the end of the chapter.

lattice site, is not to be confused with a hole, which is the absence of an electron from an atom. A somewhat less common defect, although of considerable interest in studies of radiation damage, is one in which the atom is displaced from its position in the lattice to an interstitial position; such a vacancy interstitial pair is termed a *Frenkel defect.*[2] In compound semiconductors, an antistructure defect is possible in which an atom of A occupies a B site or vice versa.

The determination of the total foreign atoms in a lattice was described in Chap. 5. However, the specific identification of a point defect as such is a very difficult task. It has been carried out in one or two cases by use of the field-ion microscope; for example, Sugata et al.[3] were able to observe the nucleation of silicon and germanium on tungsten by this technique. However, it is at present of only limited value in a few special cases, and this is outside the scope of this work. In general, information on point defects will be limited to that on impurity atoms.

6-3. DISLOCATIONS

Of more general interest is another type of defect termed the *dislocation*. Whereas the point defect is an intrinsic part of the crystal and does not alter its perfection, the dislocation is a discontinuity in the lattice. It is an area in which there are many unsatisfied bonds or, as they are often termed, *dangling bonds*. The odd electron which this bond constitutes could either pair with another to form an octet, i.e., act as an acceptor, or donate the electron to the conduction band, i.e., act as a donor. In fact, it has been shown by Gallagher[4] that, for n-type germanium, the introduction of dislocations increased the resistivity and decreased the minority-carrier lifetime. Pearson et al.[5] confirmed this and found that for p-type material the resistivity remained virtually unchanged although the lifetime was also reduced. Read[6,7] subsequently developed a theoretical treatment based on the dangling-bond idea which explained these observations on the assumption that such a bond was a deep acceptor (see Sec. 5-1). For p-type material, the reduction in minority-carrier lifetime is readily explained, as is the increase in resistivity for n-type material. The reduction in minority-carrier lifetime for n-type material is more difficult to understand but arises from the fact that, unlike point defects, these acceptors are not isolated but exist in rows. Consequently, space-charge regions are set up in the crystal which can act as deep donors. This subject is dealt with in considerable detail in a number of works.[8-10] Suffice it to say that, although their effect is less marked, the presence of dislocations is just as undesirable as the presence of impurity atoms in bulk material. For epitaxial material, as we shall see in Sec. 8-28, the presence of dislocations in the substrate can generate dislocations in the epitaxial film, and these, on a microscale, can substantially reduce the yields of microcircuits. The determination and identification of dislocations and their control is consequently an essential phase of semiconductor materials research.

6-4. EDGE DISLOCATIONS†

There are two basic types of dislocation: the edge dislocation and the screw dislocation. The edge dislocation is illustrated in Fig. 6-1. It can be envisaged as a dis-

†Adapted from Kane.[11]

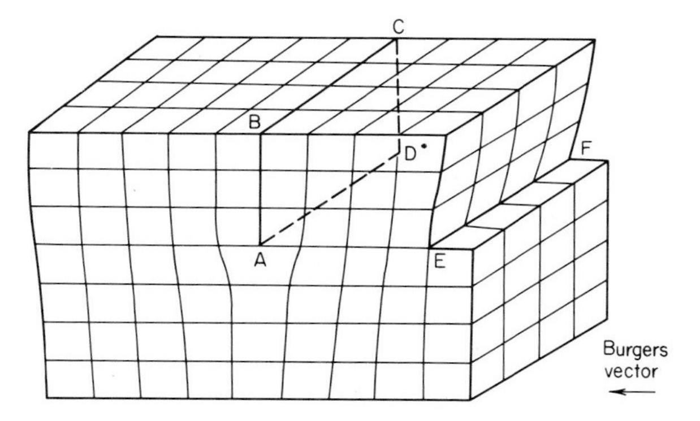

Fig. 6-1. The edge dislocation. (*From Hobstetter.*[12])

placement of part of the crystal by one atomic plane. Alternatively, one can imagine an additional layer of atoms *ABCD* being inserted between two atomic layers of the crystal. The result is a plane of atoms that terminates at *AD*, and this line *AD* is the dislocation. *ADFE* is termed the *slip plane* and can be regarded as the plane along which the movement has taken place. Figure 6-2*a* shows an edge dislocation end on in two dimensions; *A* is the dislocation viewed along its direction. The heavy line drawn around this is termed the *Burgers circuit,* and in Fig. 6-2*b* is the same circuit drawn on a perfect crystal. This circuit is of any shape but must form a closed loop around the dislocation. The element missing from the imperfect crystal is termed the *Burgers vector* **b**. It represents the mismatch of the crystal and is usually one lattice spacing. Its direction is perpendicular to the direction of the dislocation, and this is characteristic of an edge dislocation. The arrows in the circuit can be arbitrarily assigned. They have no significance unless one is considering the interaction of several dislocations, in which case they must be traversed in the same sense. In this connection, Fig. 6-2*a* represents a positive dislocation in which the extra half-plane has been inserted above the slip plane, and the Burgers circuit, if drawn with the arrows in the same direction, results in a vector which has a direction opposite to that of the negative dislocation. Interaction of a positive and negative dislocation annihilates both; that is, the two vectors cancel each other out.

It is easy to see from Figs. 6-1 and 6-2 that insertion of the extra half-plane of atoms must lead to a condition of strain extending several atoms from the dislocation, both above and below the slip plane. Above the plane (in the case of the

Fig. 6-2. The Burgers vector.

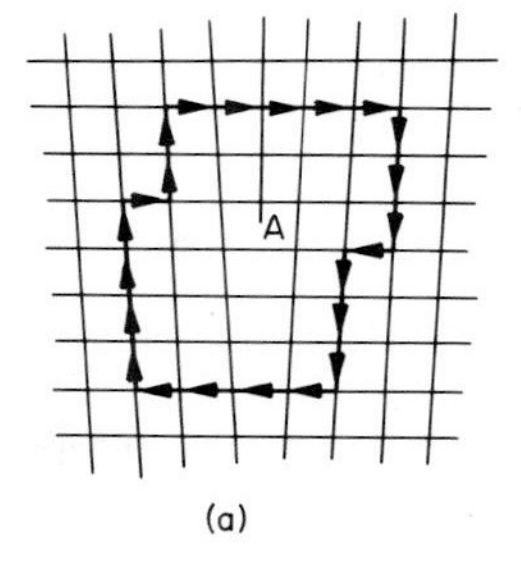

(a)

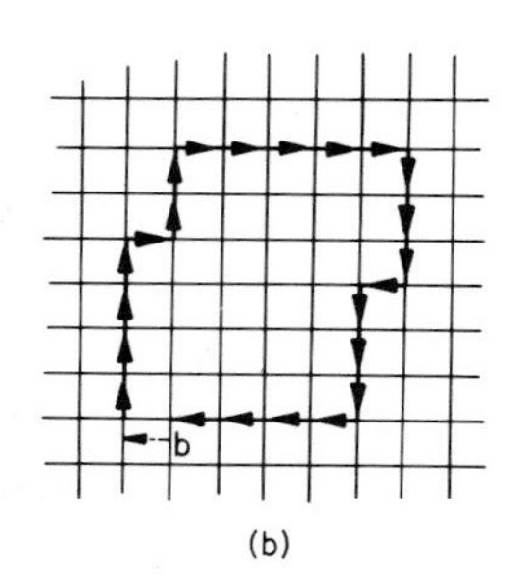

(b)

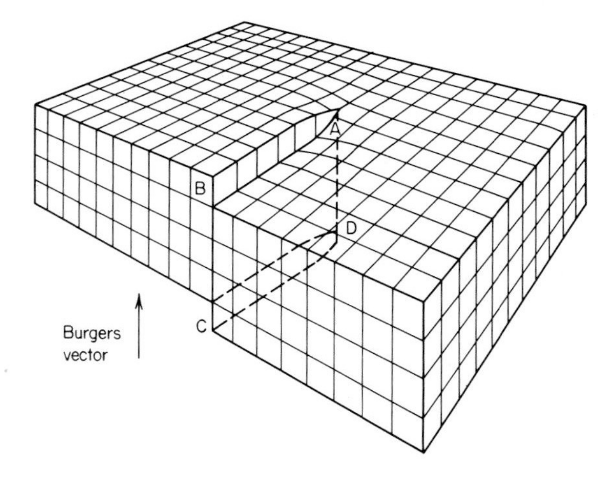

Fig. 6-3. The screw dislocation. (*After Read.*[13])

positive dislocation) atoms will be in compression, and below in tension, and it will be several atomic planes before this is relieved.

6-5. SCREW DISLOCATIONS

The other basic type of dislocation, the screw dislocation, is shown in Fig. 6-3. It is rather more difficult to visualize than the edge dislocation. The displacement is in the slip plane $ABCD$, and it can perhaps be regarded as the result of a twisting force. The Burgers vector for the edge dislocation was perpendicular to the dislocation and was the direction in which the crystal could be imagined as displaced. In the case of the screw, the Burgers vector is also in this direction of displacement, but in this case it is parallel to the dislocation AD. This important distinction characterizes the two types. As far as the dislocation is concerned, whereas the edge dislocation is the end of a half-plane of atoms, the screw dislocation has no such simple definition; it is merely a line of maximum distortion or strain.

Figure 6-4 shows the atomic pattern associated with screw dislocations. This two-dimensional representation can only show the atom layers immediately above and below the slip plane, but if one can visualize this displacement being relieved, above and below, over several planes, then the general effect will be a spiral centered around the dislocation with a pitch equal to the Burgers vector. Like the edge dislocation, it can exist as either a positive or a negative dislocation, according to whether the original mismatch was above or below the slip plane. The directions of the Burgers vectors are opposite in sign, and this results in a right-handed (positive) or left-handed (negative) strain pattern.

6-6. MOTION OF DISLOCATIONS

As we have already said, the interaction of a positive with a negative edge dislocation results in the elimination of both, and this is equally true of screw dislocations.

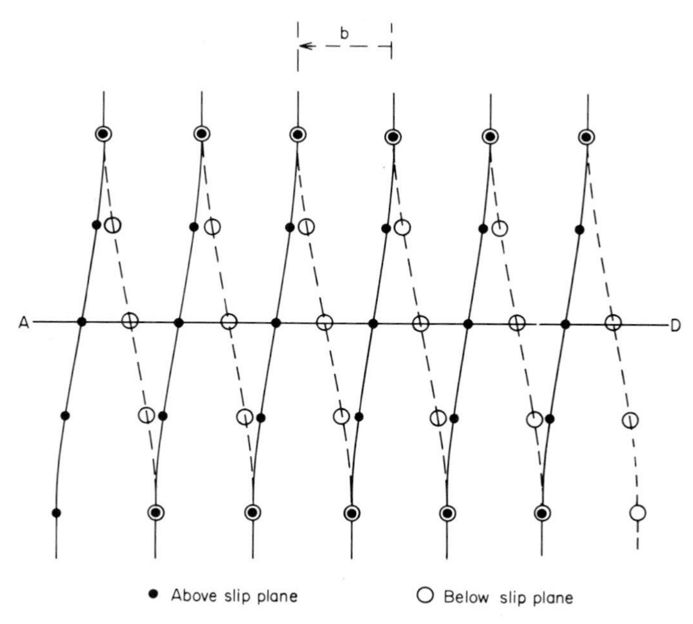

Fig. 6-4. Plane view of screw dislocation. (*After Cottrell.*[14])

This implies that dislocations are not fixed but can move through the lattice, and in fact this is true. Under strain, edge dislocations can move along their slip planes, positive in one direction, negative in the other. They are restricted to the slip plane because of their association with the half-plane of atoms. Screw dislocations do not have this restriction and can move in any plane that contains them. Like the edge dislocation, under the same shear, the positive and negative dislocations move in opposite directions.

Usually the conditions in real crystals are more complex. For one thing, dislocations are seldom of one basic type; they contain elements of both. In addition, they seldom meet on the same glide planes, so that the Burgers vectors become components of other dislocations on other glide planes. The study of the movements of dislocations is an important aspect of metallurgy and is dealt with in many standard works.[10,13-16]

For characterization of semiconductor material, the movement of dislocations is not in itself of much interest. The samples will normally be received in a static condition with the dislocations frozen in. However, care must be taken, particularly with thin samples, to avoid placing a strain on the crystal which might either generate new dislocations or cause them to move to other areas of the sample. Thermal treatments also should be avoided since these may lead to the relief of strain and hence to the movement or even elimination of dislocations.

6-7. OTHER LATTICE FAULTS

A common fault encountered in both germanium and silicon is the phenomenon of twinning. This occurs usually during growth. The lattice is redirected to a different

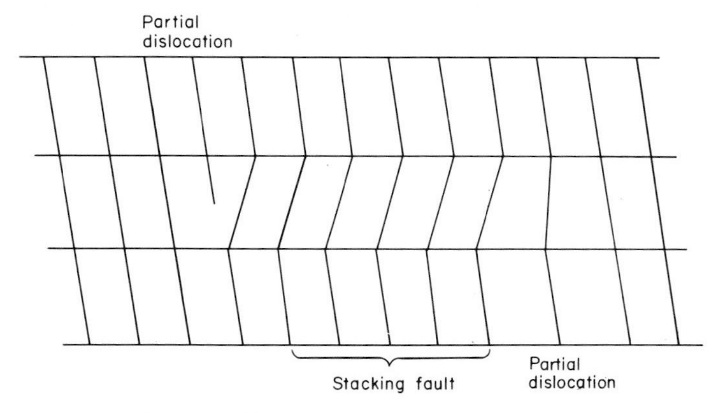

Fig. 6-5. The stacking fault. (*After Cottrell.*[14])

orientation such that the two parts of the crystal are mirror images across a boundary referred to as the *twinning plane*. In the diamond lattice, the twinning plane is the (111) and the order of the planes reverses at this boundary. A volume of the crystal bounded in parallel twinning planes and only a few lattice planes across is termed a *twin lamella*.

A stacking fault, illustrated in Fig. 6-5, is a region in the crystal which is bounded by two partial dislocations. The edge and screw dislocations described above are so-called "pure dislocations"; their Burgers vectors are lattice vectors. The displacement is equivalent to a lattice constant. However, it can be shown by using solid models that when atoms are moved from one site to the next, they cannot follow a straight path. They must move around other atoms in a zigzag path. It is possible for a position to be taken such that only half this path is traversed, i.e., the Burgers vector is only half a lattice spacing. Such a defect is called a *partial dislocation* and gives rise to strain on only one side, which is subsequently relieved by another partial dislocation. The area in between represents a part of the crystal which is out of alignment with the rest; it is not correctly stacked and is referred to as a stacking fault. This particular defect is of importance in epitaxial films (see Sec. 8-30).

6-8. CRYSTAL ORIENTATION

In Chap. 4, the methods of growing crystals from the melt were described, and it was pointed out that a seed was used to start the growth. The conditions of growth, the temperature, the pulling speed, and so on all have an effect on the distribution of dopants and impurities, and this was described fully in that chapter. They are also important in producing material of minimum dislocation density; this is defined as the number of dislocations cutting one square centimeter of the crystal.

When a crystal is grown from the melt, there is considerable evidence[10] to suggest that this occurs as a series of steps. Figure 6-6 illustrates the principle. The crystal grows along a low-index plane, such as the (111), but by laying down a series of strata growing in steps along high-index planes. The atoms find it easier to attach

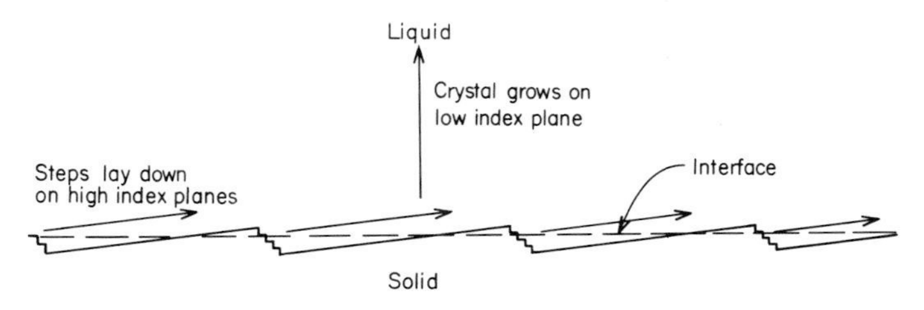

Fig. 6-6. Crystal growth from the melt. (*Adapted from Tiller.*[17])

to high-index planes since there are more bonds available. The low-index plane can be considered more perfect, and lattice sites are more difficult to find than in the broken terrain of the stepped structure. This says, in effect, that nucleation is more difficult on low-index than on high-index planes. However, this very ease of nucleation, or attachment, may give rise to mismatching which, in fast-growing directions, may be propagated. If growth is in the direction of the low-index plane, a misaligned atom may revert to the liquidus and redeposit correctly. If the growth is in the direction of the high-index planes, this may be prevented by rapid overgrowth.

The direction of growth is thus extremely important in producing material of low dislocation density, and the seed must be correctly oriented to induce this growth. Moreover, similar considerations apply to substrates for epitaxial growth. The determination of orientation therefore is a common requirement in the characterization of semiconductor material.

6-9. ORIENTATION BY X-RAYS†

The back-reflection Laue method is more generally applicable to this problem and will be described in some detail since this equipment is usually available in most laboratories. The crystal is mounted on a reference plane, usually a ceramic plate to which it is cemented, and a monochromatic x-ray beam reflected from some conveniently oriented flat surface. Generally, an experienced operator can judge a likely plane and orient fairly closely to the desired projection. The pattern resulting

†Adapted from Kane.[11]

Fig. 6-7. Laue pattern for (100) silicon crystal.

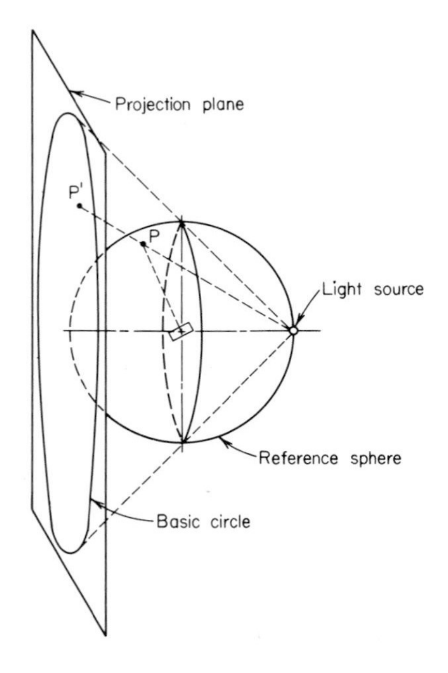

Fig. 6-8. The stereographic projection. (*From Barrett and Massalski.*[15])

from a single crystal is a series of spots, and Fig. 6-7 shows such a pattern for silicon. The next step in an orientation requires its interpretation, and this implies identification and location of the Laue spot corresponding to the direction in which the crystal must be oriented. The method used involves the application of stereograms.

Figure 6-8 illustrates the method of obtaining a stereographic projection. The crystal is imagined at the center of a sphere, and the direction of a plane can be represented by P, the latitude and longitude at which this direction cuts the sphere. The sphere is then projected onto a plane normal to one diameter of the sphere from an imaginary light source at the other end of the same diameter to give the projected point P'. If the diameter is that joining the north and south poles, then the projection is the polar projection. More commonly used is the diameter joining the intersections of the equator with the zero and 180° meridians, and this is termed the *Wulff net*, shown in Fig. 6-9. With this as a grid, it is possible to calculate from any crystal just where the various directions will appear by using any particular direction for a plane of reference. As stated in Sec. 6-1, it is fortunate that the semiconductors of interest are all cubic, so that the directions are normal to the planes and stereograms are really quite simple to apply. Figure 6-10 is the stereogram of a cubic crystal in the [001] direction, and each spot represents a particular lattice plane in the crystal. The lines represent zones, that is, families of planes with one common axis called the *zone axis*. In one direction only, they have a common angle so that they intersect the sphere on a great circle.

Returning to the Laue pattern, in Fig. 6-11 we see the geometry of this projection, which is, in fact, a section through a number of discontinuous cones, each cone the reflections from a zone. Since they intersect a plane, they appear as hyperbolas on the film. The problem is to transform this projection of the planes to the stereo-

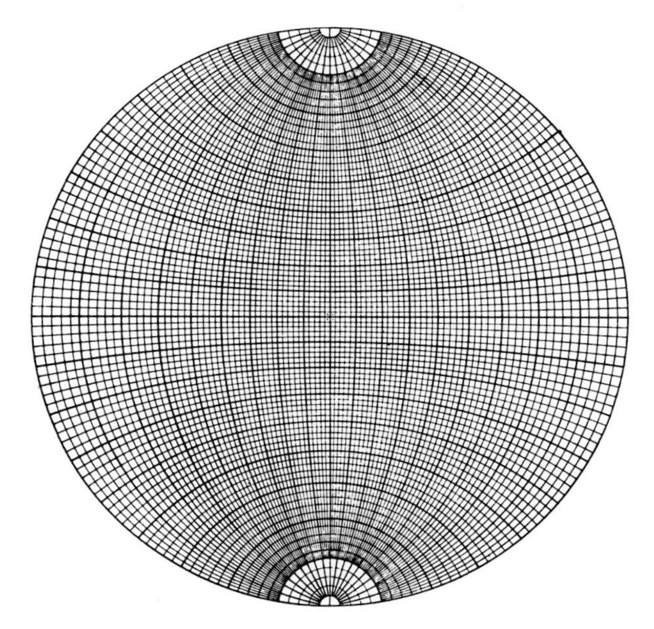

Fig. 6-9. The Wulff net.

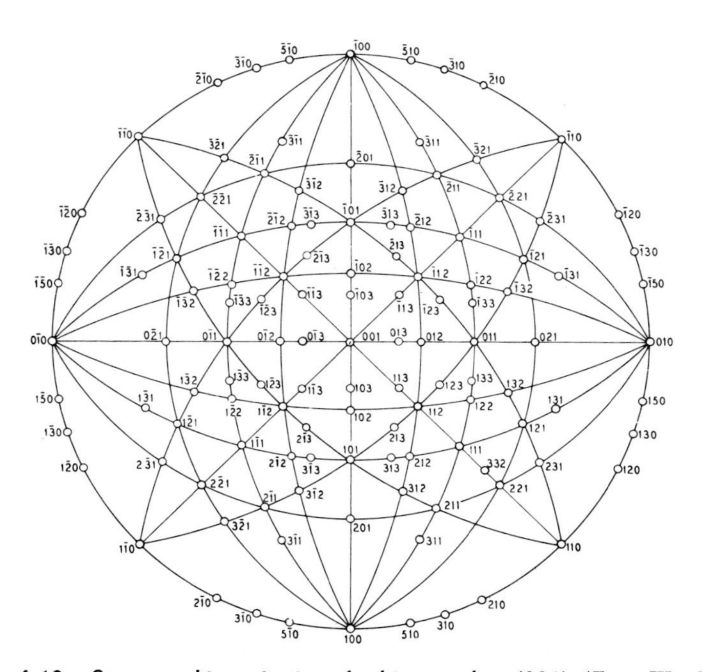

Fig. 6-10. Stereographic projection of cubic crystal on (001). (*From Wood.*[18])

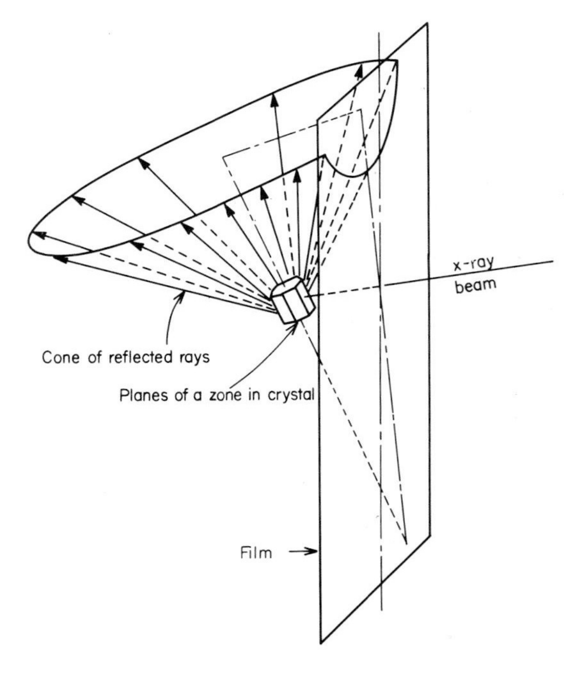

Fig. 6-11. Geometry of the Laue projection. (*After Barrett. and Massalski*[15])

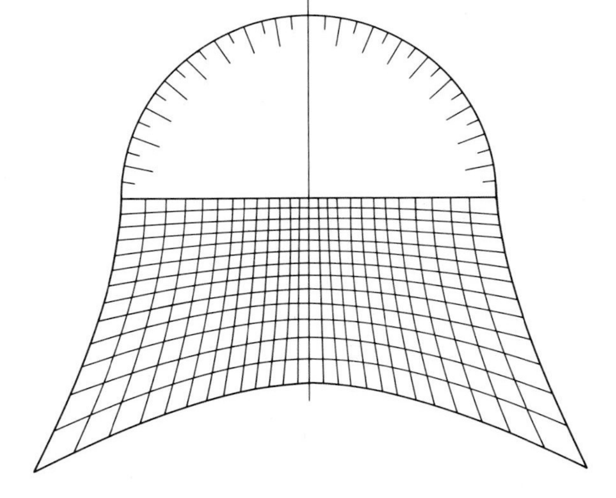

Fig. 6-12. The Greninger net.

graphic projection, and this is done by a graphical method. Figure 6-12 shows a *Greninger net,* which is a series of hyperbolas corresponding to various angular relationships at a set specimen-to-film distance, usually 3 cm. The Laue film is marked with a fiducial line corresponding to some direction of the crystal or, more commonly, the crystal mount and placed face down on the net. The center of the film corresponds to the center of the net, and it is rotated about this center until one row of spots is approximately parallel to one of the meridians of the net, as shown in

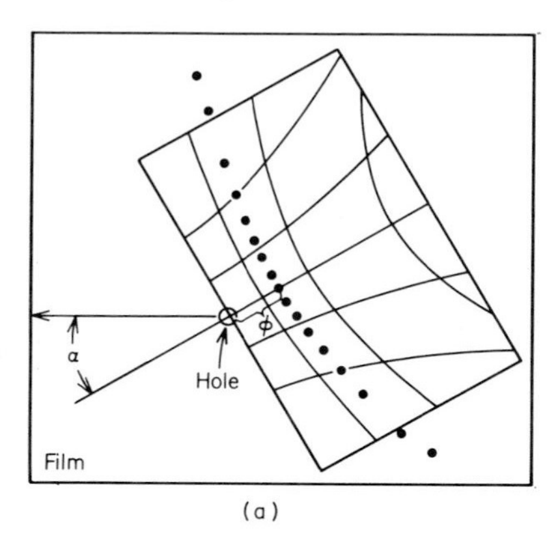

(a)

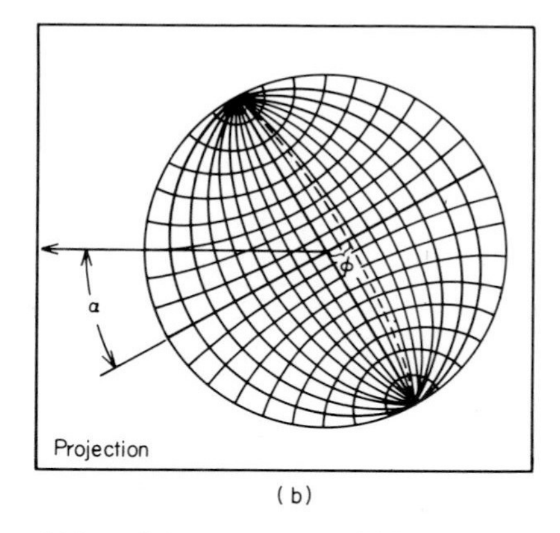

(b)

Fig. 6-13. Method of transforming Laue patterns. (*Adapted from Barrett and Massalski.*[15])

Fig. 6-13*a*. The angle of inclination ϕ of the zone axis to the film is given directly since the meridian lines are at 2° intervals. The azimuthal angle α is read from the lower scale; this is the angle between the fiducial mark and the 90° line. The zone is then transferred to the Wulff net by setting the same azimuthal angle to the equator and drawing the zone of the meridian corresponding to the angle of inclination, as shown in Fig. 6-13*b*. It should be emphasized that the zone is drawn on a transparency over the Wulff net and not on the net itself. The next zone is plotted in the same way, rotating the Wulff net under the transparency. The final result is a series of zones drawn with respect to the fiducial line. Enough are drawn to identify the projection, and the intersections of these zones will be the directions of planes common to these zones. The principal zones in face-centered-cubic materials are [100], [110], and [111], and the most important spots are their intersections, namely, the [100], [110], and [112] directions. In practice, the stereogram is placed under the projection which has been drawn and the desired orientation direction marked on the projection. This is placed again over the Wulff net, and the angles necessary to correct the direction calculated with respect to the fiducial mark.

In Texas Instruments laboratories, a simple projection device is used without transparent paper. It consists of the Greninger net mounted on one plastic wheel which is coupled by a belt to an identical plastic wheel carrying a Wulff net transparency. Both are illuminated from below. A sketch of the apparatus is shown in Fig. 6-14. The Laue film is placed face down on the plastic table over the Greninger net with the hole above the center and the fiducial mark aligned vertically. The net is rotated under the film; and as it rotates, the Wulff net also rotates to the same angle. Since in the conventional method, this is the function of the angular scale of the Greninger net, it can be dispensed with and is replaced by another set of hyperbolas. When one of the hyperbolas of Laue spots is aligned with a grid of the Greninger net, it is a simple matter to read the angle and draw the circle projection corresponding to it with a grease pencil on the plastic table over the Wulff net. This is repeated for a number of principal zones, and transparencies for various projections are placed over this stereogram until identification is made. The angles necessary to correct for misalignment can be read from the Wulff net.

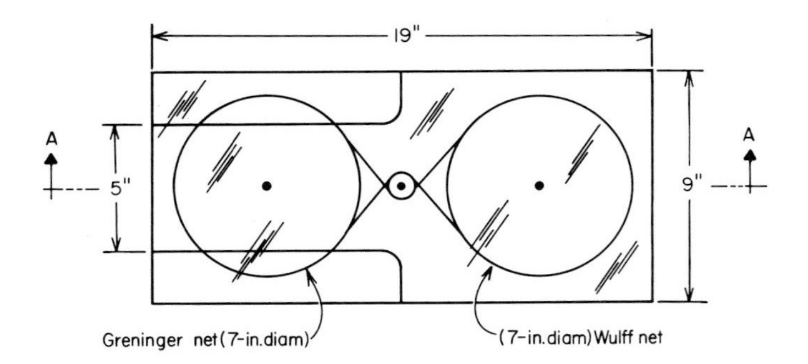

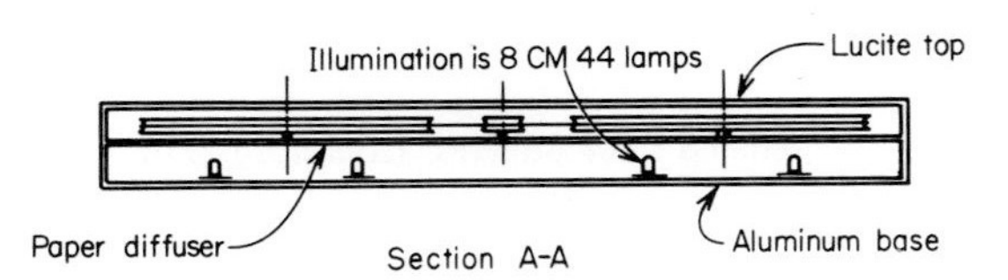

Fig. 6-14. The stereographic projector. (*From Kane.*[11])

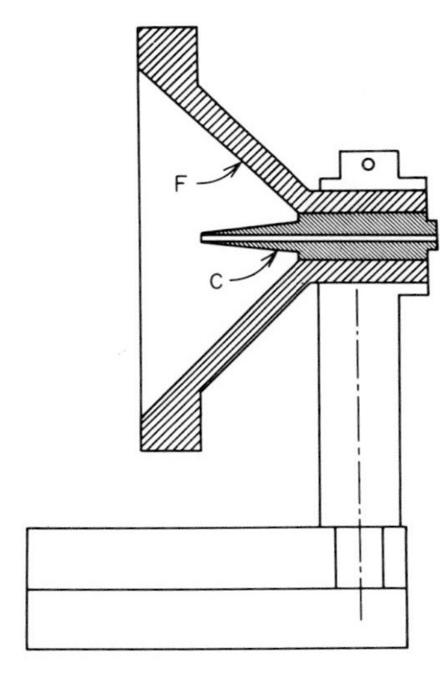

Fig. 6-15. Conical camera for crystal orientation. (*After Arguello.*[19])

This procedure sounds complicated, but with practice it can be carried out quite readily. An alternative approach has been applied by Arguello[19] in which a conical camera is used to obtain a close approximation to the stereographic projection without the use of the two nets. The camera is shown in Fig. 6-15. The film is held in the form of a cone at F, and x-rays leaving the collimator C strike the sample held in the center of the base. The cone has 45° angles and is 4 cm in radius at the base. The geometry is shown in Fig. 6-16. The normal stereographic projection of the direction of the plane shown is Y_e, and the orthographic projection of the reflection P from the same plane is Y_p. It can be shown that, if $C = 0.83R$, that is, the distance from the sample to the screen is 4.82 cm, then Y_p is never more than 2 mm from Y_e and an essentially correct stereogram can be projected. Unfortunately, the shape of

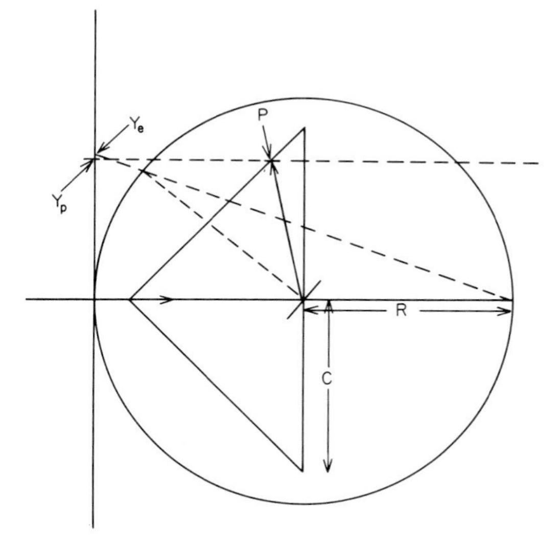

Fig. 6-16. Projection using the conical camera. (*After Arguello.*[19])

this film would make it awkward to process, and the camera itself is not currently available commercially.

The Laue method is used to give a complete orientation of the crystal. The problem above is that in which the orientation is completely unknown, but, in general, this is not the case. The orientation of the crystal is known approximately, but the crystal must be accurately aligned for cutting. The Laue method, since it uses charts marked at intervals, is accurate only to about these 2° intervals, and where the orientation is known approximately, the more accurate x-ray goniometer is used. The precision of this procedure is about ±15 minutes. The method is given

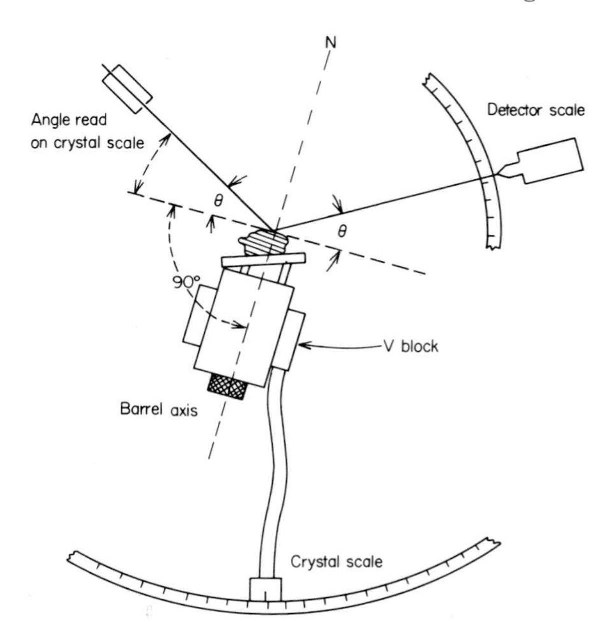

Fig. 6-17. X-ray goniometer. (*After Wood.*[18])

in some detail by Wood[18] and is outlined in ASTM Method F26-66.[20] It is based on the diffraction from one plane only, for example, the (111) in silicon. Figure 6-17 shows one typical arrangement of the apparatus used. The barrel holder rests on a V block with a stop at the back. The actual holder for the crystal can be adjusted independently of the barrel in two directions, normal to the x and y axes. The stop can be adjusted so that the surface of the crystal is at the mechanical axis of the goniometer. The detector scale is set for the 2θ angle for the particular plane of the material, as chosen from Table 6-1; e.g., for the (111) plane in silicon it is

Table 6-1. Bragg Angles θ for the X-ray Diffraction of CuK_α Radiation in Semiconductive Crystals†

Wavelength $\lambda = 1.54178$ Å

Reflecting planes h, k, l	Silicon $a = 5.43073$ Å (±0.00002 Å)	Germanium $a = 5.6575$ Å (±0.0001 Å)	Gallium arsenide $a = 5.6534$ Å (±0.0002 Å)
111	14°14′	13°39′	13°40′
220	23°40′	22°40′	22°41′
311	28°05′	26°52′	26°53′
400	34°36′	33°02′	33°03′
331	38°13′	36°26′	36°28′
422	44°04′	41°52′	41°55′

†From ASTM, Method F26-66.[20]

$2 \times 14°14' = 28°28'$. The V block is then rotated until the signal on the detector is a maximum. The crystal scale will now read $\theta + \delta$, where δ is the variation of the x axis from the true. The barrel holder is rotated on its own axis 180° and the reading repeated to obtain $\theta - \delta$. The difference thus represents twice the angle that the x axis of the (111) plane is from being correct. Two similar readings are made for the y axis. For this particular type of holder, the two adjustments can be made in the directions normal to the x and y axes to make the (111) plane exactly normal to the barrel axis. With most other holders, the cutting machine is used to adjust the crystal to the correct orientation once the angular displacements are known.

6-10. OPTICAL ORIENTATION†

An optical method, also accurate to about 15 minutes, is often used in orienting crystals for cutting since it is extremely simple and cheap. The method has been described by Schwuttke[21] and is included in ASTM Method F26-66.[20] One apparatus, shown diagrammatically in Fig. 6-18, consists merely of a beam of light which is reflected from the surface of the crystal onto a screen. The center of the beam, Z, is located by replacing the crystal with a plane mirror. The crystal surface must be preferentially etched first. More will be said about etching later, but suffice it to say

†Adapted from Kane.[11]

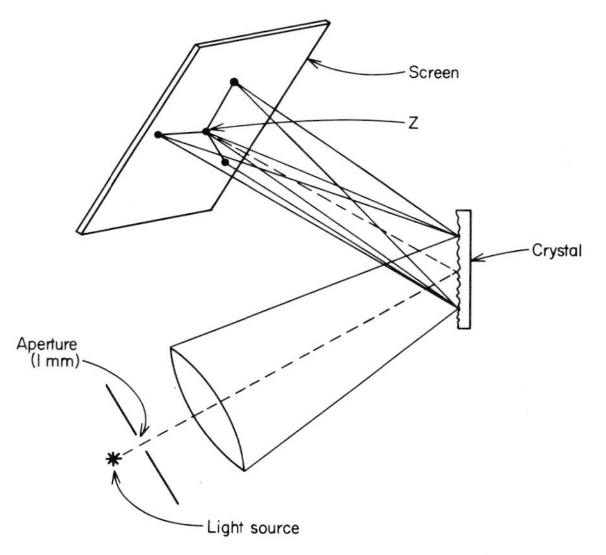

Fig. 6-18. Optical orienter. (*After Schwuttke.*[21])

at this stage that some chemical etchants will etch faster along some crystallographic axes than others. An imperfection in the surface will act as a site for this type of action, and the result is a surface containing a number of etch pits. The inside faces of these pits are actually facets parallel to some important crystal planes. They act as tiny mirrors and reflect the light to form characteristic patterns. Those for the (111), (100), and (110) planes in germanium and silicon are sketched in Fig. 6-19

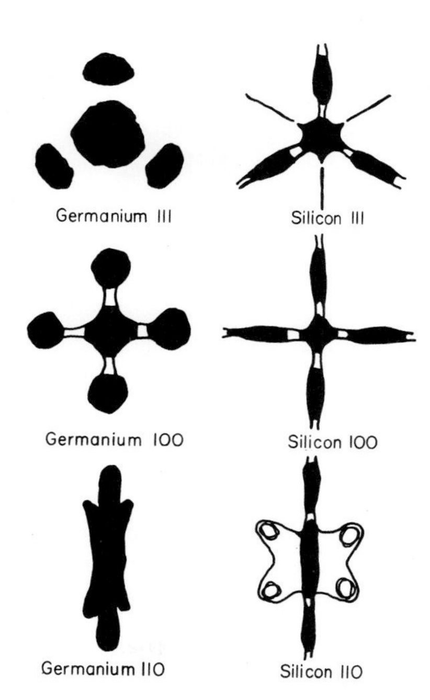

Fig. 6-19. Optical reflectograms from germanium and silicon. (*From ASTM Method F26-66.*[20])

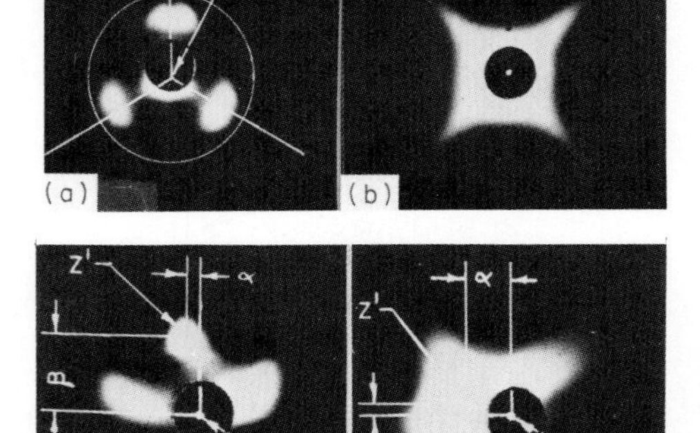

Fig. 6-20. Germanium crystal correctly oriented (*a*) on the (111) and (*b*) on the (100) plane. (*From Schwuttke.*[21])

Fig. 6-21. Germanium crystal misaligned by α and β (*a*) from the (111) and (*b*) from the (100) plane. (*From Schwuttke.*[21])

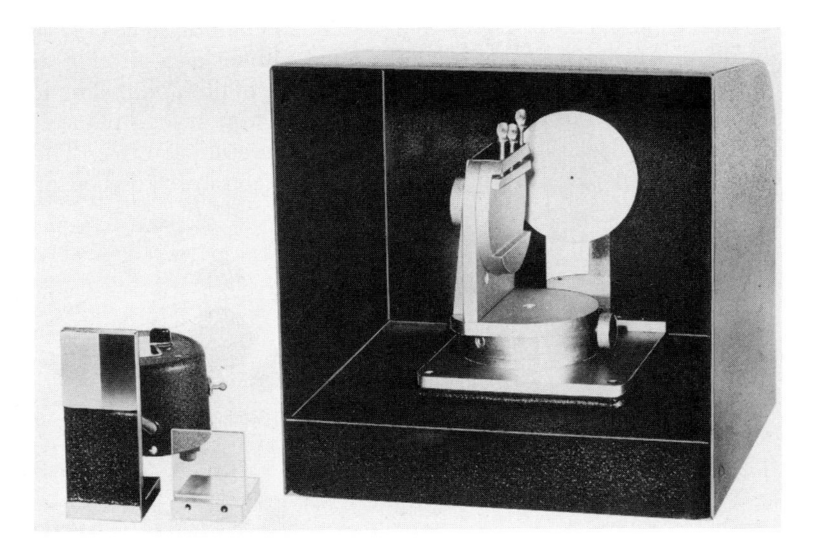

Fig. 6-22. Micromech crystal orienter. (*Micromech Mfg. Co.*)

and represent perfect alignment. In Fig. 6-20 are actual reflectograms from perfectly oriented crystals of germanium on the (111) and (100) planes. Figure 6-21 shows the same planes misaligned by α and β on the two axes of reference. The alignment can be corrected or the crystal cutter set for these two angles as in the x-ray method.

Figure 6-22 shows a commercial version of the instrument manufactured by the

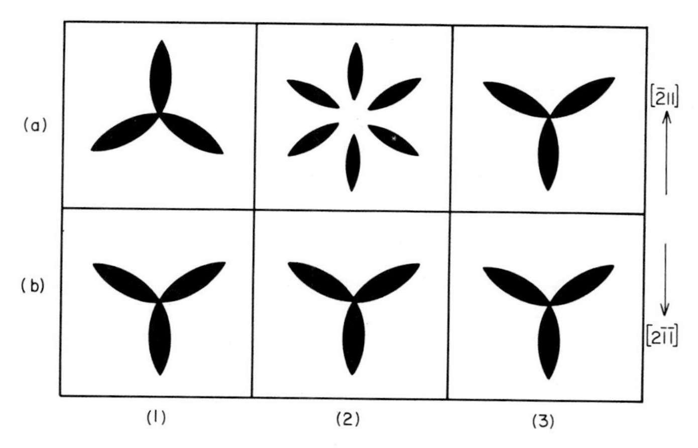

Fig. 6-23. Light figures from (*a*) **Ga surface and** (*b*) **As surface of GaAs. 1, Ground; 2, ground and slightly etched; 3, etched.** (*After Akasaki and Kobayasi.*[23])

Micromech Manufacturing Co. This is a slightly different optical system in which the light beam enters through a hole in the screen to give an angle of incidence of 0°. The sample is rotated through 180° about an axis normal to the crystal surface, as in the x-ray method, to check the readings.

The ASTM method recommends lapping the surface of the crystal with No. 600 silicon carbide and etching with the solution shown in Table 6-2. It restricts the procedure to silicon and germanium. However, it has been applied to gallium arsenide, at least for (111) wafers, by Cronin[22] and by Akasaki and Kobayasi.[23] The light figures for three types of surface preparation are shown in Fig. 6-23, and the etches used by Akasaki and Kobayasi are given in Table 6-3. It will be noticed that the polarity of the crystal can make a difference in the case of compound semiconductors. Cronin first used these light figures to differentiate the *A* and *B* faces

Table 6-2. Etching Procedure for Optical Orientation†

Material	Etchant composition‡	Etching time, min	Etch temperature, °C
Germanium	1 part (vol) hydrofluoric acid (49%) 1 part (vol) hydrogen peroxide (30%) 4 parts (vol) water	1	25
Silicon	50% sodium hydroxide (by weight) solution or 50% potassium hydroxide (by weight) solution	5	65

†From ASTM Method F26-66[20].

‡In both cases, (111)-, (100)-, and (110)-type surface planes may be prepared with these etchants.

Table 6-3. Etching Conditions for Production of Light Figures from GaAs Surfaces†

Etchant			Etching time	Surface
Notation	Composition‡	Volume ratio		
A-1	$HNO_3:H_2O$	1:1	7 min	Ga
A-2	$HNO_3:H_2O$	2:1	10 sec	Ga
B-1	$HF:H_2O_2:H_2O$	2:1:8	50 min	As
B-2	$HF:HNO_3:H_2O$§	3:1:2	7 min	As

†From Akasaki and Kobayasi.[23]
‡Concentration (wt %): HNO_3, ~60; HF, 46; H_2O_2, 30. H_2O: deionized water.
§Several drops of 1% $AgNO_3$ are added in mother solution of about 30 ml.

of a gallium arsenide wafer. Figure 6-24 shows the reflection patterns corresponding to 1*a* and 1*b* in Fig. 6-23. The patterns in this case were produced by sandblasting or lapping with No. 240 silicon carbide. The wafers were grown from the *B* end of a seed, and it is characteristic of this type of crystal that the result is a triangular cross section, bounded by (111) faces. The reflectogram from the *A* (Ga) surface shows peaks perpendicular to the faces; that from the *B* (As) face shows the peaks parallel to the faces.

6-11. DETERMINATION OF DISLOCATION DENSITY

A dislocation, as we have seen in Secs. 6-3 to 6-7, is a discontinuity in the crystal lattice and, as such, is on an atomic scale. By using the electron microscope, dislocations can be observed directly,[24] and this has proved to be a powerful research tool. As will be seen later (Sec. 6-16), x-rays and similarly electrons are attenuated less by dislocations than by perfect areas of the crystal, and an image is produced. By applying various stresses, the movement of dislocations can be studied.

For assessing the quality of a bulk material, electron microscopy is of very little value since the sample is too small to be representative. Not only is the field of view

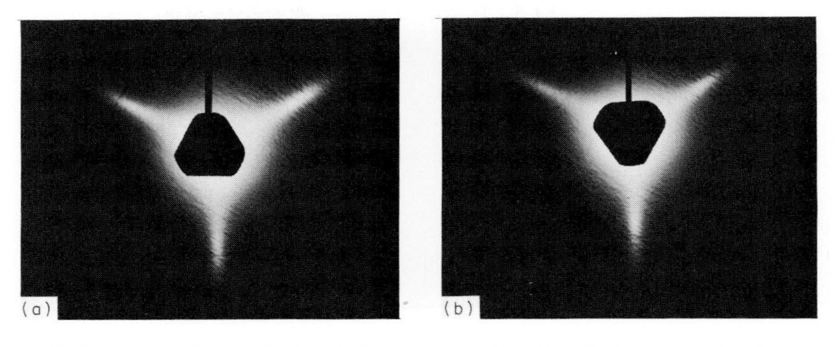

Fig. 6-24. Reflectograms from (*a*) the *A* (Ga) face and (*b*) the *B* (As) face of gallium arsenide. (*From Cronin.*[22])

very small, but the sample must also be thinned down to less than 2000 Å to allow transmission of the electrons. The total volume of the sample as observed in the microscope is something like 10^{-9} cc, and it is obviously impossible to extrapolate this to even a relatively small crystal. Of more interest in quality control are methods which can give an average value. Of these, etch-pit counts and x-ray topography are probably the most widely used.

6-12. ETCH PITS

In Sec. 6-8, crystal growth was described as being by a process of stratified deposition, and a scheme was shown in Fig. 6-6. The atoms attached more easily to high-index planes since there are more bonds available. As a corollary, there will be fewer bonds holding atoms in these planes to the lattice, so that on dissolution a similar procedure results. This implies that a low-index plane breaks up more slowly than a higher-index plane. Now a dislocation, as we saw in Sec. 6-3, is an area in the crystal in which there are many dangling bonds. In a low-index plane, it represents a point of weakness, a point at which the atoms are more loosely bound. Consequently, this point is also more easily attacked by solutions. These two effects lead to attack by a suitable etchant at points in a surface at which dislocations emerge. Figure 6-25 shows how an etch pit forms, initiating at a dislocation and dissolving in a terraced formation as the atoms are removed preferentially from the less closely packed lattice planes. V_c, V_d, and V_s are the solution velocities in different directions, where $V_s > V_d > V_c$. Microscopically, the etch pit will show a fine step structure; macroscopically, the face will be a principal lattice plane.

In using the etch-pit method, the crystal must be oriented to a low-index plane, (111), (100), or (110). When it has been cut to this plane, any mechanical damage must be removed since not only dislocations but any fault in the surface can initiate a pit. In fact, such damage is introduced by sandblasting or grinding to induce many etch pits when an optical orienter is being used (Sec. 6-10). The sawn and lapped

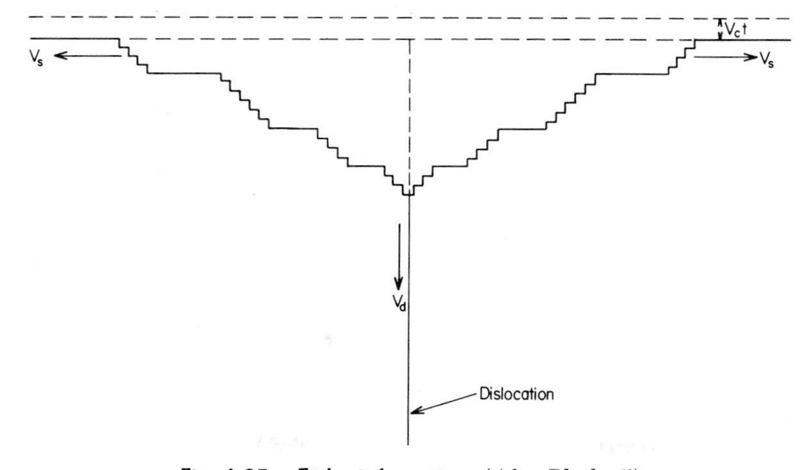

Fig. 6-25. Etch-pit formation. (*After Rhodes.*[10])

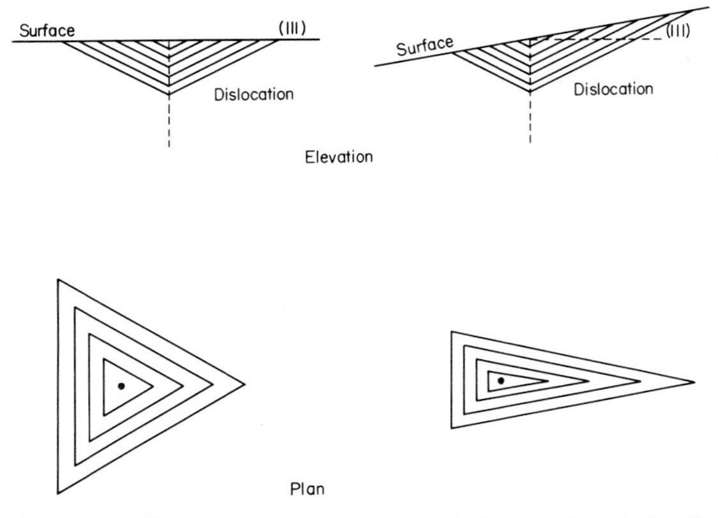

Fig. 6-26. Effect of misorientation on etch-pit shape. (*After Rhodes.*[10])

surface is chemically etched with a nonpreferential etch first to remove mechanical damage. This will usually extend to about the diameter of the grit, as was shown, for example, by Jones and Hilton[25] for gallium arsenide. The surface is then treated with a preferential etch to form etch pits similar to those shown in Fig. 6-30*a*. These preferential etches are commonly acids containing an oxidant; several are given by Rhodes[10] for silicon and germanium.

The importance of correct orientation can be seen from Fig. 6-26. The pit faces remain angled to the correct planes, hence their use in optically orienting crystals, and it follows that misorientation of the surface leads to distorted etch pits. With reference to Fig. 6-25, it was pointed out that, in etch-pit formation, $V_d > V_c$ is a necessary condition. If $V_d = V_c$, no pit will form. If the dislocation is angled to the surface, then it is not V_d which is the controlling velocity but its component along the normal. If the angle is too obtuse, then no etch pit can form. This angle can become too obtuse either because the dislocation itself is at too shallow an angle to the surface or because the surface is too misoriented. This misorientation may be as little[10] as 10°.

A tentative method for silicon has been adopted by ASTM Specification F47-64T.[20] The ingot for examination is oriented within 3° to the (111) plane, and a slice cut for examination. It is lapped with No. 600 alumina or silicon carbide and chemically polished with a mixture of hydrofluoric, nitric, and acetic acids. After rinsing, it is treated with copper etch, a hydrofluoric-nitric acid mixture containing copper nitrate and a little bromine. After rinsing and dissolving off any copper, the number of pits are counted under a low-power microscope to determine the dislocation density.

Preferential etches are usually quite specific not only for the material but also for a particular orientation. For germanium, CP-4 etch has been used quite extensively since its introduction by Vogel et al.[26] This also is a bromine etch, consisting of a

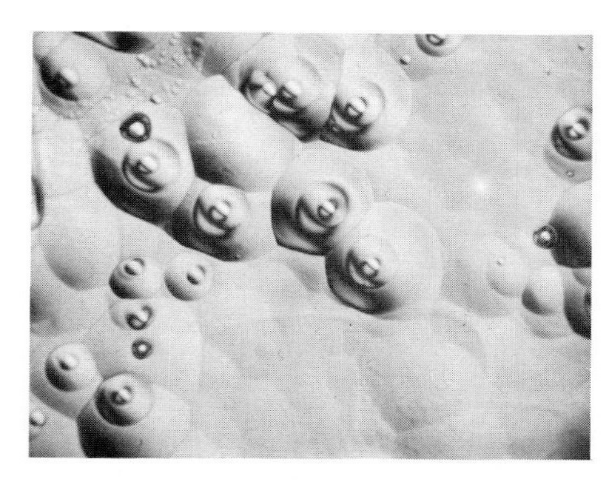

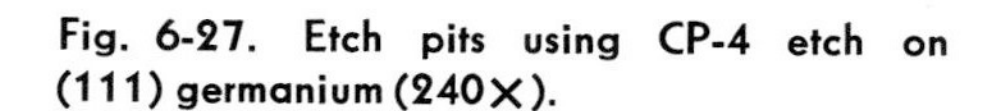

Fig. 6-27. Etch pits using CP-4 etch on (111) germanium (240×).

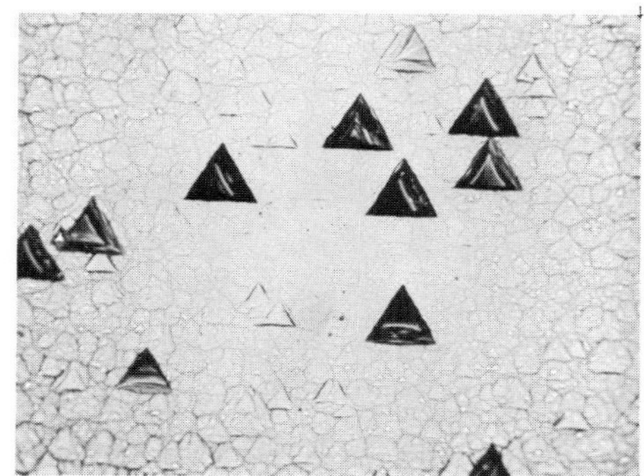

Fig. 6-28. Etch pits using WAg etch on (111) germanium (200×).

mixture of nitric, hydrofluoric, and acetic acids with a little bromine added. It has been applied to both (111) and (100) surfaces. The pits are conical in shape, as shown in Fig. 6-27. Triangular etch pits are produced by WAg (Westinghouse silver) etch,[27] a nitric-hydrofluoric acid mixture containing silver nitrate, or by a ferricyanide etch,[28] consisting of an aqueous solution of potassium ferricyanide and potassium hydroxide. Etch pits obtained by WAg etch are shown in Fig. 6-28.

For III-V compounds, there is an additional factor which can influence the etching characteristics, namely, the polarity. White and Roth[29] used a dilute aqua regia, and Richards[30] a hydrofluoric acid etch containing hydrogen peroxide to etch gallium arsenide preferentially, but both produce pits only on the *A* (Ga) face, that is, the (111) surface and not the $(\bar{1}\bar{1}\bar{1})$.† By using a silver etch, Richards and Crocker[31] were able to produce pits on both the *A* and *B* faces of (111) gallium arsenide wafers. A generally applicable etch containing silver nitrate and chromic oxide in dilute hydrofluoric acid was applied by Abrahams and Buiocchi[32] to the faces of gallium arsenide on the (111), (110), and (100) planes.

CP-4 etch has been used by Bardsley and Bell[33] on indium antimonide, and a bromine-methanol etch by Fuller and Allison[34] on gallium arsenide; this last is more valuable for indium arsenide. Etch pits on these III-V materials are typically conical in shape. Figure 6-29 shows pits similar to those obtained by Richards and Crocker[31] on the *A* and *B* faces of (111) gallium arsenide.

†This is an arbitrary convention; many European workers use the reverse Miller indices.

Fig. 6-29. Etch pits on (111) GaAs. (*a*) The *A* (Ga) face; (*b*) the *B* (As) face (220×).

6-13. ADDITIONAL INFORMATION BY ETCHING TECHNIQUES

The preferential etch will, of course, reveal the number of dislocations cutting the surface, and a dislocation density is the primary objective of this determination. However, some additional information can also be obtained, as pointed out in ASTM Specification F47-64T[20] for silicon. A line of dislocations, very close-packed, is indicative of a lineage structure, as shown in Fig. 6-30*b*. This is often met with in crystals which have been melt-grown, as are most semiconductor crystals, and is due to the aggregation of dislocations by movements in their slip planes at temperatures approaching the melting point.

The presence of several parallel lines of dislocations, as seen in Fig. 6-30*c*, is evidence of slip. In Secs. 6-4 and 6-5, the dislocation was described as due to a displacement of the lattice in the direction of the Burgers vector, and such a displacement is due to a stress which is relieved by this movement. Such a deformation is termed *plastic deformation*, as opposed to *elastic deformation*, which distorts the lattice without rupturing it. Additional stress leads to many such displacements, all

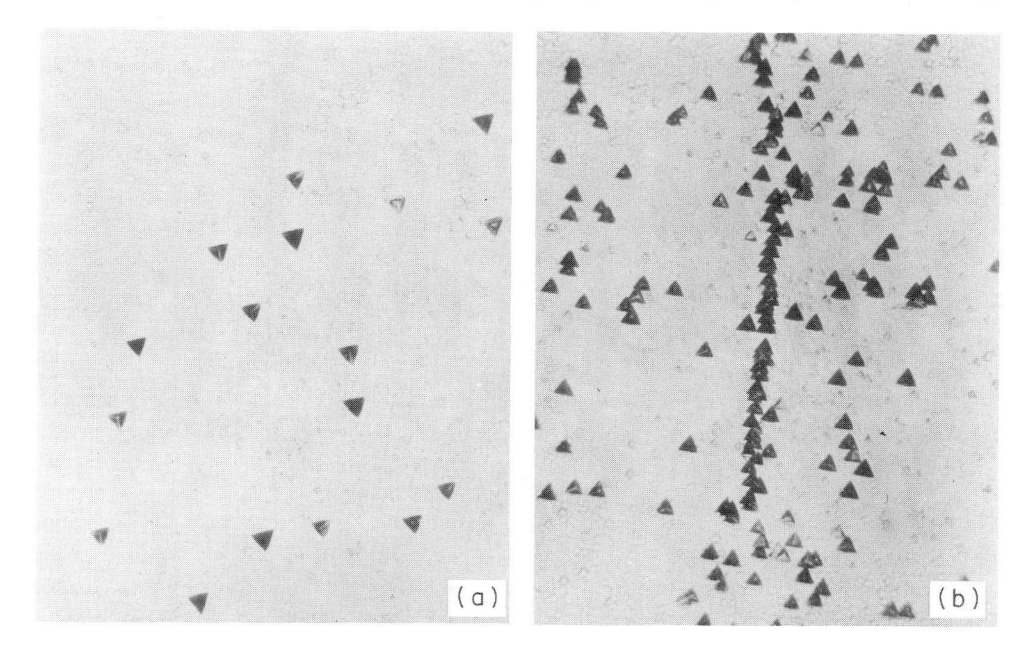

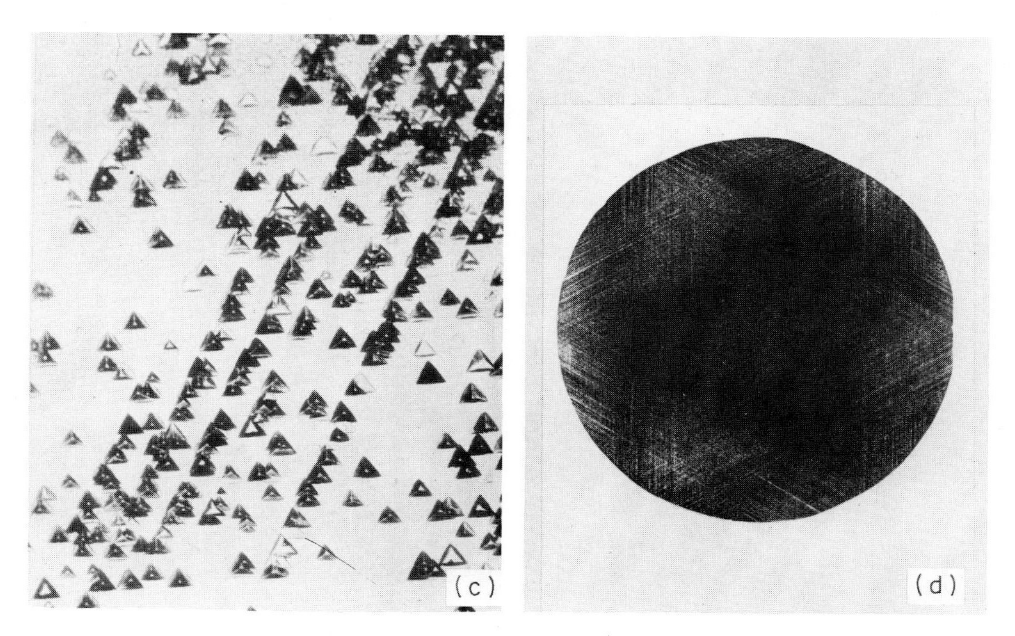

Fig. 6-30. Etch patterns on silicon. (*From ASTM Method* F47-64T.[20]) (*a*) **Pits from dislocations (200×);** (*b*) **lineage (200×);** (*c*) **slip—micro scale (200×);** (*d*) **slip—macro scale (3×).**

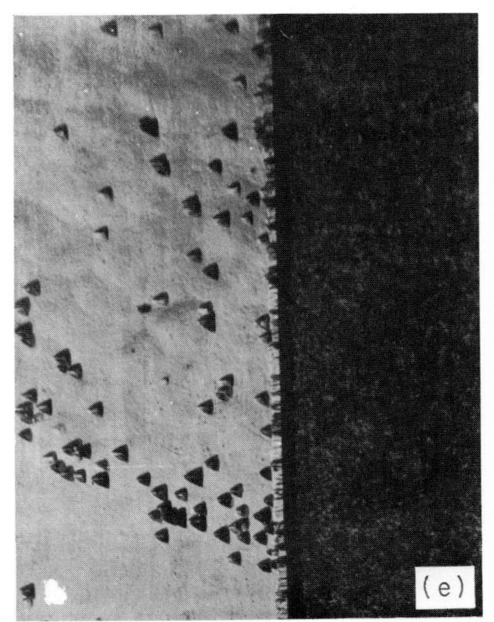

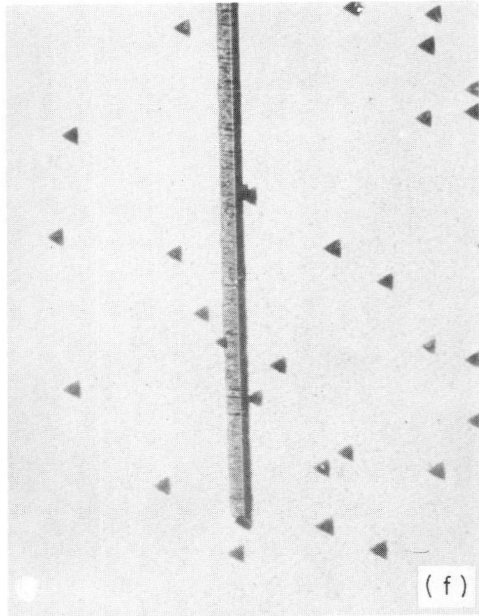

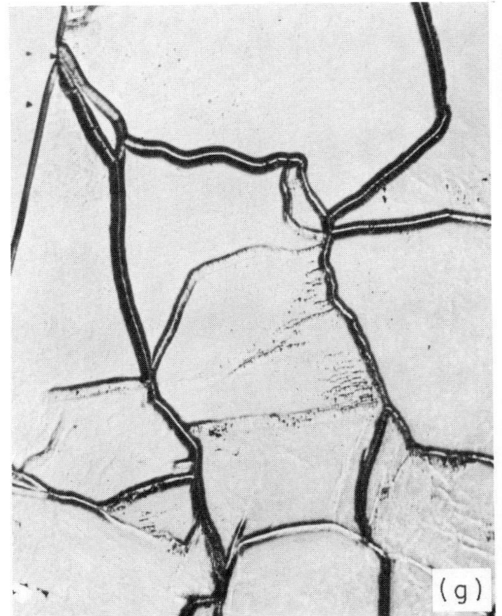

Fig. 6-30. *(continued)* **Etch patterns on silicon.** (*From ASTM Method* F47-64T.[20]) (e) **Twin boundary (200X);** (*f*) **twin lamella (200X);** (*g*) **grain boundaries (200X);** (*h*) **polycrystallinity — macro scale (3.2X).**

in the same slip plane, i.e., a whole series of dislocations in one direction. Such a phenomenon is termed *slip*. On a macroscopic scale, the appearance is as shown in Fig. 6-30*d*. The surface, after a preferential etch, shows a series of lines with an overall triangular or Star of David effect.

Since the preferential etch is sensitive to different orientations, the presence of a twin is revealed quite readily. In the diamond structure, the twinning plane is always (111), and where it cuts the surface the etch forms a straight line. The effect is shown in Fig. 6-30*e*. Twin lamellae are smaller areas of twinning, bounded on each side by twinning planes, as seen in Fig. 6-30*f*.

Grain boundaries are also etched preferentially, and a typical polycrystalline material is shown in Fig. 6-30*g*. Polycrystallinity can also be seen by using a non-preferential etch, i.e., by chemical polishing. After etching in potassium hydroxide, areas of polycrystallinity show up quite readily under reflected light. Figure 6-30*h* shows such an area in a silicon ingot.

A review of these various etching techniques has been given by Holmes,[35] and a summary of the more important ones is given in Table 6-4.

6-14. DECORATION

This is another chemical method which relies for its effect on the fact that precipitation tends to occur at dislocation sites. The method was used by Dash[45] to show the correspondence between dislocations and etch pits in silicon. After etching the sample, a drop of copper nitrate solution was placed on the specimen and the crystal heated for 1 hr in hydrogen at 900°C. Copper diffuses rapidly and is evenly distributed after this time. After cooling, the samples were examined under an infrared microscope. In Sec. 5-10, it was pointed out that the absorption edge for silicon is 1.1 μ and for gallium arsenide it is 0.92 μ; at wavelengths longer than this, it is essentially transparent. Copper, however, is not, and under this microscope the dislocations show up as lines. The images can also be recorded on infrared sensitive emulsions.

For germanium, the absorption edge is further into the infrared at 1.8 μ, and the usual detector cannot be used. However, vidicon tubes are sensitive in this region, and Schwuttke[61] has described a method using this in conjunction with a closed-circuit TV presentation.

This technique is of somewhat limited value since it is a destructive technique. The crystal cannot subsequently be used for device work. Moreover, the heating cycle may alter the dislocation pattern by thermally induced motion; and, finally, this method does not reveal screw dislocations.

6-15. X-RAY DIFFRACTION METHODS

The most useful nondestructive techniques for dislocation density are based on x-ray measurements. One method for dislocation densities in the higher ranges is line broadening. The technique uses a double-crystal spectrometer, shown in Fig. 6-31. Crystal *A* acts as a monochromator; and as crystal *B*, the sample, rotates through its Bragg angle, it generates a curve of intensity against angle of rotation,

Table 6-4. Etchants for the More Important Semiconductors†

(Figures are in milliliters of concentrated reagents unless otherwise stated)

Name of etchant	Recipe	Etching time	Principal uses, and comments	Refs.
		A. Germanium		
Iodine etch A	5 HF 10 HNO_3 11 acetic acid with 30 mg I_2 dissolved	4 min	Polishing and etching (100) and (110) surfaces. Better than CP-4 for etching (100)	36
CP-4	15 HF 25 HNO_3 15 acetic acid with 0.3 Br_2 dissolved	1½ min for etching. 2 min or more for polishing	Polishing, etching (111) and (100), revealing sharp p-n junctions,‡ and grain and twin boundaries	37 26
CP-4A CP-6 CP-8	3 HF 5 HNO_3 3 acetic acid	2 to 3 min	Chemical polish. Much slower than CP-4 at room temperature, so can be used warm (70°C). Absence of bromine means better prospects of a chemically clean surface	
Dash germanium etch	2 HF 4 HNO_3 15 acetic acid	20 sec to 1 min	Etching out of dislocations after decoration by lithium	38
White etch	1 HF 3 HNO_3	1 to 2 min warm	Chemical cleaning, and revealing p-n junctions‡	
WAg (silver nitrate etch)	2 HF 1 HNO_3 2 5% $AgNO_3$ solution	1 min	Etching of (111) planes, revealing grain boundaries. Liable to deposit silver, which must be removed chemically, preferably by a cyanide wash. Also gives dull background pitting	27
No. 1	1 HNO_3 2 HF 1 10% $Cu(NO_3)_2$ solution	1 to 2 min	Etching of (111) planes. Also used in certain etches for silicon	39

†Adapted from Holmes.[35]
‡See Sec. 9-9.

Table 6-4. Etchants for the More Important Semiconductors† (continued)

Figures are in milliliters of concentrated rengents unless otherwise stated.

Name of etchant	Recipe	Etching time	Principal uses, and comments	Refs.
Hydrogen peroxide	30% H_2O_2	Hot, or 1 hr cold	Gives a clean matte surface, revealing grain boundaries and lineage. p-n junctions shown quite sharply.‡ Good for preparing surfaces for electrical probing (the slight roughening prevents a probe from slipping)	40
No. 2 (Superoxol)	1 30%H_2O_2 1 HF 4 H_2O	1 to 3 min	Etching of (100) and (111) planes. Slow to attack polished surfaces, but otherwise a good etch for (100). Etch rate has been studied	41 42
Dilute No. 2	1 No. 2 etch 50 H_2O	2 to 16 hr	Produces etch pits, with fine structure, on (111), (100), and (110), and is one of the best etches for (110). Intermediate dilutions between this and No. 2 are also useful for orientations	43
Alkaline peroxide	8 g NaOH 100 3% H_2O_2	70°C	Controlled removal of material (etch rate falls off with age to an approximately constant value when the etch is old: 0.2 mil/min new, 0.05 mil/min at 1 hr)	
Sodium hypochlorite	1 10% NaClO 10 H_2O (i.e., 0.1 M)	40 min (40°C), or as required for thinning	Etching on (100) and (111), and for thinning slices for electron microscopy. Interesting as a single-component etchant	44
Ferricyanide	6 g KOH 4 g $K_3Fe(CN)_6$ 50 H_2O	1 min boiling	Etching of (111) planes, showing up lineage structure and grain boundaries. This is a very good etch for producing clear triangular pits on (111)	28

†Adapted from Holmes.[35]
‡See Sec. 9-9.

Table 6-4. Etchants for the More Important Semiconductors† (continued)

Figures are in milliters of concentrated reagents unless otherwise stated.

Name of etchant	Recipe	Etching time	Principal uses, and comments	Refs.
		B. Silicon		
White etch	1 HF 3 HNO_3	15 sec	Chemical polish	
Dash etch	1 HF 3 HNO_3 8–12 acetic acid	1 to 16 hr	Etching of all planes. Deep pits, following the dislocation lines into the crystal. Smaller proportions of acetic acid give a faster etch, which is often useful	45
CP-4A	3 HF 5 HNO_3 3 acetic acid	2 to 3 min	Slow chemical polishing revealing twins, twin lamellae, p-n junctions,‡ etc., and sometimes dislocations	46
	10 HF 5 HNO_3 14 acetic acid	0.5 to 3 min	Chemical polish	20
SD1	25 HF 18 HNO_3 5 acetic acid containing 0.1 Br_2 10 H_2O 1 g $Cu(NO_3)_2$ (i.e., 5 of No. 1 + 3 of CP-4)	2 to 4 min	Reveals edge and mixed dislocations on all planes. A good etch for rapid estimates of pit density, but the copper deposition is undesirable if the material is to be used again	47
	6 HF 3 HNO_3 100 H_2O with 5.5 mg Br_2 0.3 g $Cu(NO_3)_2$	4 hr	Dislocation etch for (111)	20
Hot NaOH/KOH	1–30% solution	1 to 5 min 50 to 100°C	Rapid revealing of structural details: especially large pits associated with twin lamellae on (111) surfaces. Simple to apply, but deposits traces of iron on surface, which must be removed chemically by washing in hydrochloric acid	48 49

†Adapted from Holmes.[35]
‡See Sec. 9-9.

Table 6-4. Etchants for the More Important Semiconductors† (continued)

Figures are in milliliters of concentrated reagents unless otherwise stated.

Name of etchant	Recipe	Etching time	Principal uses, and comments	Refs.
	4% NaOH Add 40% NaOCl until no hydrogen evolution on Si	As required, about 80°C	Used for thinningdown slices for electron microscopy. The specimen should float on the surface of the etch. (Too much NaOCl, or covering the dish, causes specimen to sink)	50

C. Indium antimonide (see also under *G*, below)

Name of etchant	Recipe	Etching time	Principal uses, and comments	Refs.
	1 HF 1 HNO_3	2 to 5 sec (polishing)	Polish-etch for $(\bar{1}\bar{1}\bar{1})$ and (110). No etching on (111) or (100)	51
	5 HF 5 HNO_3 2 H_2O	20 sec	Etching (100) and $(\bar{1}\bar{1}0)$	52
CP-4A	3 HF 5 HNO_3 3 acetic acid	5 to 30 sec	Chemical polish. May be further diluted with water and HNO_3. Ordinary CP-4 has also been used for etching (111)	33 53 54
111 etch	1 CP-4A 1 H_2O 1 acetic acid	1 min	Etching (111) and planes from there out to (112)	53
No. 2 etch	1 HF 1 30% H_2O_2 4 H_2O	5 to 10 sec	Etching (111)	55

D. Gallium antimonide

Name of etchant	Recipe	Etching time	Principal uses, and comments	Refs.
	1 HF 1 HNO_3 1 H_2O	1 min	Etching (111)	55

†Adapted from Holmes.[35]

Table 6-4. Etchants for the More Important Semiconductors† (continued)

Figures are in milliliters of concentrated reagents unless otherwise stated.

Name of etchant	Recipe	Etching time	Principal uses, and comments	Refs.
	1 HF 9 HNO_3	1 to 5 min	Chemical polish	56
	1 30% H_2O_2 1 HCl 2 H_2O		Etching (111) and (100), with asymmetry of pits corresponding to different rates of attack on $(\bar{1}\bar{1}\bar{1})$ and $(11\bar{1})$	55
	E. Gallium arsenide (see also under *G*, below)			
	5 5% NaOH solution 1 30% H_2O_2	5 min	Removing material at a rate of 10 to 15 μ/min	57
	2 HCl 1 HNO_3 2 H_2O	10 min	Etching (111)	29 58
	1 HF 5 HNO_3 8 to 12 1% $AgNO_3$ solution		Etching (111) and $(\bar{1}\bar{1}\bar{1})$. [The Ag inhibits attack on the normally fast-etching $(\bar{1}\bar{1}\bar{1})$]	31
	1 HF 3 HNO_3 2 H_2O		Chemical polish	31
	1 HF 1 30% H_2O_2 5 H_2O	10 min	Produces etch pits on (111)	30
	1 HF 2 H_2O with 33% Cr_2O_3, 0.3% $AgNO_3$	10 min, 65°C	Produces etch pits on (111), $(\bar{1}\bar{1}\bar{1})$, (110), (100). Also reveals dislocation lines	32
	CH_3OH with 5–20% Br_2		Polishing	34
	F. Indium arsenide (see also under *G*, below)			
	15 HF 75 HNO_3 15 acetic acid 0.06 Br_2	5 sec	Etching (111)	59
	G. Indium phosphide			
	0.4 *M* solution of ferric ion in 6 *N* HCl	1 to 5 min, 25°C	Etching (111). This solution also etches InSb, InAs, and GaAs, at various temperatures (see references)	60

†Adapted from Holmes.[35]

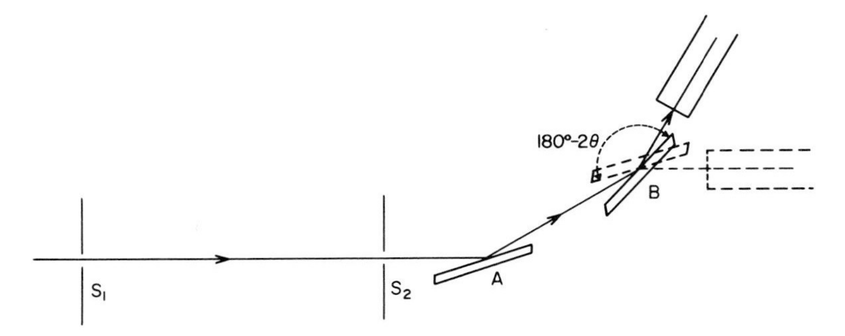

Fig. 6-31. Double-crystal spectrometer. (*After Compton and Allison.*[62])

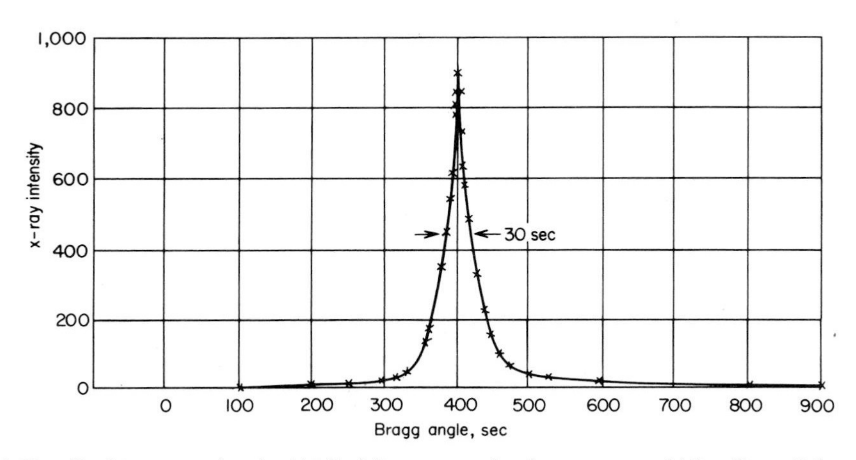

Fig. 6-32. Rocking curve for the (111) diffraction peak of germanium. (*After Kurtz;*[63] *by permission from MIT Lincoln Laboratory.*)

as shown in Fig. 6-32. It can be shown[64] that, for an ideal crystal, the width of the peak at half the height is given by the expression

$$\Delta\theta = \frac{\lambda^2 e^2 N f}{\pi m c^2 \sin(2\theta)}$$

where λ = wavelength of incident x-rays
e = charge on electron
N = number of atoms per unit volume
f = atomic scattering factor
m = mass of electron
c = velocity of light
θ = Bragg angle for particular reflecting plane

The factor f is a property of the electrons in the atom and not of the nucleus. For a (111) plane in germanium, $\Delta\theta$ has been calculated to be 15 seconds of arc. This scattering is a fundamental property of the material, and the angle represents a theoretical minimum. If the various parts of the crystal are distorted with respect to one another, then it may be regarded as a collection of small crystals almost, but not quite, fitting together; i.e., a mosaic is formed. This idea of a mosaic has been used to explain the fact that many crystals exceed the minimum value for the width at half-wave height by considerable amounts. The theory is dealt with in detail by James,[64] but briefly it can be explained as a spread in the reflected x-rays caused by the fact that the reflecting planes in different areas of the crystal are at slight angles to one another. One can assume that, in a single crystal, the mosaic is defined by the dislocations and that they are randomly distributed. In this case, the expression reduces to

$$n_D = \frac{\phi}{9b^2}$$

where ϕ = angular range
b = magnitude of Burgers vector
n_D = dislocation density

ϕ is superimposed on $\Delta\theta$, and an exact relationship is difficult since, in addition, there is a spread due also to the first crystal. If this is imperfect, this may be an even greater source of error. Batterman[65] found that substitutional impurities could also affect this value, although the levels were considerably higher than normally encountered in semiconductors. Despite these drawbacks, the method is useful for dislocation densities above about $10^5/\text{cm}^2$.

This same line broadening can be used to determine dislocation densities from back-reflection Laue patterns. These were described at length in Sec. 6-9. If the crystal is imperfect, the spots in the pattern tend to broaden and, with increasing defect structure, tend toward arcs of the hyperbolas. For a transmission Laue, the arcs may become complete circles for a polycrystalline material, and we obtain the familiar Debye-Scherrer patterns. Bell[66] applied this technique to silicon, germanium, and indium arsenide but without evidence of any spot distortion. He calculated the limit of the method to be $10^3/\text{cm}^2$, essentially that of the rocking-curve method, as might be expected.

For lower dislocation densities, the phenomenon of anomalous transmission was applied to germanium by Hunter.[67] In a perfect crystal, and with correctly oriented lattice planes, the primary x-ray beam sets up a standing-wave pattern which, when it emerges from the crystal, actually has more energy than the normal absorption coefficient for the material would forecast. The effect is termed *anomalous transmission*. Where the crystal shows nonperiodic character, e.g., at dislocation sites, this particular effect is destroyed and the beam attenuates normally. Thus dislocations in the crystal lead to a decreased energy of the total emergent beam. The theory is complicated and depends on the generation of wave fields. It is dealt with in more detail by Auleytner[68] and others. A crystal with 500 dislocations/cm^2 showed a significantly different transmitted energy from that of a dislocation-free crystal.

6-16. X-RAY TOPOGRAPHY

Of the methods described in Sec. 6-15, only the rocking-curve method is of value for evaluating material. It is applied to crystals containing over 10^5 dislocations/cm^2, but below this cannot discriminate further. Moreover, at these lower levels,

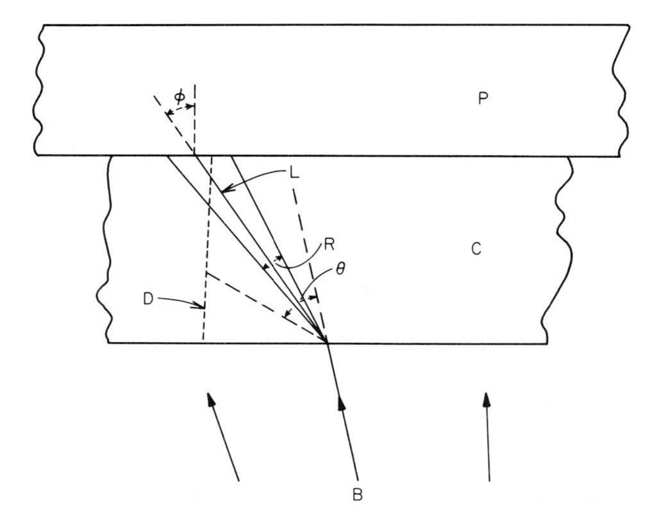

Fig. 6-33. The Borrmann method. *P*, **photographic plate;** *C*, **crystal;** *L*, **lattice plane;** *B*, **primary beam;** *θ*, **diffraction angle;** *R*, **wedge of rays with reduced absorption;** *D*, **dislocation;** *ϕ*, **angle between** *L* **and normal to surface.** (*After Borrmann et al.*[70])

which are now very common for semiconductor materials, we are faced with a distribution problem. We cannot any longer assume the distribution to be essentially uniform. At values of $10/\text{cm}^2$ or less — and much commercial material is of this level — the location, direction, and characteristics of the dislocation become important. Techniques of x-ray microscopy or, as it now more commonly termed, *x-ray topography* have been devised for this purpose and have been succinctly reviewed by Amelinckx[24] and by Webb.[69]

The phenomenon of anomalous transmission has been applied by Borrmann[70] and is illustrated in Fig. 6-33. The photographic plate is in contact with the back face of the crystal, and the primary monochromatic x-ray beam is strongly divergent. Since there are a large number of incident angles, some of these will satisfy the requirement for anomalous transmission. *L* is such a lattice plane, and the interaction of the transmitted and diffracted beams generates the wedge of anomalous transmission *R*. The thickness is such that the beam is attenuated by normal absorption, and only those areas of the plate receiving the anomalously transmitted rays are exposed. The result is a pattern of broad bands in a grid effect, generated by the correctly oriented lattice planes. However, if a dislocation occurs inside

Fig. 6-34. Borrmann topography. (*From Borrmann et al.*[70])

this wedge, the effect is destroyed and a nonexposed area occurs. A Borrmann topograph is shown in Fig. 6-34.

Variations of this technique, using a parallel beam of x-rays and setting the crystal to a suitable Bragg angle, have been described by Barth and Hosemann[71] and by Authier.[72]

A somewhat more widely used method was devised originally by Berg[73,74] and modified by Barrett.[15,75] This Berg-Barrett technique was refined by Newkirk,[76] and this modification is illustrated in Fig. 6-35. It will be noticed that, unlike the Borrmann method, only the diffracted beam is recorded by the photographic plate; the primary beam travels on through the crystal and does not strike the plate. Where a dislocation occurs, the diffracted beam is more intense than that from an area of perfect crystal. This is due to the fact that the primary beam is not attenuated as strongly in these imperfect regions; normal absorption results from the interaction of x-rays with the periodic atom centers. Consequently, when diffraction occurs it must also be stronger at this point. The image forms a uniform exposure, with dislocations showing as more heavily exposed areas. This is in contrast to the Borrmann method, where the dislocations show up as shadows.

The Berg-Barrett technique is usually employed in the reflection mode, although Newkirk[76] pointed out that it could also be used in transmission provided the crystal were thin enough. In either case, the area that can be examined by this method is relatively small, dependent on the area of the x-ray beam. Lang[77] introduced a method of transmission diffraction by which large-area topographs could be obtained, and this is shown in Fig. 6-36. A collimated monochromatic x-ray beam strikes a thin specimen set at an appropriate Bragg angle. The diffracted beam passes through a slit in a screen to the photographic plate and is recorded; the primary beam is stopped. At this point, the method resembles the Berg-Barrett transmission mode, and dislocations will be recorded as before. However, by oscillating the crystal and the film together, the whole specimen can be examined and a picture built up of a comparatively large area, e.g., a 1-in.-diameter slice.

Topographs obtained by the Lang method are shown in Fig. 6-37. The diffracting planes are selected on the basis of the Laue geometry applicable to the transmission mode and are dependent on the orientation of the crystal. For example, a crystal cut on the (001) plane might, in the absence of any previous experience, be examined in this plane first since it is relatively easy to align. The slice, thin enough to transmit, is mounted in a goniometer with a detector behind the screen slit.

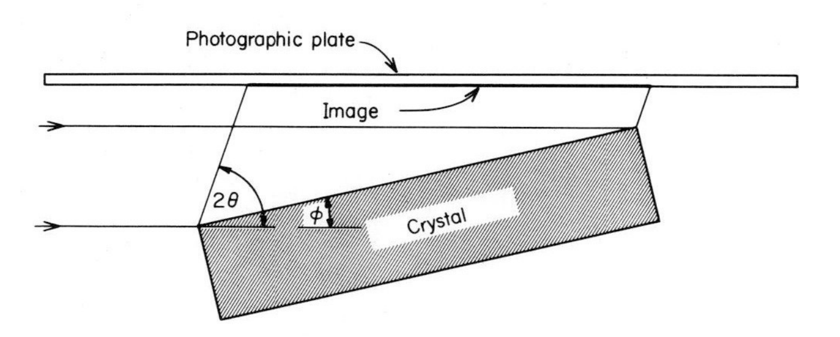

Fig. 6-35. The Berg-Barrett method. (*After Newkirk;*[76] *copyright AIME.*)

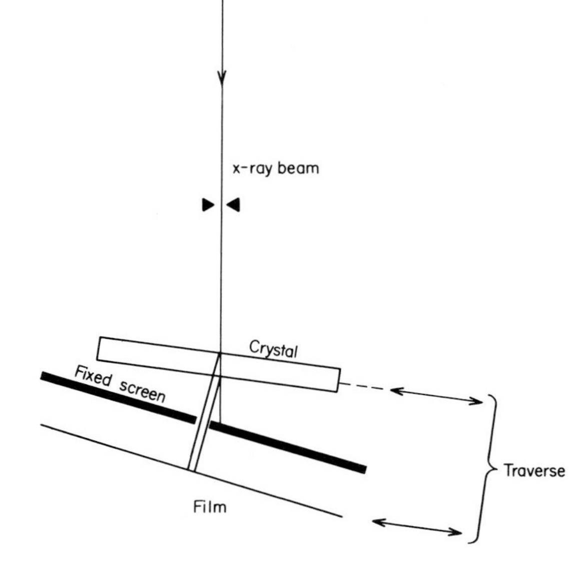

Fig. 6-36. The Lang technique. (*After Lang.*[77])

After setting approximately correctly to the calculated angle, the specimen is adjusted to a maximum signal for fine correction. This first exposure might indicate that an alternative plane would give additional information. Within the geometrical limits of the Laue arrangement, the plane can be found by reference to a stereographic projection, for example, Fig. 6-10. This would indicate that the (012) plane, for example, would be about 18° from the crystal surface, and the specimen would be aligned accordingly.

The geometrical limits can be considerably extended by adapting the Lang method to the reflection mode. Howard and Dobrott[78] modified the Berg-Barrett method with a scanning technique to obtain large-area topographs of the surface. Their method is described in more detail in Sec. 7-12, and their apparatus shown in Fig. 7-10. Since the geometry here is the Bragg reflection geometry, this allows a much larger selection of diffracting planes to be used in any one thin slice. It is also applicable to thicker slices which are to be used as epitaxial substrates (see Sec. 7-12) and to epitaxial layers (see Sec. 8-37). The selection of reflecting planes is made in the same way as for the Lang method, and this procedure is identical to that described in more detail by Newkirk[76] for the Berg-Barrett technique.

By taking topographs from three mutually perpendicular planes, it is possible to deduce the character of the dislocations. These topographs can be obtained by cutting the specimen on different orientations or, in the case of a thin wafer, a judicious use of the transmission and reflection Lang techniques. Three such transmission topographs obtained by Jenkinson[79] are shown in Fig. 6-37. Close examination will show that no dislocations show up in all three topographs; a few appear in two. Generally, a dislocation appears in only one topograph. The reason may become clearer by reference to Fig. 6-38, which is a representation of the two types of dislocation. For either case, x-rays traveling in the direction of the Burgers vector encounter misalignment in successive reflecting planes; i.e., the

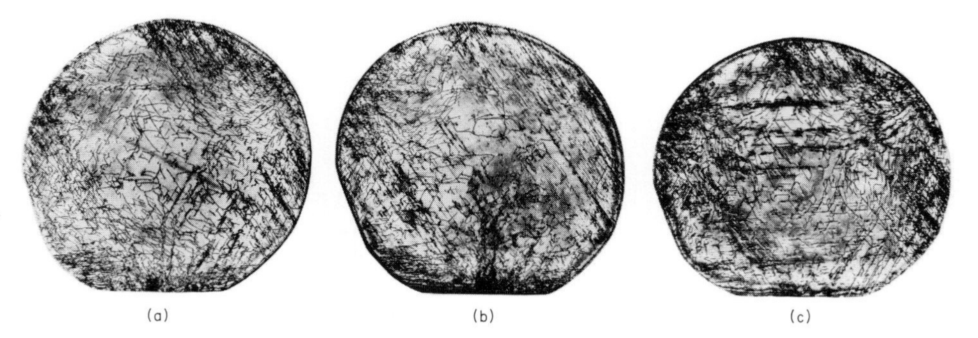

Fig. 6-37. Transmission topographs taken from (*a*) **(11$\bar{1}$),** (*b*) **(1$\bar{1}$1), and** (*c*) **($\bar{1}$11) plane of silicon slice cut (111).** (*From Jenkinson.*[79])

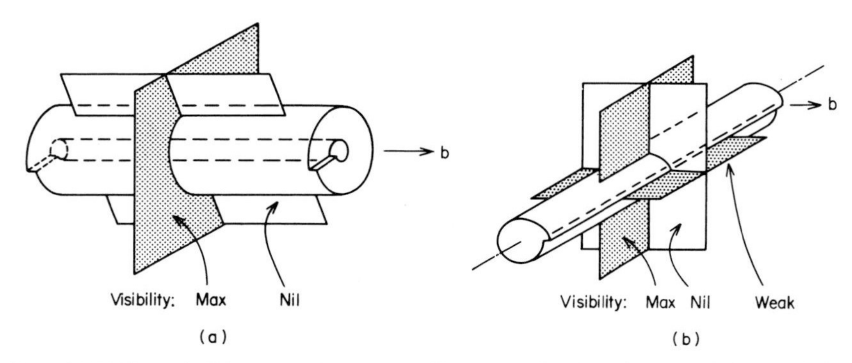

Fig. 6-38. Visibility of dislocation from mutually perpendicular planes. (*a*) **Screw dislocation;** (*b*) **edge dislocation.** (*From Jenkinson.*[79])

effect is a maximum. For x-rays traveling normal to the Burgers vector in the case of the edge dislocation, Fig. 6-38*b*, they see essentially a point defect; successive planes are in alignment. For a screw dislocation, Fig. 6-38*a*, the misalignment is confined to only one or two atom planes and is also a minimum. Returning to the topographs of Fig. 6-37, we see that the glide system of both germanium and silicon is along {111} planes in a <110> direction, that is, the slip planes are almost invariably {111} and the direction of the Burgers vector is <110>. The three topographs were obtained from {111} planes. There are six possible <110> directions (if we ignore the positive or negative sense, which is merely a question of orientation), and any {111} plane will contain three of these. If the Burgers vector of the dislocation is one of these three directions, it will be invisible. If it is invisible in a second topograph, then the Burgers vector must lie in the intersection of these two planes to satisfy both conditions. If a dislocation is visible in two topographs, then the converse is true: i.e., it cannot lie in either of these. In fact, only one possible condition is left, and it is parallel to the intersection of the third (or invisible) plane with the (111) surface.

Similar results can be obtained with III-V compounds. However, in this case the structures are more ionic, and the glide planes will prefer the electrically neutral

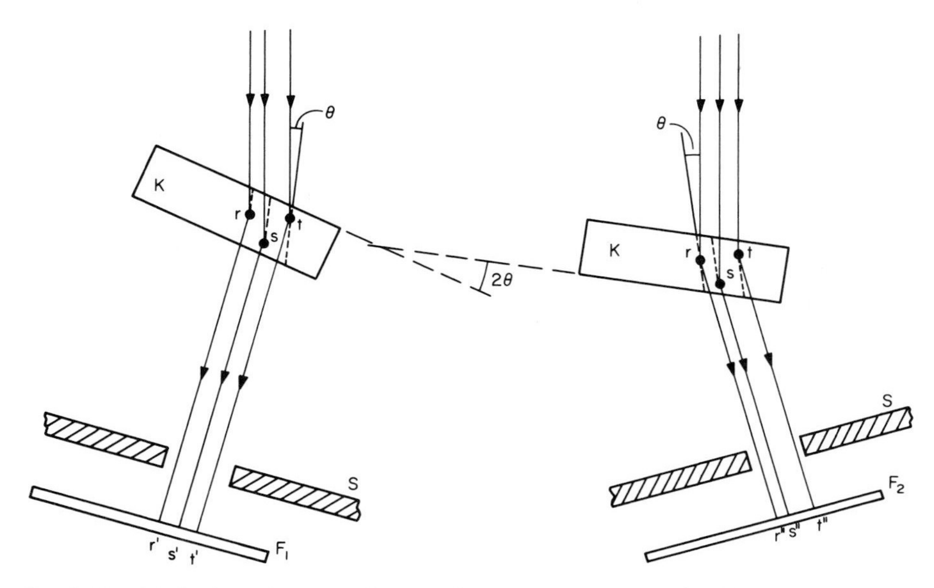

Fig. 6-39. Method for obtaining stereographic projection topographs. (*From Jenkinson.*[79])

lattice planes. The system will, in general, tend to be along a (110) plane in the [101] direction, although this is not as invariable a rule as that for the homopolar silicon and germanium structures.

For almost all work, the direction of the dislocation can be estimated reasonably closely by its appearance in the normal topographs. In a more elegant method due to Lang[77] a stereoscopic projection is obtained. The principle is shown in Fig. 6-39. Two topographs are taken, one at twice the Bragg angle to the other. This is equivalent to taking one at the (hkl) and the other at the $(\bar{h}\bar{k}\bar{l})$ reflection. Under a stereo viewer, an in-depth impression of the dislocations can be obtained. An excellent example of this technique, using red-blue spectacles and a composite red-blue print, is given by Jenkinson.[79]

Since the directions of both the Burgers vector and the dislocation can be determined, the character of the dislocation can be deduced (see Secs. 6-4 and 6-5) as either predominantly edge or screw.

REFERENCES

1. Wagner, C., and W. Schottky: *Z. Phys. Chem.*, **B11**:163 (1930).
2. Frenkel, J.: *Z. Phys.*, **35**:652 (1926).
3. Sugata, E., H. Kim, and H. Araki: in R. Uyeda (ed.), "Electron Microscopy, 1966," vol. I, p. 249, Maruzen, Tokyo, 1966.
4. Gallagher, C. J.: *Phys. Rev.*, **88**:721 (1952).
5. Pearson, G. L., W. T. Read, and F. J. Morin: *J. Phys. Rev.*, **93**:666 (1954).
6. Read, W. T.: *Phil. Mag.*, **45**:775 (1954).
7. Read, W. T.: *Phil. Mag.*, **45**:1119 (1954).

8. Van Bueren, H. G.: "Imperfections in Crystals," North Holland Publishing Company, Amsterdam, 1960.
9. Bardsley, W.: in A. F. Gibson (ed.), "Progress in Semiconductors," vol. 4, p. 155, John Wiley & Sons, Inc., New York, 1960.
10. Rhodes, R. G.: "Imperfections and Active Centers in Semiconductors," Pergamon Press, New York, 1964.
11. Kane, P. F.: *Norelco Reptr.*, **14**:47 (1967).
12. Hobstetter, J. N.: in N. B. Hannay (ed.), "Semiconductors," p. 508, Reinhold Publishing Corporation, New York, 1959.
13. Read, W. T., Jr.: "Dislocations in Crystals," McGraw-Hill Book Company, New York, 1953.
14. Cottrell, A. H.: "Dislocations and Plastic Flow in Crystals," Oxford University Press, Fair Lawn, N.J., 1956.
15. Barrett, C. S., and T.B Massalski: "Structure of Metals," 3d ed., McGraw-Hill Book Company, New York, 1966.
16. Friedel, J.: "Dislocations," Pergamon Press, New York, 1964.
17. Tiller, W. A.: *J. Appl. Phys.*, **29**:611 (1958).
18. Wood, E. A.: "Crystal Orientation Manual," Columbia University Press, New York, 1963.
19. Arguello, C.: *Rev. Sci. Instrum.*, **38**:598 (1967).
20. "1967 Book of ASTM Standards," ASTM, Philadelphia, Pa., 1967.
21. Schwuttke, G. H.: *Sylvania Technol.*, **11**:2 (1958).
22. Cronin, G. R.: *Rev. Sci. Instrum.*, **34**:1151 (1963).
23. Akasaki, I., and H. Kobayasi: *J. Electrochem. Soc.*, **112**:757 (1965).
24. Amelinckx, S.: "The Direct Observation of Dislocations," Academic Press Inc., New York, 1964.
25. Jones, C. E., and A. R. Hilton: *J. Electrochem. Soc.*, **112**:908 (1965).
26. Vogel, F. L., W. G. Pfann, H. E. Corey, and E. E. Thomas: *Phys. Rev.*, **90**:489 (1953).
27. Wynne, R. H., and C. Goldberg: *J. Metals*, **5**:436 (1953).
28. Billig, E.: *Proc. Roy. Soc. (London)*, **A235**:37 (1956).
29. White, J. G., and W. C. Roth: *J. Appl. Phys.*, **30**:946 (1959).
30. Richards, J. L.: *J. Appl. Phys.*, **31**:600 (1960).
31. Richards, J. L., and A. J. Crocker: *J. Appl. Phys.*, **31**:611 (1960).
32. Abrahams, M. S., and C. J. Buiocchi: *J. Appl. Phys.*, **36**:2855 (1965).
33. Bardsley, W., and R. L. Bell: *J. Electron. Contr.*, **3**:103 (1957).
34. Fuller, C. S., and H. W. Allison: *J. Electrochem. Soc.*, **109**:880 (1962).
35. Holmes, P. J.: in P. J. Holmes (ed.), "The Electrochemistry of Semiconductors," p. 329, Academic Press Inc., New York, 1962.
36. Wang, P.: *Sylvania Technol.*, **11**:50 (1958).
37. Heidenreich, R. D.: U.S. Patent 2619414, Nov.25, 1962.
38. Tyler, W. W., and W. C. Dash: *J. Appl. Phys.*, **28**:1221 (1957).
39. McKelvey, J. P., and R. L. Langini: *J. Appl. Phys.*, **25**:634 (1954).
40. Rösner, O.: *Z. Metallk.*, **46**:225 (1955).
41. Batterman, B. W.: *J. Appl. Phys.*, **28**:1236 (1957).
42. Camp, P. R.: *J. Electrochem. Soc.*, **102**:586 (1955).
43. Ellis, S. G.: *J. Appl. Phys.*, **26**:1140 (1955).
44. Geach, G. A., B. A. Irving, and R. Phillips: *Research*, **10**:411 (1957).
45. Dash, W. C.: *J. Appl. Phys.*, **27**:1193 (1956).
46. Holmes, P. J.: *Proc. Inst. Elec. Engrs. (Suppl. 17)*, **B106**:861 (1959).
47. Christian, S. M., and R. V. Jensen: *Bull. Amer. Phys. Soc.*, **1**:140 (1956).

48. Franks, J., G. A. Geach, and A. T. Churchman: *Proc. Phys. Soc. (London)*, **B68**:111 (1955).
49. Holmes, P. J., and R. C. Newman: *Proc. Inst. Elec. Engrs. (Suppl.* 15), **B106**:287 (1959).
50. Irving, B. A.: *Brit. J. Appl. Phys.*, **12**:92 (1961).
51. Venables, J. D., and R. M. Broudy: *J. Appl. Phys.*, **29**:1025 (1958).
52. Gatos, H. C., and M. C. Lavine: *J. Electrochem. Soc.*, **107**:433 (1960).
53. Allen, J. W.: *Phil. Mag.*, (8)**2**:1475 (1957).
54. Dewald, J. F.: *J. Electrochem. Soc.*, **104**:244 (1957).
55. Faust, J. W., and A. Sagar: *J. Appl. Phys.*, **31**:331 (1960).
56. Churchman, A. T., G. A. Geach, and J. Winton: *Proc. Roy. Soc. (London)*, **A238**:194 (1956).
57. Nasledov, D. N., A. Ya. Patrakova, and B. V. Tsatvenkov: *Zh. Tekh. Fiz.*, **28**:779 (1958).
58. Abrahams, M. S., and L. Ekstrom: in H. C. Gatos (ed.), "Properties of Elemental and Compound Semiconductors," p. 225, Interscience Publishers, Inc., New York, 1960.
59. Warekois, E. P., and P. H. Metzger: *J. Appl. Phys.*, **30**:960 (1959).
60. Gatos, H. C., and M. C. Lavine: *J. Electrochem. Soc.*, **107**:427 (1960).
61. Schwuttke, G. H.: *Sylvania Technol.*, **8**:122 (1960).
62. Compton, A. H., and S. K. Allison: "X-rays in Theory and Experiment," 2d ed., D. Van Nostrand Co., Inc., Princeton, N.J., 1935."
63. Kurtz, A. D.: *Lincoln Lab. Tech. Rep.* 85, MIT, 1955.
64. James, R. W.: "The Optical Principles of the Diffraction of X-rays," G. Bell & Sons, Ltd., London, 1962.
65. Batterman, B. W.: *J. Appl. Phys.*, **30**:508 (1959).
66. Bell, R. L.: *J. Electron. Contr.*, **3**:487 (1957).
67. Hunter, L. P.: *J. Appl. Phys.*, **30**:874 (1959).
68. Auleytner, J.: "X-ray Methods in the Study of Defects in Single Crystals," Pergamon Press, New York, 1967.
69. Webb, W. W.: in J. B. Newkirk and J. H. Wernick (eds.), "Direct Observation of Dislocations in Crystals," p. 29, Interscience Publishers, Inc., New York, 1962.
70. Borrmann, G., W. Hartwig, and H. Irmler: *Z. Naturforsch.*, **13A**:423 (1958).
71. Barth, H., and R. Hosemann: *Z. Naturforsch.*, **13A**:792 (1958).
72. Authier, A.: *J. Phys. Radium*, **21**:655 (1960).
73. Berg, W.: *Naturwissenschaften*, **19**:391 (1931).
74. Berg, W. F.: *Z. Krist.*, **89**:286 (1934).
75. Barrett, C. S.: *Trans. AIME*, **161**:15 (1945).
76. Newkirk, J. B.: *Trans. AIME*, **215**:483 (1959).
77. Lang, A. R.: *Acta Cryst.*, **12**:249 (1959).
78. Howard, J. K., and R. D. Dobrott: *Appl. Phys. Letters*, **7**:101 (1965).
79. Jenkinson, A. E.: *Philips Tech. Rev.*, **23**:82 (1962).

7

Characterization of Semiconductor Surfaces

7-1. INTRODUCTION

An important part of the processing of a semiconductor single crystal in device fabrication involves the preparation of a single-crystal slice of high surface perfection. Both physical and chemical imperfections must be considered in the surface because these slices are used for device fabrication and as substrates for the growth of epitaxial films. Any imperfections left at the surface will be deleterious to device characteristics and will contribute to imperfect epitaxial layers.

The single-crystal slices are sawed from the larger single crystal with a diamond saw. This sawing operation produces a damaged surface, which has been estimated to extend from 40 to 80 μ into the crystal.[1,†] To remove this damage and to prepare a flat surface, the sawed slice is polished with successively finer polishing grits. This mechanical polishing does remove the deeper saw work damage, but because of the brittleness of germanium and silicon these operations cause microscopic conchoidal fracture and cracking. The depth of this latter damage is of course considerably less than that obtained in the sawing operation but does extend beneath the surface. The depth of this damage appears to be approximately equal to the diameter of the abrasive material used in the lapping for germanium[1] and gallium arsenide[2] but only one-half to one-quarter as deep for silicon.[1]

To complete the preparation of the semiconductor slice it is necessary to etch the surface chemically. Chemical etching dissolves away the rest of the damaged surface left from the lapping operations. Etching also serves the vital function of cleaning the surface of chemical impurities. The effectiveness of these lapping, etching, and washing operations and methods of analyzing these surfaces will be discussed in this chapter.

†Superscript numbers indicate References listed at the end of the chapter.

7-2. PHYSICAL IMPERFECTIONS

Mechanical damage at the surface of a single-crystal semiconductor slice produced by sawing or lapping is not well understood. Clarke and Hopkins[3] reported that this damage produced a large density of acceptorlike states with two energy levels of 0.022 and 0.4 ev. These energy levels are thought to be vacancies. While there is no direct evidence for the presence of dislocations, most of the effects of work damage are explained by the presence of both vacancies and dislocations. The predominant effects observed include changes in surface conductance, surface recombination velocity, disturbance of the orderly arrangement of atoms in the single crystal, increased etch rate of the damaged part, and production of defect states that act as electron traps. All these induced imperfections lend themselves to an evaluation or analysis of the extent of surface damage.

7-3. INFRARED CHARACTERIZATION

Jones and Hilton[2] reported the use of infrared-reflection measurements in the 20- to 60-μ region to determine the depth of damage in n-type gallium arsenide surfaces caused by sawing and lapping operations. While only n-type gallium arsenide was studied in this work, the technique appears to be directly applicable to other III-V n-type intermetallic semiconductors. This technique can also be used on p-type materials since the damaged surface would cause phonon scattering and decrease the surface reflectance. However, the effect is small, and the accuracy of the method would not be nearly as good as that for n-type material.

The infrared-reflectance method is experimentally a simple technique. The sample is weighed, the reflectance measured over the 20- to 60-μ region, an increment of surface etched away, the sample reweighed, and the reflectance spectrum

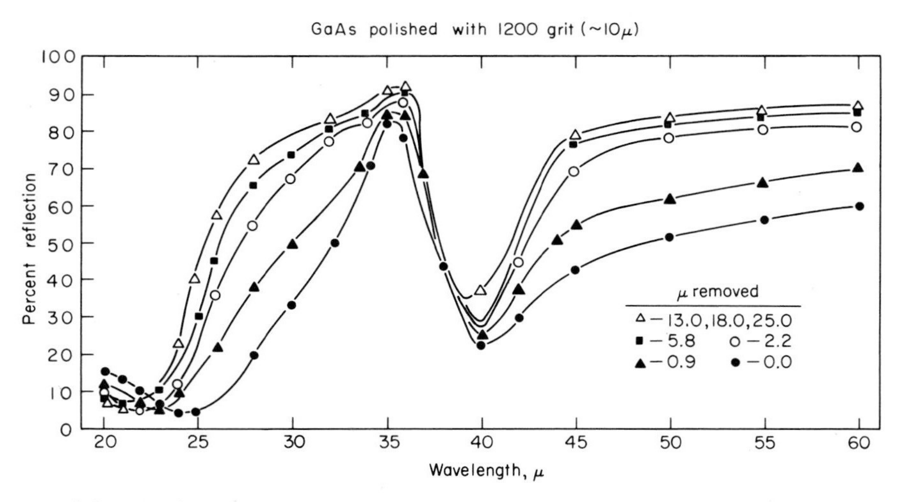

Fig. 7-1. Infrared-reflection spectra from lapped gallium arsenide surfaces, illustrating the technique used to determine the depth of damage caused by, in this case, a 1,200-grit polish. (*From Jones and Hilton.*[2])

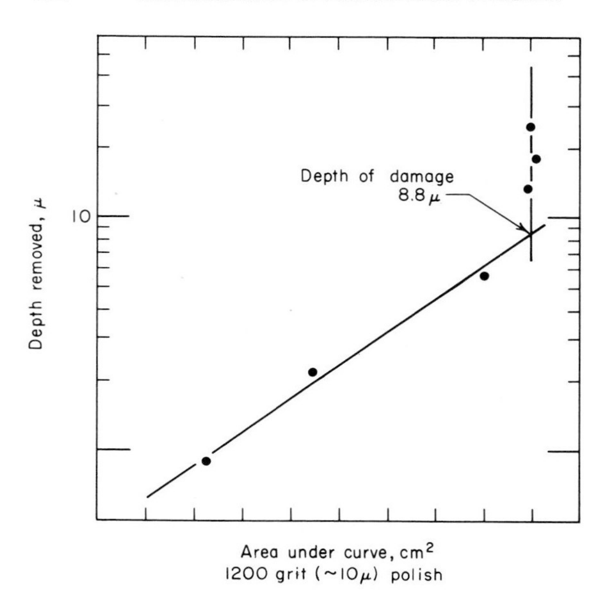

Fig. 7-2. Graphical technique used to determine the depth of damage using infrared-reflectance spectra. In this study a 1,200-grit (~10μ) polishing compound was used. (*From Jones and Hilton.*[2])

remeasured. This sequence is repeated until there is no further change in the measured spectrum after three etch cycles. A typical set of spectra for a sample of gallium arsenide that had been polished with 10-μ grit is shown in Fig. 7-1.

A plot of the area under the reflectance curve versus depth removed is shown in Fig. 7-2. As can be seen, when all damage has been removed a constant area is obtained. In this case 10-μ grit caused damage to a depth of 8.8 μ or approximated the diameter of the particles in the lapping compound.

7-4. X-RAY CHARACTERIZATION

The diffraction of x-rays from a damaged semiconductor surface will be different from the diffraction of a perfect single-crystal surface of the semiconductor. Warekois et al.[4] used the half-width of the rocking curves obtained from a double-crystal x-ray spectrometer as a measure of the depth of damage in a study of III-V intermetallic semiconductors.

In the double-crystal rocking-curve analysis of subsurface damage, radiation is diffracted and measured as shown in Fig. 6-31. Warekois et al.[4] used a germanium single crystal as a monochromator because the spacing of the planes was close to those of the III-V intermetallic semiconductors which were being studied.

A typical set of rocking curves which might be obtained from the spectrometer is shown in Fig. 7-3. As the semiconductor surface under study is rotated, the count rate of the detected radiation is recorded. If there is no surface damage, then the full width of the rocking curve at half maximum will be I_1. If the semiconductor surface were damaged, then multiple atomic planes at the Bragg angle would cause the rocking curve to broaden, as shown in Fig. 7-3, with intensity I_2. The following sequence would be used to study the depth of damage in a semiconductor. The lapped or sawed slice would be weighed, a rocking curve recorded, an increment of

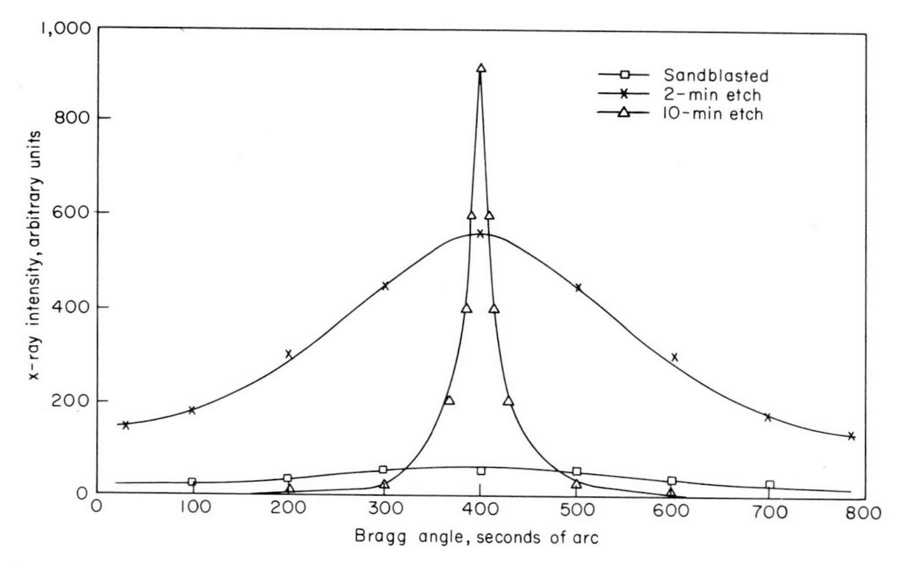

Fig. 7-3. Typical set of x-ray rocking curves showing broadening produced by damaged surfaces. *(After Kurtz;[5] by permission MIT Lincoln Laboratory.)*

the surface etched away, the slice reweighed, a rocking curve recorded, and the sequence repeated until a constant full width at half maximum for the recorded curve had been obtained. Then a plot of this full width versus depth removed would look similar to Fig. 7-2, where a constant full width would be obtained after all surface damage had been removed. The depth of damage would be obtained from the intersection of the two lines. A detailed description of this x-ray technique is given by Intrater and Weissman[6] for aluminum, but the technique is directly applicable to all semiconductor surfaces.

7-5. ETCH RATE

The etch rate of a damaged semiconductor surface will be faster than that of a perfect single-crystal surface. This increase in etch rate is probably due to a combination of factors, including increased surface area and increased dislocation density (see Sec. 6-12). When the rate of dissolution is measured as a function of distance into the crystal, a curve, as shown in Fig. 7-4, will be obtained. As the amount of damage decreases, the etch rate decreases until a constant rate of dissolution is obtained. The intersection of the two lines is the depth of damage.

The depth of damage in silicon,[7] germanium,[8,9] and gallium arsenide (*B* face)[4,5,10] is shown in Fig. 7-5 as a function of the particle size used to lap the surface. For gallium arsenide, the etch rate and x-ray rocking-curve techniques give comparable results but are in sharp disagreement with those obtained by infrared-reflection measurements.[2] This discrepancy is probably due to a change (deeper in the crystal) to a type of damage that is not detected by x-ray or etch rate. If, as suggested by Faust,[9] a deeper layer of plastically deformed material containing mechanically

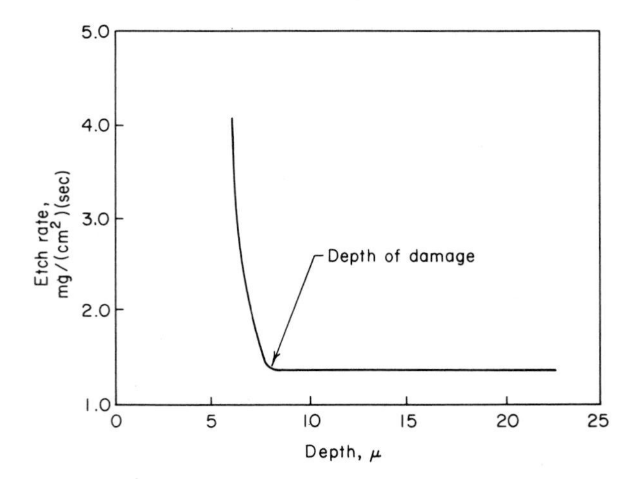

Fig. 7-4. Typical etch-rate curve used to determine the depth of damage in a semiconductor surface. *(After Gatos et al.*[10]*)*

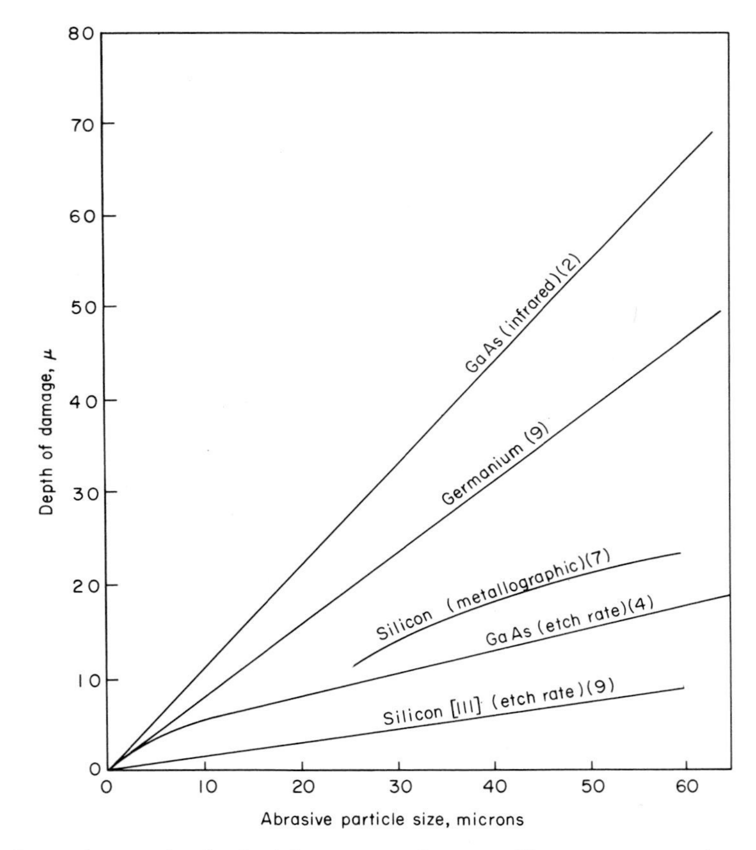

Fig. 7-5. Curves showing the depth of damage as a function of lapping-compound particle size for various semiconductors. (Note the discrepancy between etch rate and infrared data for gallium arsenide.)

induced dislocations extends into the crystal, then neither etch rate nor x-ray rocking-curve techniques can detect the damage. Since the infrared-reflection technique measures the amount of phonon absorption caused by free carriers, which in turn would be generated at dislocations, a deeper depth of damage would be obtained by this technique. This is also a more meaningful analysis since these same traps would be very detrimental to device characteristics.

7-6. ELECTRICAL METHODS

The depth of surface damage introduced by sawing or lapping can be determined by measuring some electrical parameter while incrementally removing layers of this surface until only the bulk electrical properties are observed. Intuitively it would be expected that measurements of electrical properties would be a more meaningful measure of surface damage since it is these same electrical properties that will ultimately affect the device characteristics. Buck[1] has shown that these electrical techniques, for example, the photomagnetoelectric effect and photoconductivity decay, give comparable measurements of the depth of damage when compared with x-ray and etch techniques. Generally, measurements of these electrical parameters are more difficult to perform and to interpret than the characterization techniques described earlier.

7-7. PHOTOMAGNETOELECTRIC MEASUREMENTS

The photomagnetoelectric effect (PME) can best be described as the Hall effect caused by the diffusion current of light-injected carriers. Figure 7-6 shows sche-

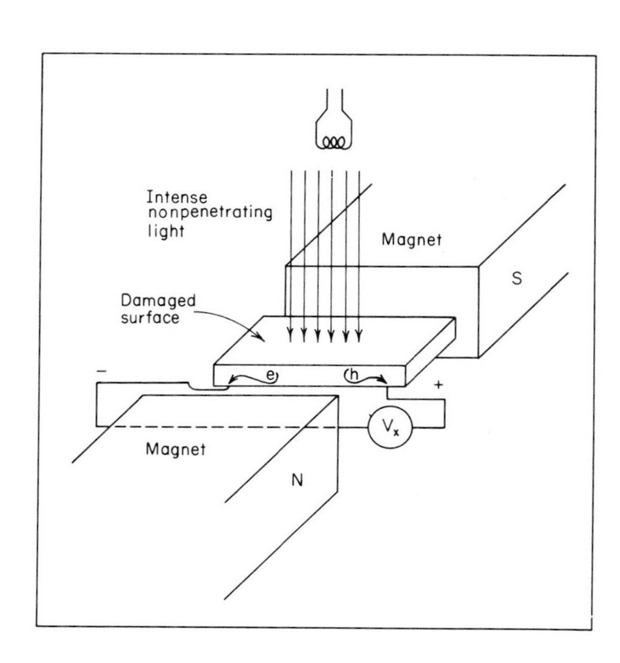

Fig. 7-6. Schematic illustrating photomagnetoelectric measurement of semiconductor surface damage.

matically how such a measurement would be made. A voltage V_x is measured, and the magnitude of the voltage is used as a measure of the amount of surface damage.

In this technique,[11] the surface under study is strongly illuminated and a magnetic field is applied perpendicular to the illuminated surface. The "photogenerated" holes and electrons are deflected in opposite directions by the magnetic field and thus set up the PME open-circuit voltage. If the surface under study has been damaged, all the carriers will recombine at the illuminated surface, and the PME voltage will be low. As the damaged surface is incrementally etched away, the PME voltage will rise until all the damaged layer has been removed and only the bulk properties control the magnitude of the signal. A typical set of curves used to measure the depth of surface damage is shown in Fig. 7-7.

7-8. CONDUCTIVITY

The bulk conductivity of a thin slice of a semiconductor will be strongly influenced by the surface conductance of the sample. A damaged surface has a large number of free carriers, and the total conductance of a thin germanium slice will be strongly affected by the conductivity. Clarke and Hopkins[3] measured the effective resistivity (surface plus volume) as a function of temperature and obtained curves as shown in Fig. 7-8. By repeating this measurement, after in-

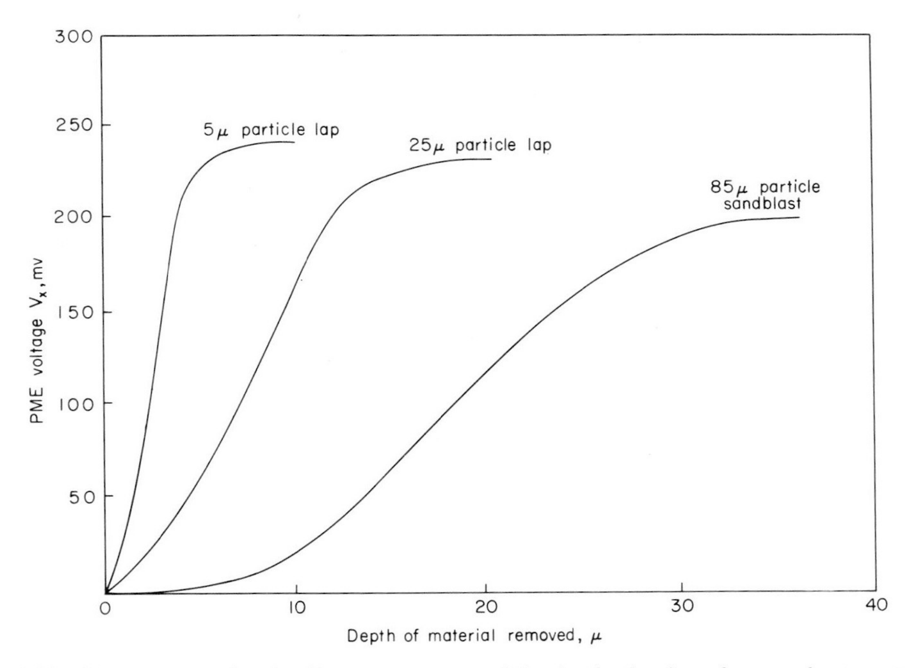

Fig. 7-7. Photomagnetoelectric effect as a measure of the depth of surface damage due to various lapping compounds on germanium. (*Adapted from Buck and McKim.*[11])

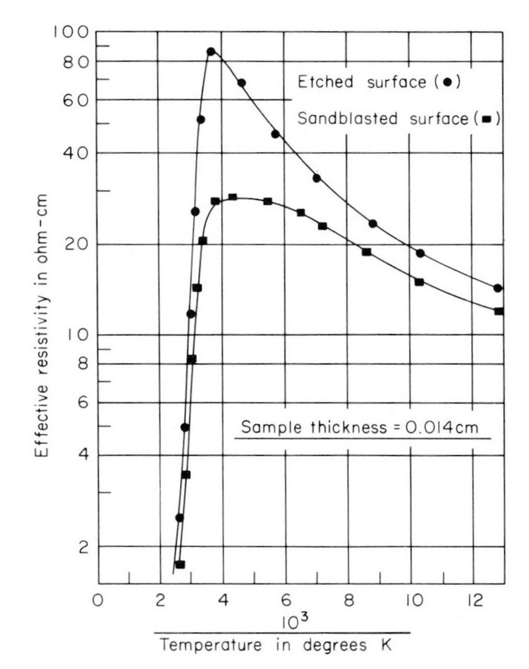

Fig. 7-8. The pronounced effect of surface damage, caused in this case by sandblasting, can be readily observed in the total resistivity (bulk plus surface) of a thin germanium single crystal. (*From Clarke and Hopkins.*[3])

crementally etching away part of the damaged surface until a constant resistivity was obtained, they were able to determine the depth of damage. They were able to identify two acceptor levels at 0.022 and 0.4 ev which were associated with the damaged surface layer.

7-9. DIODE REVERSE CURRENT

The data of Clarke and Hopkins[3] indicated that a damaged layer of a semiconductor surface would cause a decrease in effective resistivity because the damaged layer would act as a leakage path. Buck and McKim[11] used this increased surface conductance as a measure of the depth of damage. Large-area p-n diodes were lapped or polished on all four sides. The reverse leakage current I_R was measured. The diode surfaces were then incrementally etched away and the reverse current measured after each etch. The results are shown in Fig. 7-9, and the depths of damage obtained by this method agreed with values obtained by other techniques for the same size of grit.

7-10. MISCELLANEOUS TECHNIQUES FOR PHYSICAL IMPERFECTIONS

Buck[1] has used photoconductivity decay (PCD) to determine the depth of damage in germanium surfaces and obtained reasonable agreement with the photomagnetoelectric effect. There are significant random variations and experimental difficulties that make the PCD method less attractive than the other techniques discussed. Walters[12] reviewed the use of magnetic resonance techniques

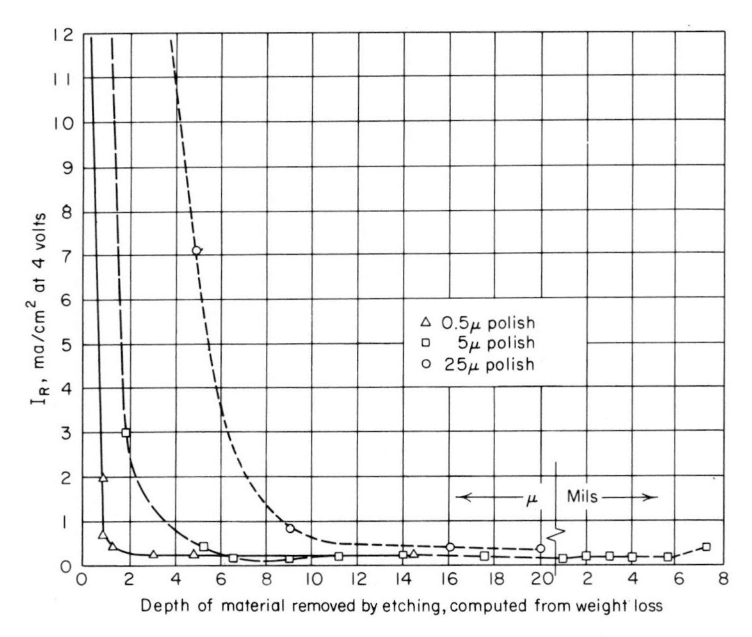

Fig. 7-9. Measurement of the depth of surface damage using reverse dark current on a germanium grown-junction p-n diode. (*From Buck and McKim.*[11])

in the study of semiconductor surfaces. The difficulty with this latter technique is the need for very high surface areas, necessitating finely crushed and powdered samples.

7-11. ETCH PITS CAUSED BY SURFACE DAMAGE

The prime reason for chemically etching a semiconductor surface is to complete the removal of surface damage from previous cutting and lapping operations. Generally a polished, flat surface is desired for further slice processing, and planar or nonpreferential etches are available for this purpose. However, there are other etches available for each semiconductor that are used to produce pits (see Sec. 6-13), and these etch pits are related to the dislocation density of the bulk semiconductor. These same etch pits will form at dislocations produced by mechanical lapping or polishing of a semiconductor surface. The real difficulty in this method of analyzing for surface damage is differentiating the causes of pit formation. Frequently, it is not possible to decide whether an etch-pit pattern was caused by mechanical damage or a dislocation line in the original semiconductor bulk material.

7-12. X-RAY TOPOGRAPHY

The technique of x-ray topography (see Sec. 6-16) appears to offer the only nondestructive method for examining physical imperfections in semiconductor surfaces. The x-ray topograph can produce a photographic image of the imperfections at the surface of the single-crystal semiconductor slice. The ultimate resolution of the technique is controlled by the type of photographic emulsion used to record the image. Emulsions with resolutions near 1 μ are available, e.g., Ilford L-4, but as much as 20 hr is required to generate the topograph. Generally resolutions in the order of 10 μ are adequate for the study of surface damage, and Eastman Kodak Type A autoradiographic or similar plates are used. With these plates more reasonable exposure times of 3 to 5 hr are employed to record the topograph.

While this technique yields detailed images of the damage in the semiconductor slice, it does not record the depth of damage. If the depth of damage is required, it is necessary to remove layers of the surface incrementally and record an x-ray topograph after each etch. Considering the length of time required to record each x-ray topograph, it would appear that this is a needlessly time-consuming approach to the problem. However, the high sensitivity of this technique to subsurface dislocations allows the use of this method after all the others have reached their ultimate limit of sensitivity. Techniques such as etch rate, x-ray rocking curves, PME, and other electrical techniques are useful only where gross damage is present. These are only relative techniques in that they record the point at which that technique can no longer distinguish between the bulk parameter and surface-damage change in the bulk parameter. X-ray topography, on the other hand, can record individual dislocation lines and can follow the removal of these imperfections through incremental etching. No other technique has this inherent sensitivity for studying semiconductor surfaces.

The experimental x-ray topographic technique which has proved most useful in surface study is the scanning-reflection method developed by Howard and Dobrott[13] (see Sec. 6-16). This was a modification of the stationary Berg-Barrett back-reflection technique,[14] and it allows the sampling or analysis of the entire surface of a slice up to 1 in. in diameter.

The experimental arrangement is shown in Fig. 7-10. The sample-to-film distance is in the order of 15 mm, which is large compared with the 0.1 mm used in the stationary Berg-Barrett technique. The semiconductor slice under study is mounted on the goniometer head in the micrographic camera and the beam stops opened. Copper radiation ($K\alpha_1$) from a micro-focus x-ray tube is used for this surface work. The crystal is angularly rotated until the desired diffracting planes are in exact orientation (see Sec. 6-16). The sample (c) in proper diffracting position is coupled to a film holder mounted perpendicular to the diffracted beam (D).

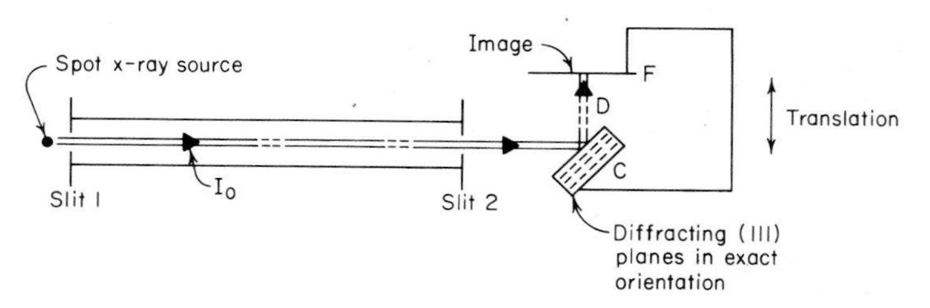

Fig. 7-10. Experimental arrangement used for scanning-reflection x-ray topography. (*From Howard and Dobrott.*[13])

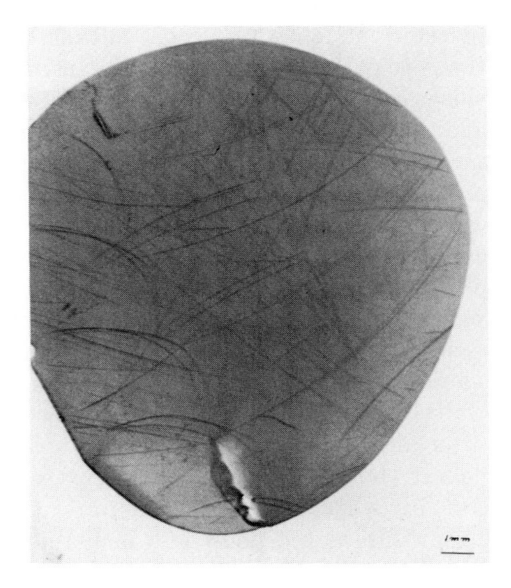

Fig. 7-11. A (440) x-ray topograph showing polishing damage below the surface. This slice appeared optically to have a mirror finish with no evidence of surface damage.

The sample and film are translated perpendicular to the incident beam (I_0). When the x-rays strike the crystal, the surface diffracts coherently and is registered on the photographic plate (F). The resulting image is a photograph of the distribution of flaws in the semiconductor slice.

As an illustration of the sensitivity of this technique, Fig. 7-11 shows a (440) x-ray topograph of the $(\bar{1}\bar{1}\bar{1})$ face of a gallium arsenide slice that had been chemically polished to a mirror finish on the surface. There was no visible evidence of any surface damage. This topograph revealed polishing scratches over the entire surface. The surface was further chemically etched to remove 4.8 μ, and a

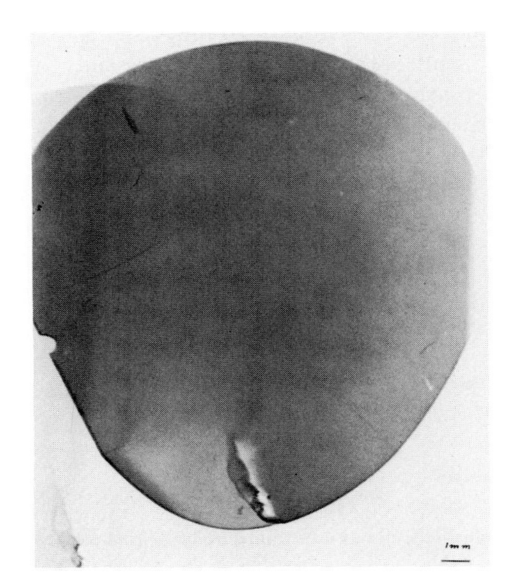

Fig. 7-12. A (440) x-ray topograph of the slice shown in Fig. 7-11 after 4.8μ had been etched away, removing almost all the surface damage.

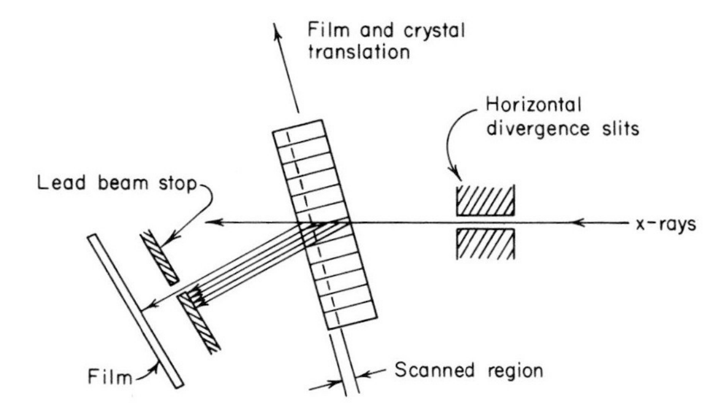

Fig. 7-13. Schematic diagram of the transmission-surface topography apparatus used by Blech et al.[15]

second (440) topograph was run. This topograph (Fig. 7-12) showed that almost all the scratches had been removed. Similar studies of other chemically polished surfaces often have shown damage depths of 1.1 to 8.1 μ.

This technique has been most useful in the study and correlation of surface preparation and epitaxial deposition (Chap. 8).

It is possible to examine surface imperfections by transmission x-ray topography on very thin semiconductor slices. Blech et al.[15] used a modified translating Lang method, shown in Fig. 7-13, to examine diffusion-generated dislocations in silicon surfaces. The positioning of the slits (100 μ) allows either the surface or various depths below the surface of the crystal to be examined. In this case the diffusion-induced dislocations were all at the surface. This technique is of course also applicable to the study of all surfaces and circumvents the usual repeated etching-analysis sequence.

7-13. ELECTRON MICROSCOPY

The high resolution of the electron microscope has made it a valuable instrument for the examination of surface features. For these types of studies, it is necessary to prepare replicas of the surface for examination in a transmission electron microscope. The direct-carbon-replica technique of Bradley[16] gives the highest fidelity and is the easiest to prepare. Stickler[17] reported a highly successful nondestructive direct-carbon-replica method for the examination of silicon surfaces. This technique made use of the very thin (10 to 50 Å) native oxide film that is always present on a silicon surface. Following the vacuum evaporation of an approximately 300 Å carbon film on the silicon surface, the sample was lowered into HF–H_2O (1:10), and the thin native oxide film immediately dissolved away and the freed carbon replica floated to the surface. The silicon sample was quickly removed, rinsed, and dried to prevent surface stains. A new thin oxide layer immediately grew on the silicon surface, and the entire process could be repeated many times without any deleterious effect on the original surface. For higher-contrast work it was found possible to preshadow the surface at an oblique angle with a platinum-palladium alloy and follow this with carbon deposited at normal incidence. The replica was easily floated away from the surface by the HF–H_2O treatment.

Koehler and Mattern[18] and Savanick[19] have reported techniques for preparing replicas in such a manner that correlations between light and electron microscopic examinations are possible. Both techniques involve attaching an electron-microscope grid to the replica in such a manner that the area of interest is located in a mesh window.

7-14. CHEMICAL IMPERFECTIONS

The final step in the preparation of a semiconductor slice is to complete the removal of work damage and to clean the surface. This is accomplished by chemically etching the surface with a nonpreferential polishing etch such as those shown in Table 6-4. The etch must produce a smooth, uniformly etched surface and must be used long enough to dissolve away all the damaged surface. The effectiveness of the etch in removing physical imperfections can be determined by any of the techniques described earlier.

7-15. CHEMICAL IMPURITIES DEPOSITED FROM SOLUTION

The effectiveness of the etch in cleaning the surface of the semiconductor is considerably more difficult to evaluate. Atalla et al.[20] have pointed out that as little as one ten-thousandth of a monolayer of an ionic impurity is sufficient to invert (cause to change type, for example, p to n) the surface of 1 ohm-cm silicon and cause device instability. On a (111) oriented silicon surface one ten-thousandth of a monolayer is only 4.8×10^{10} atoms/cm^2. For an impurity with an average atomic weight of 100 this represents 8 pg/cm^2 of surface area. Any analytical technique would be hard pressed even to approach this detection limit. Morrison[21] studied the effect of trace amounts of copper (as low as 0.05 ppm) in the etch and subsequent rinse solutions used to treat germanium surfaces. By measuring the field-effect mobility and surface recombination velocity, Morrison was able to cause the surface of n-type germanium to become p type. Figure 7-14 shows the results of Morrison's work for copper contamination in rinse solutions and, as can be seen, as little as 0.05 ppm affected the surface. The same result was observed for copper contamination from etch solutions, but it was necessary to use about one hundred times more copper to achieve the same effect. Frankl,[25] and Boddy and Brattain,[26] observed similar effects for copper contamination on germanium. While surface contamination is foreign to a semiconductor, these contaminants may be applied in a controlled manner to produce some desired effect in or on devices.[22-24] Sullivan and Scheiner[27] reported some work where the addition of the complexing agent tetrasodium ethylenediaminetetraacetate to a hydrogen peroxide etch solution resulted in germanium transistors that degraded some 2,000 hr later than those etched in regular hydrogen peroxide etch solution. An excellent discussion of some effects of semiconductor surface treatment on device operation was given by deMars.[28] As early as 1958, Bemski and Struthers[29] reported that traces of gold from reagents used in etching deposited on silicon and degraded the lifetime of the semiconductor after heating. Carlson[30] reported that fast-diffusing elements such as iron, copper, manganese, and zinc could act in a similar manner. Copper

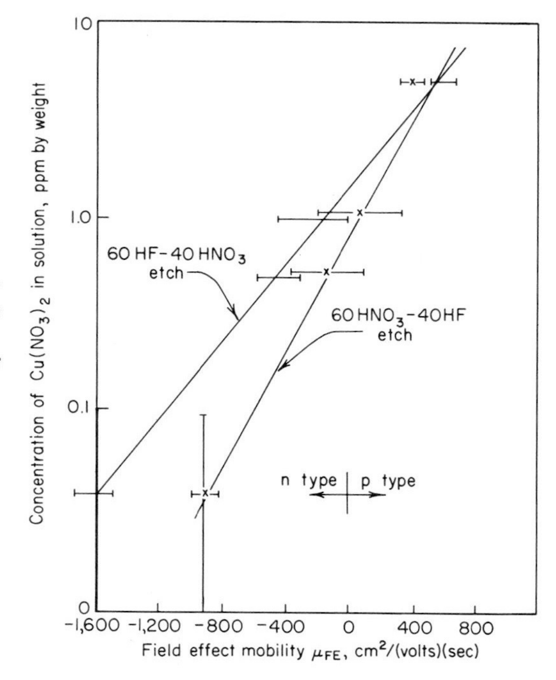

Fig. 7-14. The detrimental effect of copper ions depositing on a germanium surface as measured by the field-effect mobility. The surface becomes p type at higher copper concentrations. (*From Morrison.*[21])

is generally accepted as the element that causes thermal conversion (change from n to p type on heating) in most semiconductors. This is particularly difficult to control if present, because of its very fast diffusion coefficient.

The most fruitful approach for studying the contamination of semiconductor surfaces from solution has been by radiotracer techniques.[31-34] Radiotracers offer the most sensitive and specific means that can be used for directly measuring the amount of surface contamination after an etching or washing operation. Since the mechanism of adsorption or amount of surface contamination is a function of the species presented to the surface, the various contaminants will be discussed as cations, anions, and organics.

7-16. CATION CONTAMINATION

The contamination of semiconductor surfaces by cations from etches and subsequent washing solutions can be adequately described either by irreversible electrochemical reaction at the semiconductor-liquid interface or by reversible physical adsorption. If the electrochemical potential of the contaminating metal-cation couple lies above that of the semiconductor (Latimer's convention), then physical adsorption will occur. If the electrochemical potential lies below, then an electrochemical reaction will occur. Table 7-1 lists some of the more important electrochemical potentials in acid and alkaline solution. In acid media, gold and copper, for example, will electrochemically deposit as metal on silicon and germanium. On the other hand, iron will deposit only from alkaline solutions.

Table 7-1. Selected Examples of Electrochemical Potentials of Metal–Metal Ion Couples in Acid and Alkaline Solution

Metal-metal ion couple	E_0, *volts*
In acid solution:	
$Na \rightleftarrows Na^+ + e^-$	−2.714
$Al \rightleftarrows Al^{+++} + 3e^-$	−1.662
$Si + 2H_2O \rightleftarrows SiO_2 + 4H^+ + 4e^-$	−0.869
$Ge + 2H_2O \rightleftarrows GeO_2 + 4H^+ + 4e^-$	−0.15
$Cu \rightleftarrows Cu^{++} + 2e^-$	+0.337
$Au \rightleftarrows Au^+ + e^-$	+1.691
In alkaline solution:	
$Ca + 2OH^- \rightleftarrows Ca(OH)_2 + 2e^-$	−3.02
$Al + 3OH^- \rightleftarrows Al(OH)_3 + 3e^-$	−2.30
$Si + 6OH^- \rightleftarrows SiO_3^{--} + 3H_2O + 4e^-$	−1.697
$Ge + 5OH^- \rightleftarrows HGeO_3^- + 2H_2O + 4e^-$	−1.03
$Fe + 2OH^- \rightleftarrows Fe(OH)_2 + 2e^-$	−0.877
$Pt + 2OH^- \rightleftarrows Pt(OH)_2 + 2e^-$	+0.15

Holmes and Newman,[35] using electron diffraction techniques, showed that silver-ion contamination in acid solution on silicon and germanium was deposited as microcrystallites of silver metal over the surface of the semiconductor. Larrabee[32] showed that the amount of surface contamination was linearly related to the amount of cation in the solution according to the following relationship:

$$\log [M^0] = n \log [M^{+n}] + \log k \tag{7-1}$$

Figure 7-15 shows a typical set of curves for various cations deposited on indium antimonide from solution. Electrochemical deposition of these cations from solution is an irreversible process. Table 7-2 shows the results of repeated attempts to wash gold contamination from germanium surfaces with hot-water washes.

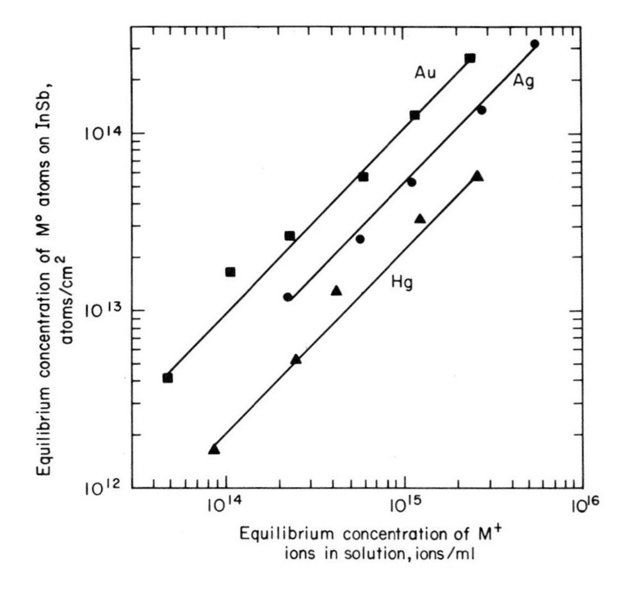

Fig. 7-15. Electrochemical deposition of metal ions on indium antimonide from solutions containing ppm levels of impurities. (*From Larrabee.*[32])

Table 7-2. Adsorption of Gold from Solution on Germanium Surfaces

ppm Au presented	First H_2O wash, atoms/cm²	Second H_2O wash, atoms/cm²	Third H_2O wash, atoms/cm²	Fourth H_2O wash, atoms/cm²
0.03	8.54×10^{11}	7.07×10^{11}	5.76×10^{11}	5.34×10^{11}
0.06	1.29×10^{12}	1.14×10^{12}	1.00×10^{12}	9.22×10^{11}
0.30	4.54×10^{12}	4.03×10^{12}	3.73×10^{12}	3.58×10^{12}
0.60	8.57×10^{12}	7.75×10^{12}	7.05×10^{12}	6.80×10^{12}

Table 7-3. Washing of Germanium Surfaces with 1% KCN Solution to Remove Gold Contamination

ppm Au presented	First KCN wash, atoms/cm²	Second KCN wash, atoms/cm²	Third KCN wash, atoms/cm²
0.03	3.76×10^{11}	2.77×10^{11}	2.18×10^{11}
0.06	6.00×10^{11}	3.88×10^{11}	2.99×10^{11}
0.30	2.31×10^{12}	1.48×10^{12}	9.35×10^{11}
0.60	4.44×10^{12}	2.82×10^{12}	1.43×10^{12}

Subsequent washing of these same surfaces with 1% KCN solution (Table 7-3) did not remove the gold, which illustrates how difficult it is to eliminate this type of surface contamination. Its removal must involve a two-step process: dissolution, immediately followed by complexing of the dissolved species to prevent redeposition.

Those cations whose electrochemical potentials lie above the semiconductor are adsorbed by a physical adsorption mechanism. This reversible mechanism is

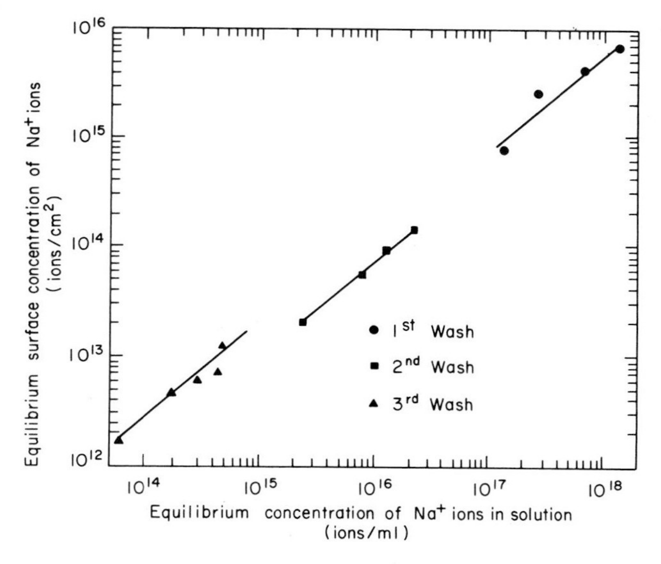

Fig. 7-16. Washing of sodium-ion contamination from gallium arsenide surfaces. (*From Larrabee.*[32])

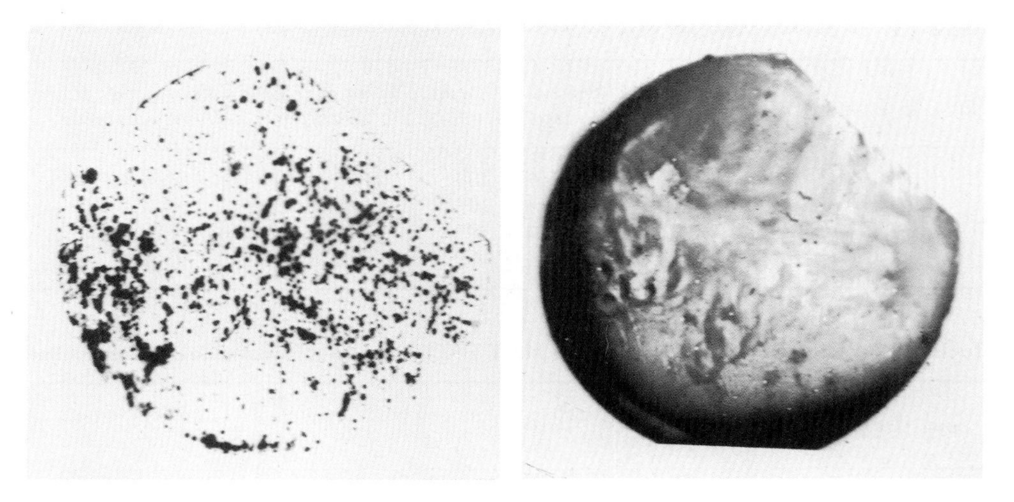

Fig. 7-17. Autoradiograms showing silver-ion distribution on nitric acid-boiled *(left)* **and freshly etched** *(right)* **silicon surfaces.**

characterized by a Freundlich isotherm, where a log plot of equilibrium surface concentration vs. equilibrium solution concentration is a straight line. Figure 7-16 shows this straight-line relationship for sodium ion on gallium arsenide. As can be seen, after each wash a new equilibrium is established with that wash solution. This behavior demonstrates the need for ultraclean rinsing media and preferably flowing rinse solutions.

The impact of radiotracers is clear from these examples. Only radiotracers have sufficient sensitivity to allow the entire process to be followed from etching through final wash. Radiotracers have the added property that the surface distribution can be readily followed by using autoradiography. In a study of the contamination of silicon transistor bars by silver ion from solution, it was observed that freshly etched bars retained more impurity from solution than nitric acid–boiled bars. When this experiment was repeated on silicon slices and autoradiograms obtained (Fig. 7-17), it was immediately apparent that the nitric acid–boiled silicon surface had grown an oxide that protected the surface from the solution. The silver ion deposited only where there were breaks or imperfections in the oxide film. The freshly etched surface showed a more uniform distribution of silver because the surface was essentially oxide-free.

7-17. ANION CONTAMINATION

As in the case of cation interaction with semiconductor surfaces, anion radioactive-tracer techniques offer the most sensitive and easiest method of studying surface contamination from solutions. The interaction of the various types of anion with semiconductor surfaces is not as well understood as for cations. The adsorption of anions from aqueous solution on silicon surfaces is probably best understood if it is remembered that all silicon surfaces are covered with a porous

15 to 25 Å native oxide film. As a result the anion is presented with the possibility of physical or chemical adsorption on either or both the silicon surface and the silicon dioxide film.

Some work in the Texas Instruments Incorporated laboratories[36] on the adsorption of phosphate using radio-phosphorus-32 from solution points to the silicon dioxide film as the adsorption site. Table 7-4 shows the effect of pH on the amount of phosphate retained on silicon. In these studies reject, low-breakdown, silicon transistor bars were used, and flowing distilled-water rinses were employed. As can be seen, most adsorption was obtained from the higher pH solutions and least from fluoride-containing etches, where all oxide would have been dissolved away.

Table 7-4. Effect of pH on the Amount of Phosphate Ion Adsorbed from Solution

Solution pH	*Phosphate surface concentration, atoms/cm²*
10	2.49×10^{14}
5	1.09×10^{14}
1	3.24×10^{13}
0 (1 N HNO_3)	3.28×10^{13}
Etch No. 2†	4.56×10^{12}
Etch No. 4‡	2.12×10^{12}

†Etch No. 2: 120 ml HF, 180 ml HNO_3, 150 ml CH_3COOH, 6 ml 2% Na_2HPO_4.

‡Etch No. 4: 80 ml HF, 120 ml HNO_3, 300 ml CH_3COOH, 4 ml 2% Na_2HPO_4.

Further evidence of the role of the native oxide film in the adsorption of anions was obtained by using radiosulfur-35 to study sulfate-ion adsorption. In one experiment silicon slices were treated for 20 min at 200°C in $H_2{}^{35}SO_4$ and then treated in various ways. Table 7-5 shows the effect of water wash followed by HF, which dissolves the SiO_2 film. These results suggest that the sulfate anion was entrapped or chemically adsorbed on or in the SiO_2 film.

Table 7-5. Adsorption of Sulfate Ion on Silicon Surfaces from $H_2{}^{35}SO_4$ at 200°C for 20 min

Slice	Treatment, $SO_4{}^{--}/cm^2$		
	Boiled distilled H_2O, 20 min	Hydrofluoric acid, 5 min, blotted dry	Second 5 min hydrofluoric acid, blotted dry
1	1.21×10^{14}	2.54×10^{12}	1.88×10^{12}
2	7.59×10^{14}	9.12×10^{12}	4.86×10^{12}
3	2.72×10^{14}	3.56×10^{12}	2.04×10^{12}

The retention of $Cr_2O_7^{--}$ ion from solution on silicon surfaces is readily studied by using radiochromium-51. The study of the effect of $Cr_2O_7^{--}$ solution concentration and repeated washing is shown in Table 7-6. As can be seen, the adsorption

Table 7-6. Adsorption of $Cr_2O_7^{--}$ Ion from Solution on Silicon Surfaces

ppm $K_2Cr_2O_7$	Atoms Cr presented	Atoms Cr adsorbed/cm^2		
		First wash	Second wash	Third wash
2	8.21×10^{16}	8.44×10^{13}	6.06×10^{13}	5.44×10^{13}
2	8.21×10^{16}	6.98×10^{13}	5.08×10^{13}	4.52×10^{13}
1	4.11×10^{16}	4.38×10^{13}	2.26×10^{13}	1.85×10^{13}
0.2	8.21×10^{15}	1.17×10^{13}	6.00×10^{12}	4.94×10^{12}
0.2	8.21×10^{15}	8.74×10^{12}	6.68×10^{12}	5.92×10^{12}
0.1	4.11×10^{15}	2.06×10^{12}	1.07×10^{12}	8.96×10^{11}
0.05	1.64×10^{15}	1.18×10^{12}	8.54×10^{11}	7.68×10^{11}
0.02	8.21×10^{14}	6.18×10^{11}	4.94×20^{11}	4.50×10^{11}

is irreversible and concentration-dependent. Since physical adsorption is a reversible process, it must be concluded that the $Cr_2O_7^{--}$ either is chemically adsorbed on the oxide or electrochemically reacts with the silicon surface, according to Eq. (7-2),

$$2Cr_2O_7^{--} + 3Si + 16H^+ \rightleftarrows 3SiO_2 + 8H_2O + 4Cr^{+++} \qquad (7\text{-}2)$$

at the same time becoming entrapped in the newly formed SiO_2. To determine which mechanism was operative, adsorption on freshly etched silicon and nitric acid–boiled silicon was studied as a function of $Cr_2O_7^{--}$ concentration. These results are shown in Fig. 7-18, where it can be seen that the freshly etched surfaces retained more radiochromium-51 than the nitric acid–boiled surfaces, which had a thicker protective film of SiO_2. From this behavior, the adsorption process would appear to involve electrochemical reaction with the silicon surface.

Radiotracers have also been used to study the retention of halide ions by silicon surfaces in the Texas Instruments Incorporated laboratories.[37] By using radioactive iodine-131, it has been demonstrated that the iodide ion is not retained ($< 10^{10}$ atoms/cm^2), while the fluoride ion (fluorine-18) is irreversibly and chemically adsorbed[38] on both silicon and silicon dioxide films at 10^{14} atoms/cm^2 concentrations. This behavior can be explained on the basis of Si—X and Si—O bond strengths. Only the Si—F bond is stronger (136 kcal/mole) than the Si—O bond (106 kcal/mole). All other Si—X bonds are weaker and therefore are readily hydrolyzed.

While the iodide ion does not interact with the silicon surface, elemental iodine in solution does react[37] and becomes entrapped in freshly grown SiO_2 film:

$$Si + 2I_2 + 2H_2O \rightleftarrows 4I^- + SiO_2 + 4H^+ \qquad (7\text{-}3)$$

The physically entrapped iodine was found to be readily removed with an HF etch. This reaction is analogous to the $Cr_2O_7^{--}$ reaction with silicon surfaces.

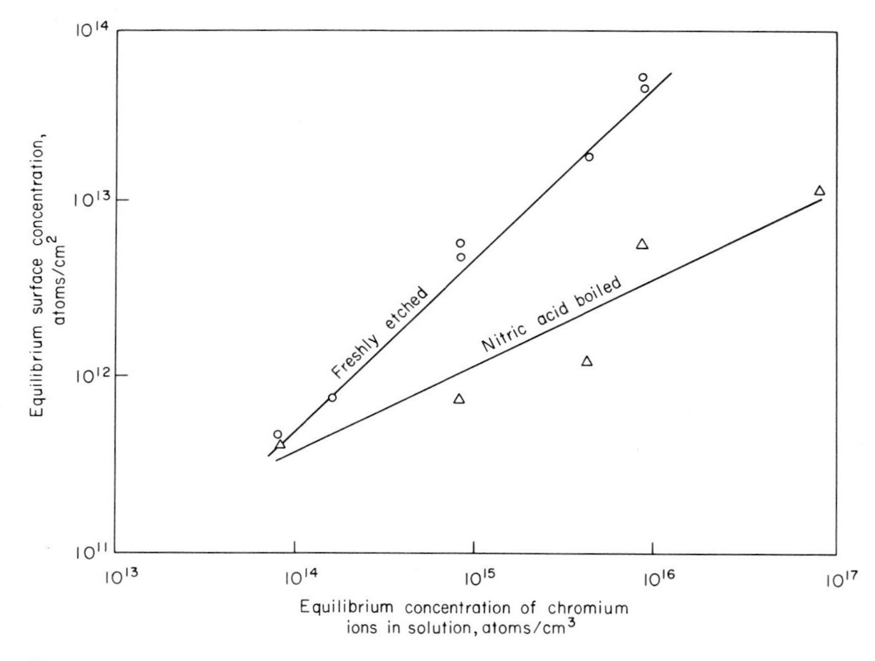

Fig. 7-18. Adsorption of chromium ions from solution on two types of silicon surface.

7-18. ORGANIC CONTAMINATION

Very little work has been carried out on the adsorption of organic molecules on semiconductor surfaces. Cunningham et al.[39] have described a method for the formation of chemisorbed monomolecular films of organo-substituted silanes on the natural oxide of silicon. Cullen et al.[40] carried out similar work on germanium. No work has been carried out to measure directly the amount of organic material adsorbed by using radiotracer carbon-14 tagged organic molecules. Anderson[41] has published a rather interesting technique using carbon-14 labeled solvents to detect preexisting organic surface contaminants. In this technique, the rate of evaporation of a volatile material from a surface was found to be an inverse function of the amount of preexisting contaminant. With an experimental arrangement as shown in Fig. 7-19, the rate of evaporation of a tagged material such as tetrabromo-

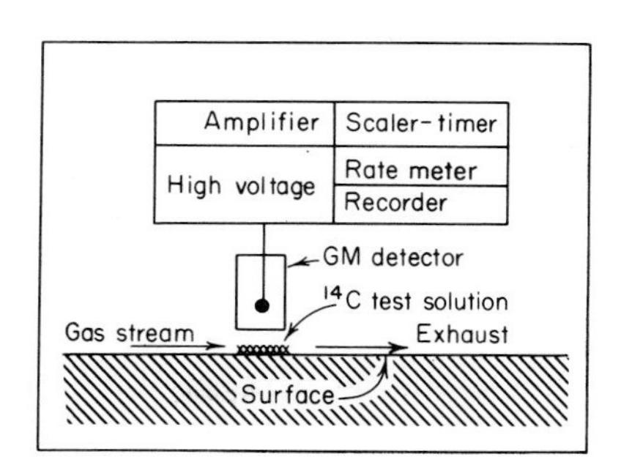

Fig. 7-19. Apparatus used to measure surface contamination using a carbon-14 tagged organic test solution. (*After Anderson.*[41])

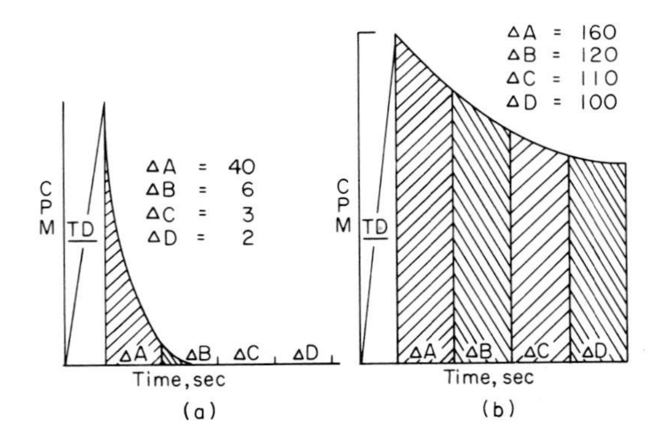

Fig. 7-20. **Evaporation curves (*A* clean, *B* contaminated) used to measure the amount of surface contamination. Data from apparatus shown in Fig. 7-19.** (*Courtesy Dr. John Anderson.*[41])

ethane-C14 from a clean surface and from a contaminated surface would appear as shown in Fig. 7-20. Anderson feels this technique can detect 0.1 $\mu g/cm^2$ of organic contaminate. This level of contamination is still rather high for semiconductor surface contamination and probably represents in the order of 10^{15} molecules/cm^2, depending on the molecular weight of the organic compound.

Some recent work in the Texas Instruments Incorporated laboratories[42] has evolved a method whereby iodine labeled with iodine-131 is reacted with surface organics (e.g., KMER†) and appears to have high sensitivity ($\approx$ 20 ng/cm^2). This technique is discussed in Sec. 10-41 on films.

7-19. ELECTRON MICROPROBE

The electron-probe microanalyzer offers one of the most useful tools for the analysis of semiconductor surfaces for chemical imperfections. The technique answers many apparently divergent criteria at the same time. Both qualitative and quantitative analysis are possible on surface areas ranging from 100 to 160,000 μ^2. The volume of material analyzed can be controlled by the energy of the electron beam, but the depth of penetration is typically 1 to 2 μ for a 50-kv beam. Figure 7-21 schematically illustrates the operation of the instrument. In fact, it is a combination of the electron microscope and the fluorescent x-ray spectrograph. Birks[43] gives an excellent discussion of the microprobe, its operation, and its applications.

The specimen whose surface is to be analyzed is mounted in an evacuated chamber, and the electron beam is focused on the surface. The 5- to 50-kv beam of electrons generates characteristic x-ray spectra of the chemical elements contained in the area being analyzed. A system of x-ray optics is used to analyze the emitted x-rays and thus yield both qualitative and quantitative chemical analysis of the micron-size area under study.

The forms of readout available from today's commercial instruments provide a wealth of information which otherwise would not be attainable. The x-rays which are generated by the high-energy electron beam are analyzed by diffracting them

†KMER is Kodak Metal Etch Resist, a light-sensitive polymer used in the photolithographic process (Sec. 1-5).

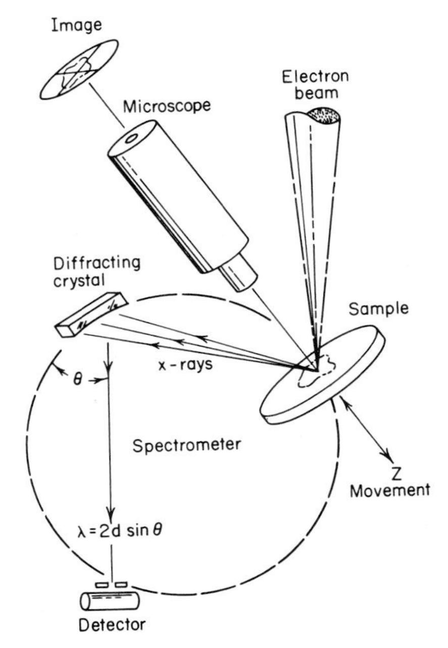

Fig. 7-21. An electron microprobe showing the light, electron, and x-ray optical systems all focused to the point of analysis.

with a crystal and measuring them with an appropriate detector. The output of the detector, which then consists of voltage pulses of varying amplitude, is analyzed with any of several types of electronic signal-processing equipment. These analyses can be carried out with the electron beam stationary or with the electron beam scanning the surface over a small area of 10×10 to 400×400 μ.

Electron Beam Stationary. With the electron beam stationary, it can be focused on an area on the surface of the semiconductor. The diameter of the spot can be varied from 0.5 to 500 μ by defocusing the electron beam. Under these operating conditions the instrument is a micro x-ray fluorescence spectrometer. By scanning the x-ray diffracting crystal and feeding the output of the detector through an amplifier and count rate meter to a recorder, a standard x-ray fluorescence spectrum will be obtained. This will give a qualitative or semiquantitative analysis for all elements present in the spot or area under analysis.

For quantitative analysis, the output of the detector is fed from the amplifier-pulse-height analyzer to a scaler-timer. The precision of any measurement should be controlled by normal counting statistics, where the standard deviation is equal to the square root of the total number of counts accumulated. Smith[44] has carried out a careful evaluation of the entire microprobe analytical system and has observed that the true standard deviation can be 1.5 times that predicted by Poisson statistics of counting data alone. Causes of this larger standard deviation are not clear but probably include accuracy of electron-beam focus and fluctuations in the detector system (gas flow rate, high voltage, preamplifier gain).

This experimentally observed decrease in precision will of course affect the ultimate detection limit for any impurity in a matrix. Typical detection limits are in the 300- to 1,000-ppm range. This impurity level, when segregrated in a $10 \times 10\mu$

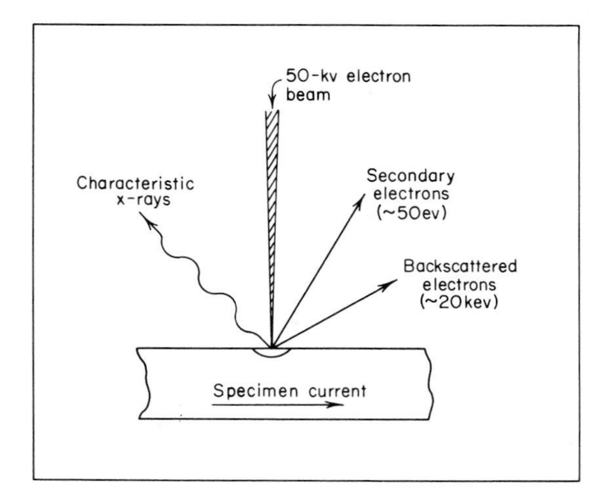

Fig. 7-22. Behavior of high-energy electrons on interaction with matter.

area, with 2 μ beam penetration, represents 7.5×10^{-10} to 2.5×10^{-9} μg total detectable element for a germanium or gallium arsenide matrix. Detection levels of this order are very useful when semiconductor-device surfaces are being analyzed.

Electron Beam Scanning. When a beam of 50-kv electrons strikes the surface of a sample, a number of other phenomena occur as well as x-ray generation. Figure 7-22 shows this interaction, and, as can be seen, both backscattered electrons and secondary electrons will be emitted from the surface. The number of backscattered electrons will depend on the atomic number of the elements in the specimen. Figure 7-23 shows the fraction of electrons that will be backscattered as a function of atomic number. Conversely, the number of collected electrons in the specimen is equal to $(1 - R)$.

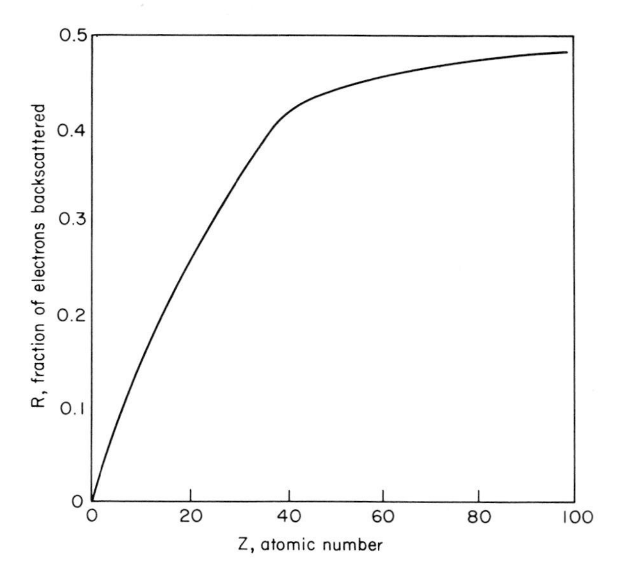

Fig. 7-23. Fraction of electrons backscattered as a function of atomic number. (*After Birks.*[43])

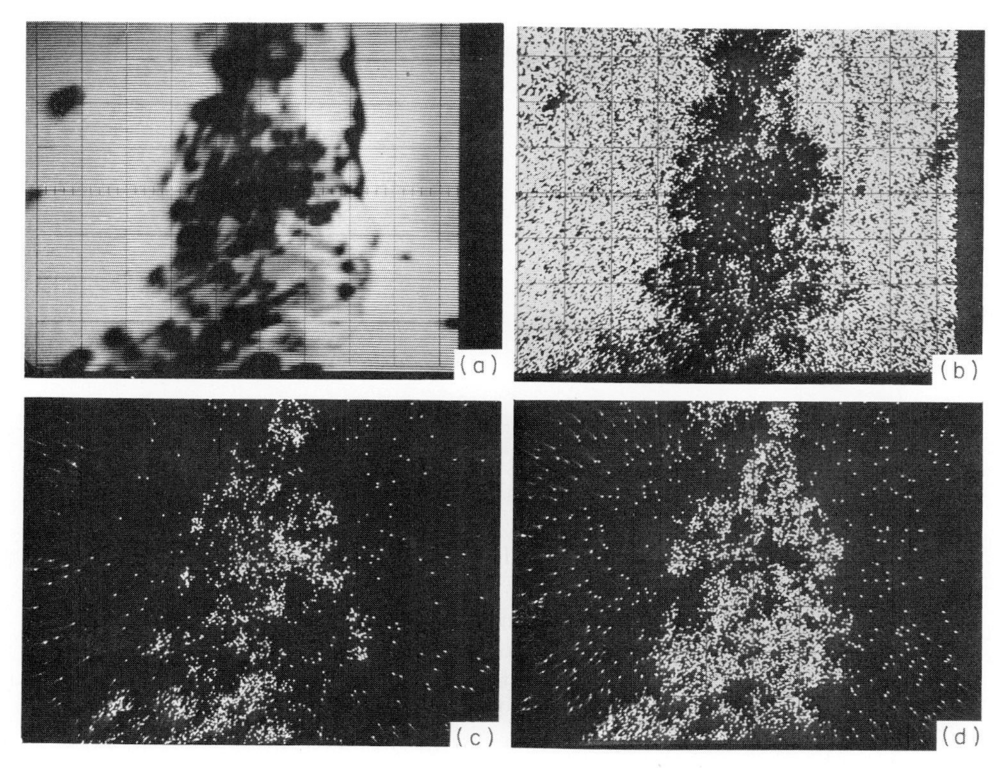

Fig. 7-24. Electron-probe microanalysis of a fault in a gold film on silicon. (*a*) **Specimen-current picture;** (*b*) **gold distribution;** (*c*) **aluminum distribution;** (*d*) **copper distribution.**

By using a scanning electron beam and measuring either the backscattered electrons or the specimen current, it is possible to see differences as small as one atomic number. When the two parameters, electron-beam position and specimen current density, are correlated and displayed on the raster of an oscilloscope, a picture is obtained of the elemental distribution over the scanned area (Fig. 7-24*a*). Similarly, a picture can be obtained from the backscattered electrons. While it is not feasible to identify elements from these displays, it is possible to identify areas of interest and to carry out subsequent electron-microprobe x-ray analyses of these areas on the surface. This x-ray analysis of an area for a specific element is accomplished by setting the diffracting crystal and electronics so that only the x-ray of the element of interest is detected. As the electron beam is scanned over the area of interest, the density of that particular x-ray emission is recorded on the raster of the oscilloscope. Figure 7-24*b*, *c*, and *d* show the distribution of Au, Al, and Cu on a silicon substrate determined by this scanning technique.

The electron microprobe appears to be one of the more powerful tools available for the study of semiconductor surfaces. The full potential of the technique has yet to be exploited on purely surface studies since most work to date has been on the analysis of finished electron devices, particularly integrated circuits.

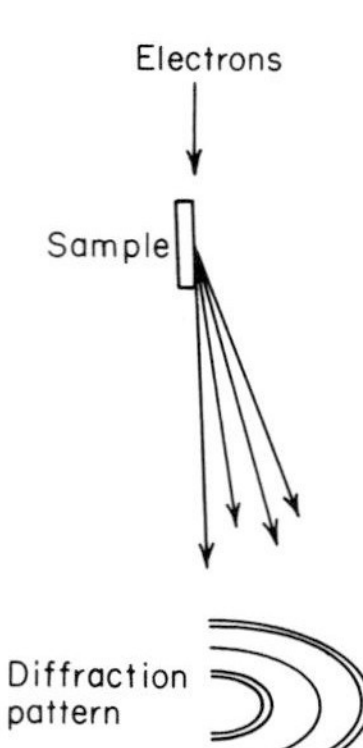

Fig. 7-25. Schematic illustrating reflection-electron diffraction technique.

7-20. ELECTRON DIFFRACTION

Electron diffraction is directly applicable to the analysis of small areas on the surface of a semiconductor. The deposit or area under study must be crystalline, or diffraction will not occur. Schematically the technique of selected-area reflection-electron diffraction is shown in Fig. 7-25. A beam of 20-kv electrons is focused on the area of interest on the surface at a small grazing angle, typically 1°. The electrons will diverge because of interaction with bound electrons in the lattice and will travel in the proper direction to satisfy the Bragg cones of diffraction. A series of rings will result and be recorded on a photographic plate. The ring diameters correspond to interplanar spacings in the crystal, and identification of the crystal under study is made by carefully measuring the spacings of these rings and comparing them with known patterns. Compilations of these patterns are available, arranged in a systematic manner to aid in the identification. The technique is not quantitative but does give more information than a simple elemental analysis since it is possible to identify a compound from the ring spacings.

Holmes and Newman[35] used electron diffraction to study the state of etched silicon and germanium surfaces and particularly metallic deposits from impurities in the etching solutions. They observed that when these impurity metals electrochemically deposited on silicon or germanium, they formed small three-dimensional islands of 40 to 60 Å across rather than flat, coherent films. Table 7-7 lists the results of the work of Holmes and Newman and gives a good picture of the sensitivity of this technique.

7-21. MASS SPECTROSCOPY

Mass spectroscopy has found only limited use in the investigation of semiconductor surfaces. Kozlovskaya[45] has studied the amount and nature of gases evolved from silicon and germanium on heating to 800°C, using mass spectroscopy. It was observed that nitrogen, carbon monoxide, carbon dioxide, hydrogen, and water were evolved from the semiconductor surfaces. The relative amounts of these gases were shown to be a function of the surface treatment, and the total amount of gas represented two to four monolayers' coverage.

Table 7-7. Summary of Quantitative Data on the Sensitivity of Electron Diffraction[35]

Substrate	Deposit material	Atomic number of deposit	Orientation of deposit	Crystal size of deposit, Å	Thickness when deposit pattern appears, Å	Thickness when substrate pattern disappears, Å
Evaporated silver single crystal on mica (111) surface atomically smooth at room temperature	Copper	29	Parallel	60 across, 12 high	0.8	5
Electropolished (111) face of massive silver crystal at room temperature	Copper	29	Parallel	About 50 across	1.5	?
Electroetched (111) face of silver at room temperature	Copper	29	Parallel and random	About 50 across	3.5	>12
Evaporated silver single crystal on mica (111) surface atomically smooth at room temperature	Surface attacked with bromine gas to give silver bromide	41	Parallel	70 across, 40 high	0.6	10
As above, but at 200°C	Copper	29	Parallel	>200	0.25	?
Sodium chloride cleavage face at room temperature	Copper	29	Random	Very small crystallites, almost amorphous	1.5 Very high background with weak substrate pattern	7.0
Silicon carbide cleavage face at room temperature	Silica	. .	. . . .	Amorphous	3	10

Ahearn[46] has used this technique to examine semiconductor surfaces for adsorbed impurities such as metals. Pairs of silicon electrodes were fabricated and sparked in the source of a solids mass spectrometer. The experimental arrangement was such that a pointed electrode was scanned over the surface of the counter electrode. Ahearn was able to show significant surface contamination of silicon by sodium hydroxide etching.

The problems with spark-source solids mass spectrometry are the quantitation and interpretation of the results. The depth of penetration of the spark varies with its energy but probably penetrates as much as 20 μ. For true surface-contamination studies a penetration depth in the order of 20 Å would be preferable. Further, the true area covered by the scanning spark can only be roughly estimated, which makes quantitative results very difficult. However, the judicial use of control samples does make it possible to show the effect of some process on the amount of surface contamination, and the qualitative identification of surface impurities has been demonstrated by Ahearn's work.

7-22. ELECTROCHEMICAL POTENTIAL AND CAPACITANCE

Brattain and Boddy[26] measured the differential capacitance of the interface between germanium and a pH 7.4 buffered electrolyte to determine the effect of certain trace-metal ions in solution on the electrical properties of the semiconductor. The capacitance of an electrode-electrolyte interface is inversely proportional to the charge separation distance. This charge is made up of the Helmholtz double layer (a surface charge formed by adsorbed ions) and a diffuse electrochemical double layer or space-charge region.

The capacitance of the germanium-electrode solution interface was measured by a current pulse technique, and the potential of the semiconductor was followed with an oscilloscope through a platinum reference electrode.

Since minute quantities of impurities in solution interact with the semiconductor surface, it was necessary for Brattain and Boddy to purify the electrolyte. This was accomplished by a simple technique whereby germanium powder was produced in situ by rapidly stirring a solution containing a few pieces of germanium with a magnetic stirrer. This stirring action gradually produced a dense cloud of fine particles with, of course, a very high surface area. With this purified electrolyte, Brattain and Boddy's capacitance measurements showed only the presence of the semiconductor space-charge region. The surface recombination velocity was found to be close to zero, which, coupled with the absence of the Helmholtz layer, indicated the absence of fast surface states.

In later work, Boddy and Brattain[47] systematically added trace amounts of copper ion to the electrolyte and observed the presence of another capacitance in parallel with the semiconductor space-charge region. Surface recombination velocity was observed to increase to significant levels. Boddy and Brattain attribute this change in electrochemical capacitance and the presence of surface states to the copper ion in solution. They also observed similar states produced by ions of gold and silver.

This electrochemical technique appears to offer a unique and powerful method of producing nearly perfect semiconductor surfaces (i.e., no fast states) and then systematically observing the effect of deliberately added impurity ions on the semiconductor.

7-23. MISCELLANEOUS PHYSICAL TECHNIQUES FOR CHEMICAL IMPERFECTIONS

The analytical chemist will generally find physical methods involving microscopic examination of semiconductor surface for contaminants of only limited value. A visual microscopic examination is at best qualitative and then can only detect agglomerates of material. If the contaminant is uniformly distributed, the visual method will be quite ineffective. Occasionally, if the contaminant is organic, it will fluoresce under ultraviolet light. However, even this method, when applicable, is also quite insensitive.

An indirect method used by Atalla et al.,[20] based on the hydrophobic character of a freshly etched silicon surface, appears to be capable of yielding a qualitative measure of the cleanliness of a surface. The surface to be examined is dipped in liquid nitrogen for about 10 sec and then observed at 400 magnifications in a closed chamber in which wet nitrogen is circulated. A thin sheet of ice forms immediately on the surface; and as the semiconductor warms up, a uniform layer of fine water droplets forms over the surface if the surface is clean. By observing the shape and size of the water droplets and the way the water evaporates, a qualitative measure of the cleanliness of the surface can be obtained. This technique offers no information on the amount of contamination and is probably not capable of detecting inorganic contaminants.

These techniques are of more use as qualitative tests in a production operation where they can be applied in quality control of some step in the process.

REFERENCES

1. Buck, T. M.: in H. C. Gatos (ed.), "The Surface Chemistry of Metals and Semiconductors," p. 107, John Wiley & Sons, Inc., New York, 1960.
2. Jones, C. E., and A. R. Hilton: *J. Electrochem. Soc.*, **112**:908 (1965).
3. Clarke, E. N., and R. L. Hopkins: *Phys. Rev.*, **91**:1566 (1953).
4. Warekois, E. P., M. C. Lavine, and H. C. Gatos: *J. Appl. Phys.*, **31**:1302 (1960).
5. Kurtz, A. D.: *Lincoln Lab. Tech. Rep.* 85, MIT, 1955.
6. Intrater, J., and S. Weissman: *Acta Cryst.*, **7**:729 (1954).
7. Stickler, R., and G. R. Booker: *Phil. Mag.*, **8**:859 (1963).
8. Camp, P. R.: *J. Electrochem. Soc.*, **102**:586 (1955).
9. Faust, J. W.: in Symposium on Cleaning of Electronic Device Components and Materials, 1966, *ASTM Spec. Tech. Publ.* 246, Philadelphia, 1959.
10. Gatos, H. C., M. C. Lavine, and E. P. Warekois: *J. Electrochem. Soc.*, **108**:645 (1961).
11. Buck, T. M., and F. S. McKim: *J. Electrochem. Soc.*, **103**:593 (1956).
12. Walters, G. K.: *Phys. Chem. Solids*, **14**:43 (1960).
13. Howard, J. K., and R. D. Dobrott: *Appl. Phys. Letters*, **7**:101 (1965).
14. Barrett, C. S.: *Trans. AIME*, **161**:15 (1965).
15. Blech, I. A., S. Meieran, and H. Sello: *Appl. Phys. Letters*, **7**:176 (1965).
16. Bradley, D. E.: *J. Appl. Phys.*, **27**:1399 (1956).
17. Stickler, R.: *J. Sci. Instrum.*, **41**:523 (1964).

18. Koehler, H., and D. Mattern: *Rev. Sci. Instrum.*, **36**:1602 (1965).
19. Savanick, G. A.: *Rev. Sci. Instrum.*, **38**:43 (1967).
20. Atalla, M. M., E. Tannenbaum, and E. J. Scheibner: *Bell System Tech. J.*, **39**:749 (1959).
21. Morrison, S. R.: *Phys. Chem. Solids*, **14**:214 (1960).
22. Sanders, D.: U.S. Patent 3,032,484, May 1, 1962.
23. Schnable, G. L., and J. G. Javes: U.S. Patent 3,034,970, May 15, 1962.
24. Schnable, G. L., and J. G. Javes: U.S. Patent 3,017,332, June 16, 1962.
25. Frankl, D. R.: *J. Electrochem. Soc.*, **109**:238 (1962).
26. Boddy, P. J., and W. H. Brattain: *J. Electrochem. Soc.*, **109**:812 (1962).
27. Sullivan, H. J., and L. L. Scheiner: *Semicond. Prod. Solid State Tech.*, **4**(4):48 (1961).
28. deMars, G.: *Semicond. Prod. Solid State Tech.*, **2**(4):24 (1959).
29. Bemski, G., and J. D. Struthers: *J. Electrochem. Soc.*, **105**:588 (1958).
30. Carlson, R. O.: *J. Appl. Phys.*, **29**:1001 (1958).
31. Sotnikov, V. S., and A. S. Belanovskii: *Zh. Fiz. Khim.*, **34**:2110 (1960).
32. Larrabee, G. B.: *J. Electrochem. Soc.*, **108**:1139 (1961).
33. Irving, B. A.: in P. J. Holmes (ed.), "The Electrochemistry of Semiconductors," p. 256, Academic Press Inc., New York, 1962.
34. Keiler, D.: *Solid-State Electron.*, **6**:605 (1963).
35. Holmes, P. J., and R. C. Newman: *Proc. Inst. Elec. Engrs.* (*Suppl.* 15), **B106**:287 (1959).
36. Larrabee, G. B., and J. F. Osborne (Texas Instruments Incorporated): Unpublished work, 1963.
37. Larrabee, G. B. (Texas Instruments Incorporated): Unpublished work, 1967.
38. Larrabee, G. B., K. G. Heinen, and S. Harrell: *J. Electrochem. Soc.*, **114**:867 (1967).
39. Cunningham, J. A., L. E. Sharif, and S. S. Baird: *Electrochem. Tech.*, **1**:242 (1962).
40. Cullen, G. W., A. J. Ameck, and D. Gerlich: *J. Electrochem. Soc.*, **109**:124 (1962).
41. Anderson, J. L.: *J. Am. Assoc. Contam. Contr.*, **2**(6):9 (1963).
42. Heinen, K. G., and G. B. Larrabee: *Solid State Tech.*, **12** (4): 44 (1969).
43. Birks, L. S.: "Electron Probe Microanalysis," Interscience Publishers, Inc., New York, 1963.
44. Smith, J. P. and J. E. Pedigo: *Anal. Chem.*, **40**: 2028 (1968).
45. Kozlovskaya, V. M.: *Fiz. Tverd. Tela*, **1**:1027 (1959).
46. Ahearn, A. J.: in C. R. Meissner (ed.), "1959 Sixth National Symposium on Vacuum Technology, Transactions," p. 1, Pergamon Press, New York, 1960.
47. Brattain, W. H., and P. J. Boddy: *J. Electrochem. Soc.*, **109**:574 (1962).

8

Characterization of Epitaxial Films

8-1. INTRODUCTION

In 1928 Royer[1,†] introduced the term *epitaxy* to denote the phenomenon of oriented growth of one crystal upon another. The word *epitaxy* is derived from the Greek *epi,* meaning "on" or "upon," and the past tense of the verb *teinein,* meaning "arranged," and hence "arrangement on." Royer's work was applied to the orientation of crystal layers on a substrate of different material and structure. Dash[2] extended the term to describe the fresh growth of silicon on a silicon seed in melt-grown crystals. The present-day usage of epitaxy and epitaxial crystal growth describes a process where a thin single-crystal film is deposited on a substrate. In the case of semiconductors this now usually describes the growth of the thin single-crystal film from the vapor on an oriented melt-grown single-crystal surface of the same material. This vapor growth can be carried out in a closed-tube system or in a flow system. While much of the early work on semiconductor epitaxial film growth was carried out in closed-tube systems,[3,4] the flow systems are now the preferred methods.

8-2. GROWTH IN CLOSED-TUBE SYSTEMS

The epitaxial growth of single-crystal semiconductor films in closed-tube systems is accomplished as shown in Fig. 8-1. In this technique, a quartz tube is loaded with the polished semiconductor substrate seeds, a large source of the semiconductor, e.g., 50 to 100 g, and some element such as iodine or hydrogen which will combine with and transport the feed semiconductor. Iodine was used in the earlier germanium[3] and silicon[4] work carried out at the IBM laboratories, and the chemical reaction proceeded via the disproportionation of the diiodide for both elemental semiconductors; e.g.,

$$2GeI_2(g) \rightleftarrows Ge(s) + GeI_4(g) \qquad (8\text{-}1)$$

†Superscript numbers indicate References listed at the end of the chapter.

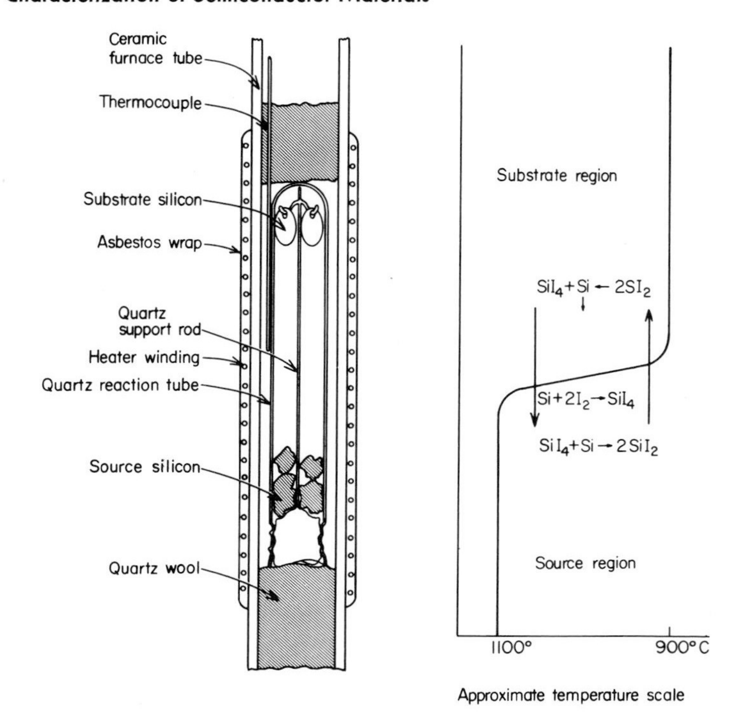

Fig. 8-1. Schematic illustrating epitaxial film growth in a closed-tube system. (*From Wajada et al.*[4])

Deposition rates of 10 μ/hr were obtained for germanium[3] and silicon.[5] Later work by May[6] showed the deposition rate to be directly proportional to the iodine pressure and roughly proportional to the reciprocal of the source-to-substrate distance.

The closed-tube system for epitaxial growth has received comparatively little attention, and generally the quality and uniformity of the deposits are inferior to those produced by other vapor techniques. The lack of flexibility in system control and substrate doping has severely limited the usefulness of the closed-tube system.

8-3. GROWTH BY SUBLIMATION AND EVAPORATION

The techniques of sublimation, evaporation, and sputtering have not been widely used in the semiconductor industry for the production growth of epitaxial layers. These three techniques are basically the same in that silicon atoms are energized by some method and, under high vacuum, migrate to the substrate, where they impinge on the substrate and align themselves. The overall quality of films formed by this technique is comparable to that produced by other vapor-transport methods. However, as with the closed-tube system, these techniques are not amenable to mass production and lack versatility in system control and layer doping.

Sublimation. Handelman and Povilonis[7] reported the epitaxial growth of silicon by vacuum sublimation at pressures of 10^{-9} to 10^{-8} torr in the apparatus shown in Fig. 8-2. As can be seen, the experimental arrangements required for sublimation are exacting with a specially designed, metal and organic-free ultrahigh-vacuum system a prime prerequisite. These workers successfully grew

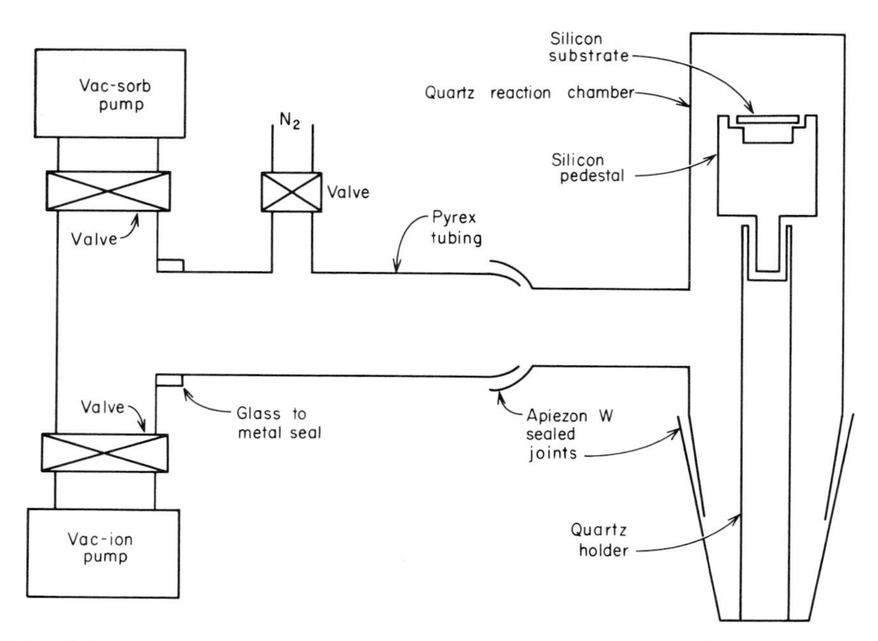

Fig. 8-2. Schematic of a sublimation apparatus for the growth of epitaxial films. (*From Handelman and Povilonis.*[7])

n-type silicon layers where previous workers had obtained only p-type layers. They also demonstrated the feasibility of vacuum-sublimation growth of epitaxial germanium layers.

Evaporation. In sublimation techniques for the growth of epitaxial films, the source or feed is not heated to the point of melting as it is in the evaporation methods. The use of a melted source requires higher temperatures and more intimate contact of the molten source with the container and implies higher risk of contamination for the epitaxial film. Newman,[8] in an excellent review of the growth and structure of epitaxial films of germanium and silicon, has pointed out that residual oxygen and nitrogen in vacuum systems are major sources of contamination. Refractory oxide boats[9] and refractory metals such as tungsten and tantalum[10,11] are reported to cause contamination of evaporated films. The use of electron-beam bombardment to melt the silicon, as shown in Fig. 8-3, results in considerably cleaner films.

Doping of epitaxial films during epitaxial growth can be accomplished by evaporation of the dopant in a separate boat at the same time as the film is being grown. This doping technique is difficult to control and lacks the flexibility required for industrial production.

8-4. GROWTH IN FLOW SYSTEMS

The gas flow system for epitaxial film growth has become the accepted industrial technique for the mass production of epitaxial films. This system is best described as "single-crystal growth from the vapor," where there can be simultaneous dep-

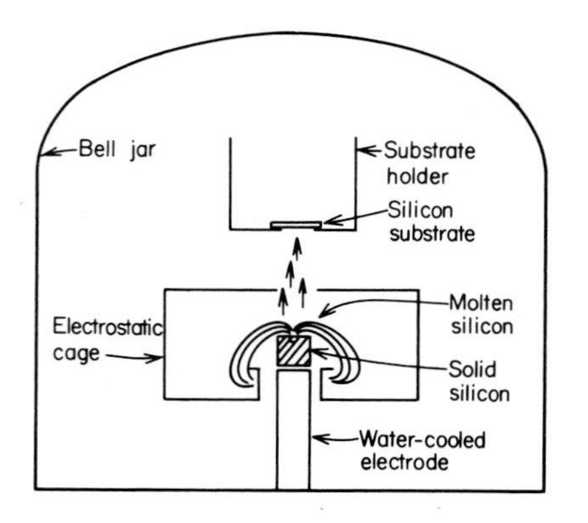

Fig. 8-3. Schematic illustrating electron-beam evaporation of silicon for epitaxial film growth.

osition of silicon and a doping impurity into a single-crystal film of high perfection. The deposition rate of the silicon or desired doping impurity can be rapidly changed simply by changing the composition of the gases flowing over the substrate. This results in a highly versatile system.

A schematic of a typical gas flow system is shown in Fig. 8-4. In this type of system, Teal-Little grown substrates, which have been carefully polished and cleaned, are mounted in the reaction chamber of a susceptor. Usually the susceptor is inductively heated with RF in production areas because of the rapid turnaround time between runs. Resistance heating is more frequently encountered in research work, where speed of operation is not a consideration.

Although every effort is made to ensure that the substrate is free of crystallographic and chemical imperfections, additional vapor etching is performed in the reactor immediately prior to epitaxial deposition. This etching can be performed

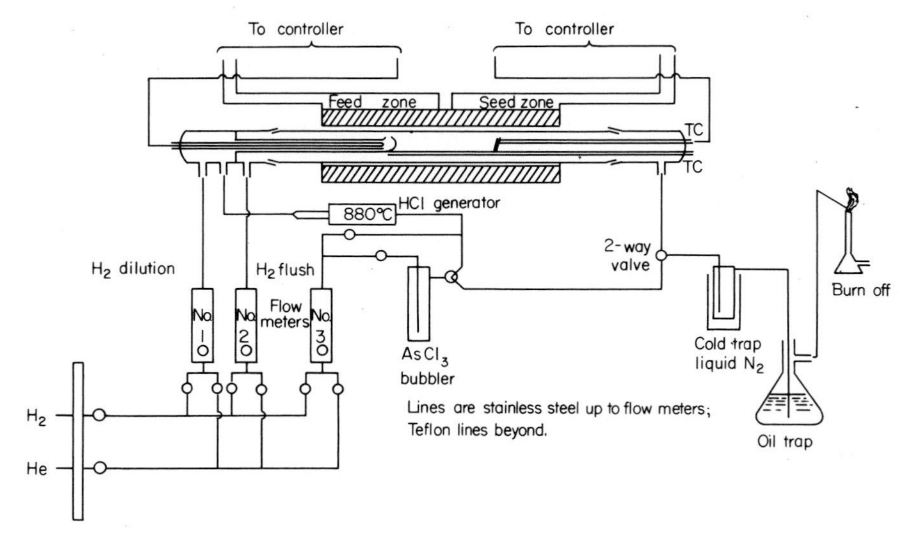

Fig. 8-4. Schematic of a typical flow system used for the growth of epitaxial films. This system was used for indium arsenide films. (*From Mehal and Cronin.*[13])

with high-purity HCl, chlorine, bromine, or water vapor. The HCl system has been most extensively studied[12] and utilizes the reversibility of the reaction

$$SiCl_4 + 2H_2 \rightleftarrows 4HCl + Si \tag{8-2}$$

where an excess of HCl in the hydrogen–$SiCl_4$ gas stream will force the reaction to the left. After sufficient silicon has been etched from the surface, the HCl flow is stopped, the reaction immediately reverses, and epitaxial silicon growth begins.

The doping or incorporation of desired impurities into the epitaxial layer is also accomplished through the gas phase. The two techniques used to incorporate the dopant into the flowing gas phase are solution doping and gas doping. Generally, solution doping is used for the more heavily doped epitaxial layers and is carried out by bubbling the carrier gas through a solution of the dopant. In silicon and germanium epitaxy, common solution dopants are PCl_5 or $SbCl_5$ dissolved in $SiCl_4$ or $GeCl_4$.

Gas doping is more flexible in that gases such as B_2H_6, AsH_3, and PH_3 are diluted with hydrogen to the ppm level and then are simply introduced into the H_2–$SiCl_4$ stream by appropriate valving. Both n- and p-type layers can be grown with impurity concentrations ranging from 5×10^{14} to 10^{20} atoms/cm^3 by using gas doping.

The epitaxial growth of the III-V compounds was reviewed by Mehal and Cronin.[13] A slightly more complex flow system is used with these compound semiconductors than with silicon and germanium. The versatility of this flow system, in which Mehal and Cronin grew mixed III-V epitaxial layers of GaInAs, is shown in Fig. 8-4. Doping was accomplished by introducing H_2S, zinc, or cadmium vapor into the gas stream. Rubenstein and Meyer[14] grew gallium arsenide epitaxial layers by using $GaCl_3$ and arsenic in a hydrogen carrier gas where reaction and single-crystal growth occurred at the substrate surface. Conrad et al.,[15] using elemental gallium, indium, and arsenic metals as feed, grew $Ga_xIn_{(1-x)}As$ epitaxial layers on GaAs substrates. These epitaxial mixed III-V semiconductor layers are an ideal medium for tailoring electrical properties to produce the desired band gap, mobility, or graded structure required for given device parameters.

8-5. CHARACTERIZATION PROBLEMS

The evaluation of epitaxial films presents many problems that are absent in the analysis of bulk semiconductors. Since these single-crystal layers are very thin (1 to 25 μ), there is only a limited amount of sample available for analysis. On a 1-cm^2 surface of a 10-μ silicon film there would be a total sample of 2.3×10^{-3} g available for analysis. Typical analytical requirements for the analysis of this film would include film thickness, electrical properties, composition if it were a mixed III-V such as $Ga_xIn_{(1-x)}As$, chemical impurities in the film, and physical imperfections such as dislocation count, stacking faults, and other damage in the epitaxial layer. Most of these measurements would be expected to be nondestructive.

8-6. FILM THICKNESS

There are both destructive and nondestructive methods available for the accurate measurement of the thickness of an epitaxial layer. Unfortunately, the nondestructive techniques are not applicable to every epitaxial layer, and therefore in such cases the destructive techniques must be utilized.

8-7. ANGLE LAP AND STAIN WITH INTERFEROMETRY

The angle lap and stain is a destructive technique and is much like that in Sec. 9-9 for the determination of diffusion depth. Bond and Smits[16] at Bell Telephone Laboratories described the use of an interference microscope to measure the thickness of extremely thin surface layers. In this technique the sample was mounted on a steel cylinder at a shallow angle of 0.5 to 1.0°, as shown in Fig. 8-5. The angle-lapped sample was then transferred to an interferometer, as shown in

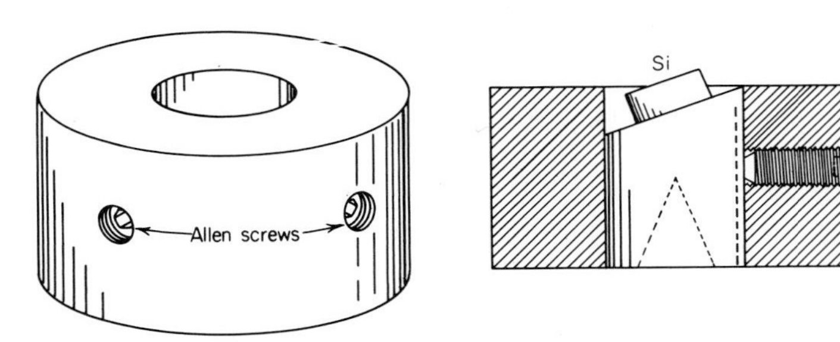

Fig. 8-5. Fixture used for lapping a bevel at a small angle. *(From Bond and Smits,[16] copyright 1956 by the American Telephone and Telegraph Co., reprinted by permission.)*

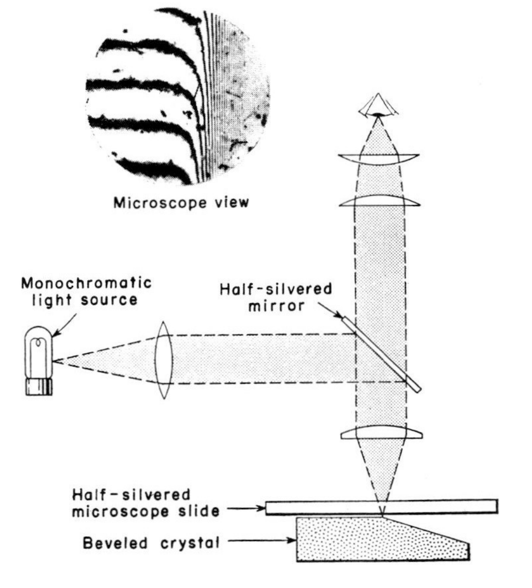

Fig. 8-6. Diagrammatic view of the light path in an interferometer. *(From Bond and Smits,[16] copyright 1956 by the American Telephone and Telegraph Co., reprinted by permission.)*

Fig. 8-6. The monochromatic light normally used in this work is sodium and has a wavelength of 5.89×10^{-5} cm, or 0.589 μ. A fringe pattern is observed, and the number of fringes between the epitaxial surface and the substrate surface are counted. Since the fringe spacings correspond to $\lambda/2$, for sodium light each fringe equals 0.2945 μ. For the case where 20 fringes were observed, the layer thickness would be

$$\Delta n \times \frac{\lambda}{2} = 20 \times 0.2945$$
$$= 5.89\ \mu$$

Bond and Smits feel that an accuracy of $\pm$ 5 percent is easily obtained by using half-silvered microscope slides for the mirror. A detailed procedure for angle lapping, staining, and interferometrically determining the thickness of epitaxial layers was published by Monsanto Company.[17] These workers report the precision of the measurement to be $\pm$ one fringe or $\pm 0.2945\ \mu$. When an interferometer is not available, the angle-lap-and-stain technique is still employed, but the thickness of the epitaxial layer is measured with a calibrated microscope. The thickness is then calculated as described in Sec. 9-9 for diffused-junction depth measurement.

While this angle-lap technique is in fact a destructive technique, it requires only a small edge of the epitaxial slice. Frequently it is possible to cleave a small piece from the slice and measure the epitaxial-layer thickness on this chip. In either case it is very seldom necessary to sacrifice the entire slice for the thickness measurement.

8-8. STACKING-FAULT DEFECTS

Frequently during the growth of epitaxial films stacking faults will grow into the film. A schematic drawing of a stacking fault grown in an epitaxial layer on a (111) substrate is shown in Fig. 8-7. This stacking fault is shown originating at the substrate surface, as most do, and is the result of the intersecting of the three stacking-fault planes $(\bar{1}11)$, $(1\bar{1}1)$, and $(11\bar{1})$. By measuring the length L of the side of the equilateral triangle formed at the surface, the thickness D can be calculated:

$$D = L\sqrt{2/3} = 0.816L \tag{8-3}$$

Dash[18] reported excellent agreement between the thickness found by this technique and the angle-lap-and-stain technique. Since stacking faults can originate in the growing layer, care must be taken to measure only the largest faults. This technique has declined in usefulness since the quality of epitaxial films has pro-

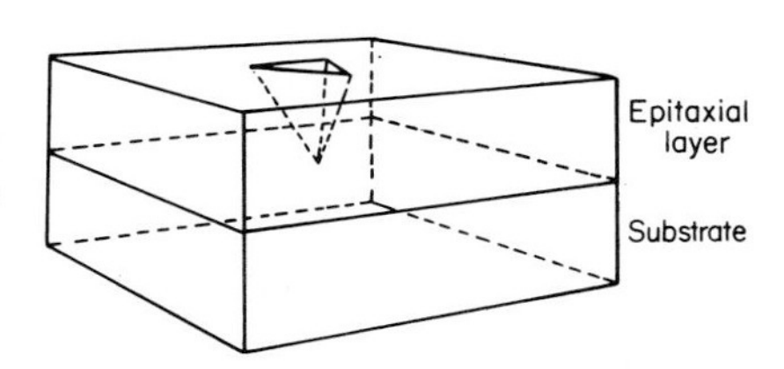

Fig. 8-7. Illustration of a stacking fault that originated at the substrate and was propagated through the epitaxial layer during film growth.

gressed to the point where there are few, if any, stacking faults. Since it is necessary to etch the surface to delineate these stacking faults, the technique is best considered a destructive technique.

8-9. INFRARED INTERFERENCE

The most widely used technique for the measurement of the thickness of epitaxial films is infrared (IR) interferometry. This is a nondestructive technique first reported by Spitzer and Tanenbaum[19] for epitaxial-film studies. Albert and Combs[20] subsequently made a detailed study of the technique.

The interaction of infrared radiation with an epitaxial film on a reflecting substrate is shown in Fig. 8-8. The incident ray I, at angle ϕ, is partly reflected, as ray 1 at the epitaxial-layer surface, and partly refracted, at angle ϕ', to the substrate, where it is then reflected. The reflected ray emerges from the epitaxial-layer surface as ray 2. In the analysis, when the wavelength of incoming ray I is varied, the two reflected rays will interfere at integral multiples of half-wavelengths, resulting in alternate bright and dark interference fringes. An infrared detector looking at the reflected radiation would record the interference fringes as shown in Fig. 8-9. This figure is an experimental recording of an infrared scan of an epitaxial layer.

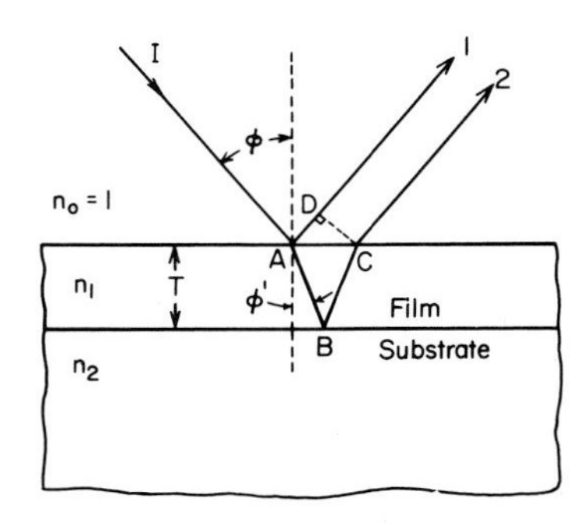

Fig. 8-8. Interaction and reflection of infrared radiation from an epitaxial layer on a reflecting substrate.

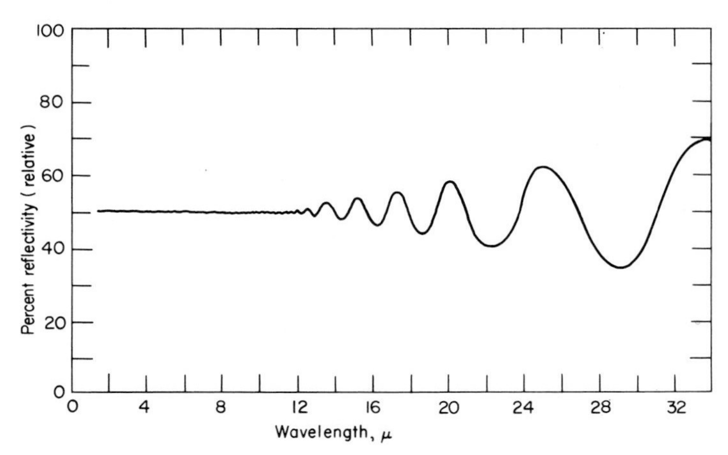

Fig. 8-9. Typical infrared-reflection spectrum of an epitaxial layer.

Since this measurement technique is restricted to infrared transparent epitaxial layers on heavily doped reflecting substrates, it can be assumed that the refractive index n_2 is lower than the film index n_1, and that the extinction coefficient k_1 of the film is negligibly small. By referring to Fig. 8-8 it can be seen that the optical path difference δ for the two interacting rays is

$$\delta = n_1(AB + BC) - AD \tag{8-4}$$

$$\delta = 2T \cos \phi' \tag{8-5}$$

where T is the film thickness. For any given spectrometer or measurement system, the incident angle is fixed, so that $\cos \phi'$ in Eq. (8-5) is constant. Therefore, to calculate the thickness it is necessary only to obtain the best value for δ, the optical path difference, by using the infrared interference scan shown in Fig. 8-9.

If the order m of a fringe maximum occurring at wavelength λ_m is known, then $\delta = m\lambda_m$ and can be substituted in Eq. (8-5). To obtain the value of m for the λ_m maximum, it is necessary to count x fringes and record λ_{m+x}, where

$$m\lambda_m = (m + x)\lambda_{m+x} \tag{8-6}$$

and

$$m = \frac{x\lambda_{m+x}}{\lambda_m - \lambda_{m+x}} \tag{8-7}$$

The thickness can then be calculated by substituting these values in Eq. (8-5) and solving for T.

$$T = \frac{x\lambda_m\lambda_{m+x}}{2n_1 \cos \phi' \, (\lambda_m - \lambda_{m+x})} \tag{8-8}$$

These calculations are simple since for a given spectrometer and material $2n_1 \cos \phi'$ is a constant. Albert and Combs[20] developed a fringe chart which gives reliable thickness measurements with only three well-defined fringes. Walsh[21] reports a circular slide rule that quickly does the same calculations.

The infrared interferometric method is the preferred method for epitaxial-film-thickness measurement. It is fast (1 to 2 min), accurate, and nondestructive. Unfortunately, it is not applicable to all films. The technique depends on a precise set of optical constants for the two silicon layers. The epitaxial layer must transmit and the substrate must reflect radiation. These optical constants are controlled by the free-carrier concentration of the silicon. Figure 8-10 shows the reflectivity of silicon for several dopant concentrations. As can be seen, high-resistivity or lightly doped semiconductors show small but constant reflectance over the wavelength range of interest. On the other hand, the heavily doped material shows a dip and then a rise in reflectance over the same wavelength range. Spitzer and Fan[22] determined the optical constants of silicon, germanium, and indium antimonide in the 5- to 35-μ region as a function of free-carrier concentration. They reported reflectance curves similar to those of Albert and Combs.[20] In general, the resistivity of the substrate must be less than 0.01 to 0.02 ohm-cm to obtain satisfactory reflectance. The resistivity of the epitaxial layer must be greater than 1 ohm-cm to allow sufficient radiation to penetrate to the substrate for reflection and subsequent interferometric interaction with the radiation reflected at the

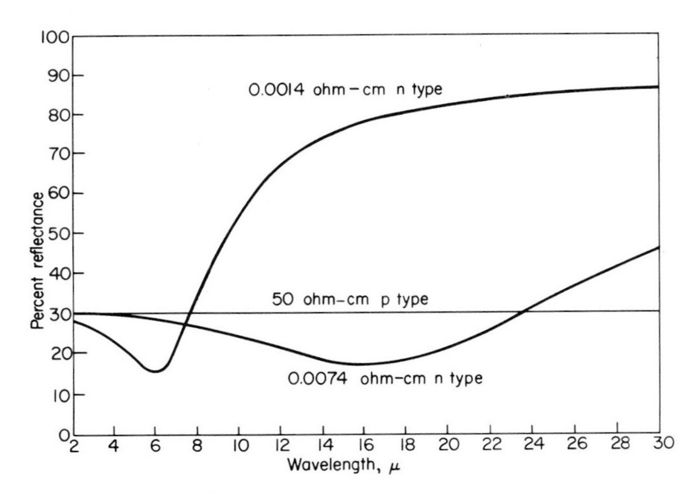

Fig. 8-10. Reflectance spectra of silicon samples showing the effects of carrier concentration on the optical constants. Decreasing reflectance, below 30 percent, indicates decreasing refractive index, and large increasing reflectance indicates large increasing extinction coefficient. (*From Albert and Combs.*[20])

epitaxial surface. Fortunately, most production epitaxial material meets these requirements of a lightly doped epitaxial layer on a heavily doped substrate.†

The two reflecting surfaces must be nearly parallel, or fringes from regions of varying thickness will cancel out.[23] Variations in layer thickness must be less than 1 to 1.5 μ. Good interference fringes will not be obtained if outdiffusion of dopant from the heavily doped substrate occurs.[24] Very thin films are difficult to measure by this technique because only one or two fringes are obtained, and films less than 2 μ are more accurately measured by other techniques.

Albert and Combs[20] reported good agreement between the IR and angle-lap-and-stain methods. Schumann et al.[25] report that the infrared interference technique can be in error under certain conditions of wavelength and substrate doping concentration. This error occurs because a significant phase shift of the radiation occurs when wavelengths greater than those employed by Albert and Combs are used to measure the layer thickness. The effects of this phase shift can be seen when successive peaks or valleys do not yield the same layer thickness by using Eq. (8-8). Schumann et al.[25] developed a set of equations to correct for this phase-shift error, and some typical results of corrected data are shown in Table 8-1. This correction improves the agreement between thickness measurements techniques and also eliminates the variation in calculated thickness when successive peaks are used. The ASTM proposed method[26] for epitaxial-film-thickness measurement includes these corrections and lists an excellent table of these phase shifts as a function of substrate resistivity.

†Heavily doped material is referred to as either n^+ or p^+.

Table 8-1. Results of Phase-shift Correction on Epitaxial-layer-thickness Measurements[25]

Sample	Corrected, μ	Uncorrected, μ
n/n^+ Si	7.01 ± 0.091	7.62 ± 0.52
n/n^+ Si	2.90 ± 0.016	3.32 ± 0.073
p/p^+ Si	3.10 ± 0.033	3.64 ± 0.12

8-10. INFRARED ELLIPSOMETRY

The theory and technique of visible-light ellipsometry is discussed in detail in Sec. 10-15. Infrared ellipsometry is the extension of this technique for the specific purpose of measuring the thickness of epitaxial layers which are themselves transparent to infrared radiation and are on reflecting substrates.

Hilton and Jones[27] developed the infrared ellipsometer shown in Fig. 8-11, which is an infrared version of the visible ellipsometer. The infrared ellipsometer is operated at 54.6 μ because this wavelength satisfies all the requirements of the optics in the instrument and maximizes system energy transmission. At 54.6 μ

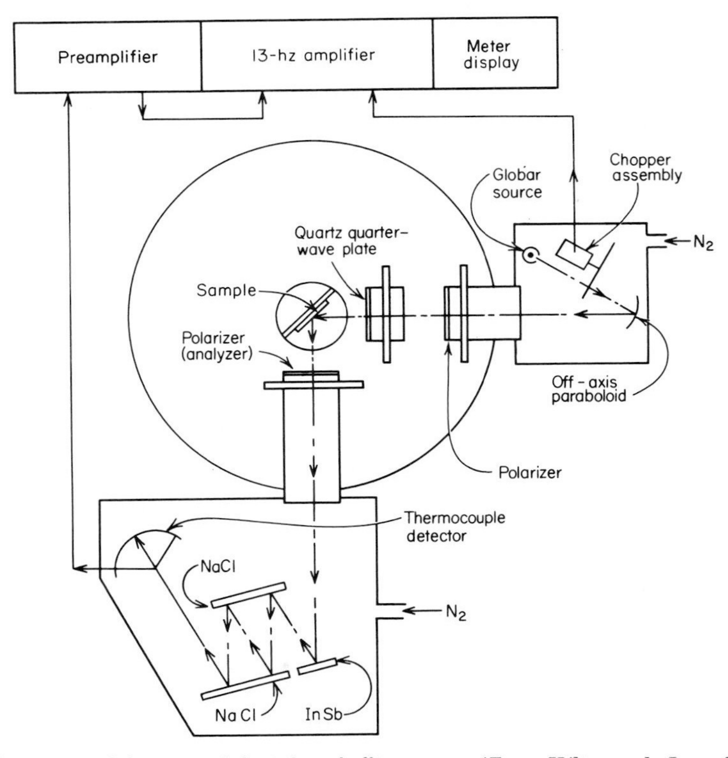

Fig. 8-11. Schematic of the infrared ellipsometer. *(From Hilton and Jones.[27])*

there is good optical transmission through the quartz quarter-wave plate and maximum reflection from the NaCl and InSb optics.

Subsequent work by these authors[28] has improved the system. A new 13,000 line/in. aluminized polyethylene polarizer was fabricated and resulted in an increase from 26 to 70 percent transmitted 54.6-μ radiation. Simultaneously a Golay detector[29] was installed. This detector is the most sensitive room-temperature detector available for this wavelength range. The Golay cell substantially increased the output signal over the thermocouple detector. With these improvements the two angle measurements on the polarizer and azimuth can be read to $\pm 1°$.

A typical ellipsometer curve for germanium epitaxial layers, p on p^+, is shown in Fig. 8-12. The thickness of the epitaxial layer is shown for whole microns with numbers and arrows, and the half-micron values are shown by arrows only. The first order of thickness is read inside the curve (1.0 to 6.8 μ), and the second order (6.8 to 13.6 μ) outside. The experimental points shown on the curve are for over 40 samples. The infrared ellipsometer performed well over the entire thickness range and gave good reproducibility on films below 2 μ, where infrared interferometry cannot be used. An average reproducibility of ± 2.8 percent and an average difference between IR ellipsometry and interferometry of ± 9.1 percent were obtained.

This technique has also been used to measure the thickness of silicon epitaxial layers covered with as much as 14,000 Å of SiO_2. The approximate thickness of the oxide layer must be known, and this can be readily determined with a visible-light ellipsometer.

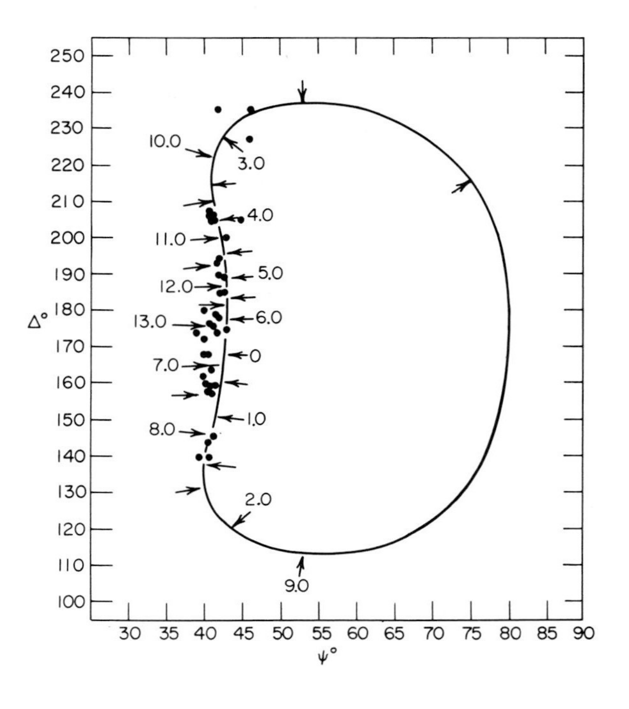

Fig. 8-12. Ellipsometer curve for p-type germanium epitaxial layers on a p^+ germanium substrate. $\phi = 50°$, $\lambda = 54.6\ \mu$. (*From Hilton and Jones.*[27])

8-11. FILM COMPOSITION

It is obvious that the chemical composition of silicon and germanium epitaxial layers will not be considered. Also the stoichiometry of the III-V epitaxial compounds cannot be determined with sufficient accuracy by using existing analytical techniques. It is generally felt that if stoichiometry deviates by more than 1 part in 10^6, the electrical properties of the deposit will be seriously degraded. The third class of semiconductor materials is the mixed III-V compounds (for example, GaInAs and GaAsP) where the desired electrical properties, such as band gap, can be controlled by the composition of the epitaxial single-crystal alloy semiconductor. In these cases it is necessary to know the chemical concentration of each of the constituents of the semiconductor alloy. Here, as with other epitaxial layers, the amount of total sample available for analysis is very small ($\sim$5 to 10 mg/cm^2), and nondestructive analytical techniques are vital since in the exploratory research area each slice is different. X-ray diffraction and ultraviolet reflection techniques are used for these analyses.

8-12. X-RAY DIFFRACTION

Vegard's law states that the axial parameter of a parent lattice is very nearly a linear function of the atomic concentration of the dissolved atoms. That is, there will be a shrinkage of the host crystal lattice when small-radius atoms are introduced substitutionally and expansion of the lattice for larger atoms. These changes in the lattice parameters are measured by x-ray diffraction techniques. It is estimated that the unit-cell parameter d can be measured to an accuracy of 1 part in 5×10^4 with a relative error of $\pm 2 \times 10^{-3}$ percent of the unit cell.[30]

Rubenstein[31] used powder x-ray techniques to show that Vegard's law did hold, and there was a linear relationship between the lattice constant and the composition for the $GaP_{(1-x)}As_x$ alloy system over the entire concentration range from $x = 0$ to $x = 1$. The samples used in Rubenstein's work were powdered bulk melt-grown crystals, and the technique was not directly applicable to thin epitaxial layers.

Williams et al.[32] have described in detail the use of x-ray diffraction to determine the composition of epitaxial Ga(As,P) alloys. They used a Supper-type goniometer head which was adapted to a Philips wide-range goniometer. With copper Kα x-radiation, the sample was carefully oriented by using the GaAs substrate (333) diffraction maximum. Then a diffraction trace was made, as shown in Fig. 8-13. The $K\alpha_1$ and $K\alpha_2$ peaks are clearly defined in homogeneous films. The shape and intensity of the region between the GaAsP and GaAs peaks give qualitative information on the concentration gradient in the film. This is due to the fact that the x-radiation penetrates in the order of microns into the film and gives the average concentration of that part of the epitaxial film. Vegard's law was found to be obeyed over the concentration range GaAs to $GaAs_{0.5}P_{0.5}$. Tietjen and Amick[33] reported a linear relationship with lattice constant over the entire concentration range from $x = 0$ to $x = 1$ for epitaxial $GaAs_{(1-x)}P_x$. Reproducibility in the order of ± 1 percent was reported by both workers.[32,33] This x-ray diffraction analysis technique has also been used for $Ga_xIn_{(1-x)}As$.[15]

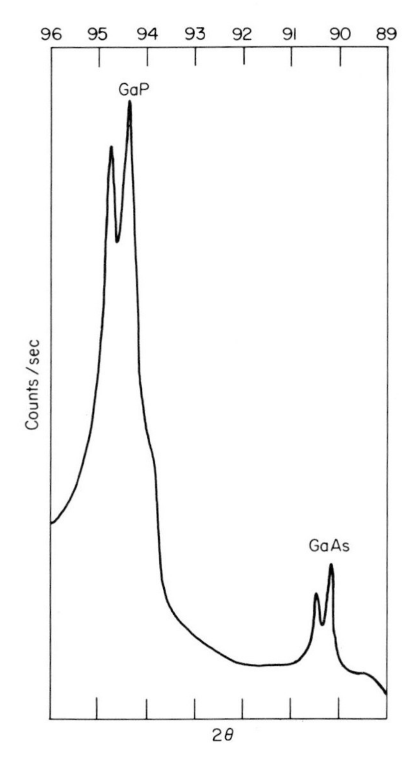

Fig. 8-13. Diffractometer trace of a GaP film on a GaAs substrate. (*From Williams et al.*[32])

8-13. OPTICAL REFLECTIVITY

The use of optical reflectance spectroscopy and its analytical applications have been reviewed by Wendlandt and Hecht.[34] Peaks in the reflectance spectra of semiconductors, at wavelengths shorter than the band edge, are due to electronic interband transitions caused by phonon interaction of the light with the semiconductor. Woolley and Blazey[35] reported a linear relationship between the wavelength of the energy associated with the reflectance peak and the composition of polycrystalline (GaIn)Sb and (GaIn)As. Jones et al.[32,36-38] applied this technique to the analysis of epitaxial (GaIn)As and Ga(AsP) alloy films.

The specular reflectance technique is rapid, and since measurements are made directly on the solid, it is nondestructive. Unlike the x-ray diffraction techniques described above, optical reflectance examines only the surface of the epitaxial layer. This makes it a valuable complementary technique and, when used with incremental etching, will give a concentration profile through the epitaxial layer.

The measurements are made with a double-beam spectrophotometer equipped with a specular reflectance attachment in the 350- to 750-mμ range (3.54 to 1.65 ev). In the region of the peaks (Fig. 8-14), a scale expansion of 20 is used to obtain a more accurate position for the principal reflection peak. The relationship between the energy of the reflectance peak and the composition of the GaAs–InAs is not linear. A linear relationship was observed for the Ga(As,P) alloys of composition 0 to 50 mole percent GaP (Fig. 8-15). Jones[36] reported a standard deviation of ± 1 percent at the 90 percent confidence level for GaAs and InAs.

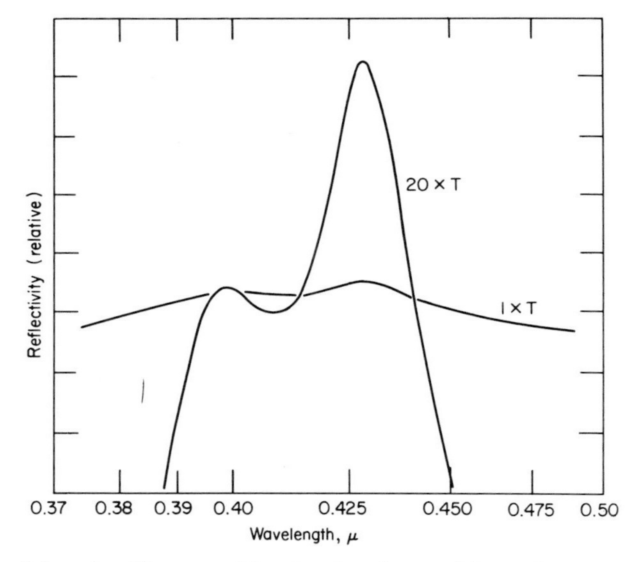

Fig. 8-14. Reflectivity of gallium arsenide, showing the need for a 20 times expansion on the reflectivity scale. (*From Jones.*[36])

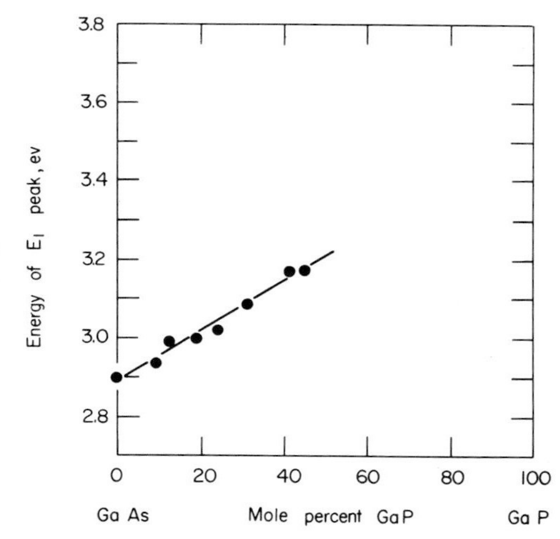

Fig. 8-15. Variation of the E_1 reflectivity peak at room temperature for etched GaAsP alloys. (*From Williams et al.*[32])

8-14. COMPOSITIONAL X-RAY TOPOGRAPHY

The technique of x-ray topography was described in Sec. 7-12 for the analysis of surfaces for physical imperfections. The scanning-reflection method was used to examine an entire surface (maximum 1 in.^{2}) for defects. Compositional x-ray topography was developed by Howard and Dobrott.[39] It is an application of the scanning-reflection x-ray topographic method and is used in conjunction with the x-ray diffraction method described above. These two techniques can yield the composition, alloy homogeneity, perfection, and crystallinity of the epitaxial deposit.

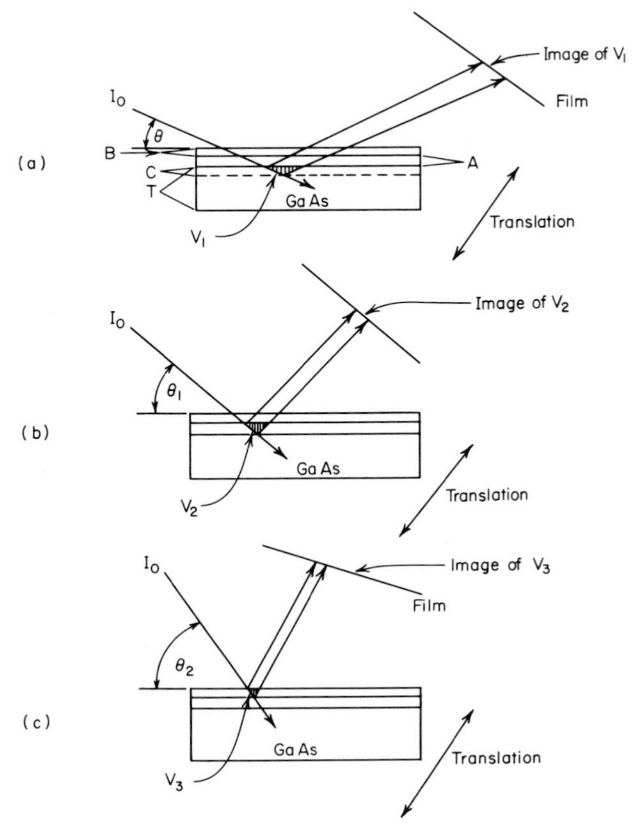

Fig. 8-16. Conceptual diagram for compositional x-ray topography. Each layer can be diffracted separately by utilizing the diffraction angle which corresponds to the lattice parameter (composition) of that layer. (*From Howard and Dobrott.*[39])

Since Vegard's law holds for these alloy systems, the diffraction angle depends on the lattice constant, which is unique for each composition of the alloy. Once this diffraction angle is known or determined, the epitaxial layer can be examined topographically for *that* composition. Then, by changing the diffraction angle, the substrate or some other layer can be examined. This technique of selecting the layer to be examined is shown schematically in Fig. 8-16, where two epitaxial layers A and B have been deposited epitaxially on a GaAs substrate. In case *a*, the incident beam I_0 impinges at angle θ and the shaded volume V_1 in the GaAs substrate diffracts to form a topographic image. Then the slice-film combination is translated to examine the entire slice. Changing the angle of incidence of the beam I_0 to θ_1 and θ_2, where θ_1 and θ_2 are the diffraction angles for the particular composition of epitaxial layers A and B, makes it possible to obtain topographs of each layer.

To illustrate the applicability of this technique, a GaInAs epitaxial layer yielded the diffraction scan shown in Fig. 8-17, where the GaAs substrate diffraction angle

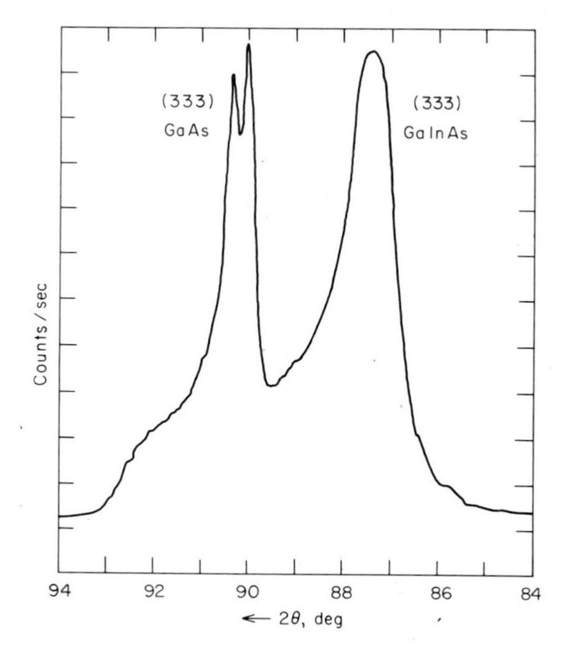

Fig. 8-17. Diffraction scan of an inhomogeneous (111) deposit of GaInAs. The alloy composition was 35% InAs. (*From Howard and Dobrott.*[39])

was 66.2° and the GaInAs layer was 64.2°. This diffraction angle for the layer corresponds to $Ga_{0.5}In_{0.5}As$. With these diffraction angles, (400) x-ray topographs were made (Fig. 8-18) and show compositionally a nonuniform layer. Changing the diffraction angle made it possible to show that the GaAs layer was incorporated into the alloy.

This technique is applicable to any mixed III-V alloy epitaxial system. Howard and Cox[40] report that this method clearly differentiated two regions in a Ga(As,P) layer with a compositional difference of only 3% GaP.

This technique was used by Howard and Dobrott[39] to examine the composition

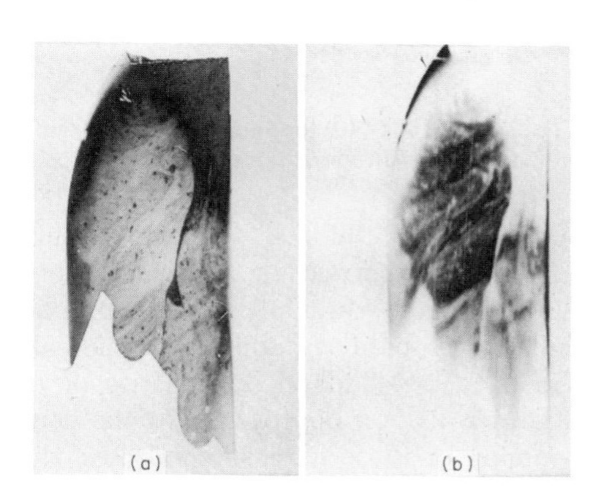

Fig. 8-18. (*a*) **(400) topograph reveals inhomogeneous regions in the (100) deposit of GaInAs; the alloy reflection was used to form the topographic image.** (*b*) **The relative contrast was reversed when the GaAs diffraction angle was employed to obtain the (400) topograph.** (*From Howard and Dobrott.*[39])

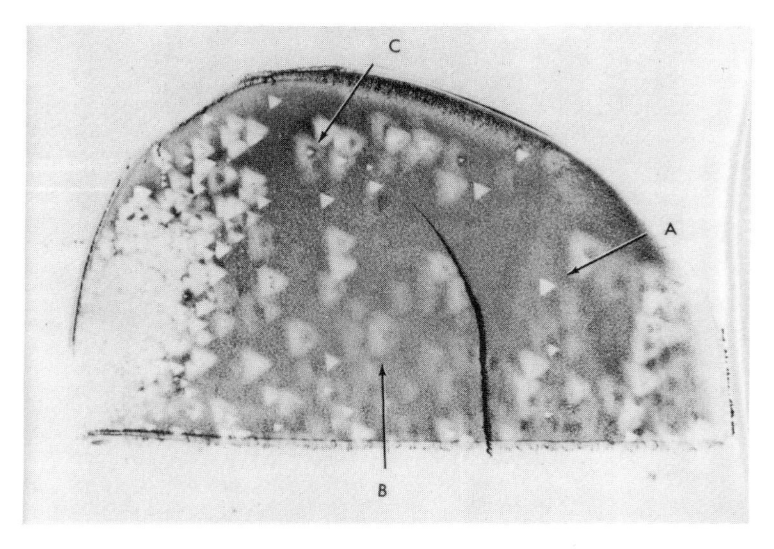

Fig. 8-19. The (440) topograph of the GaAs substrate reveals the hillocks as triangular regions of null contrast. The dark line traversing the image is a spurious reflection. *(From Howard and Dobrott.[39])*

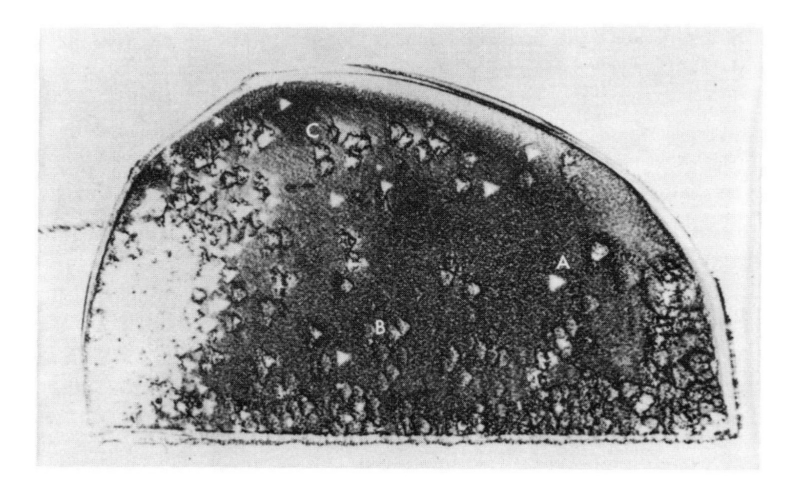

Fig. 8-20. The (440) topograph of the $GaAs_{0.67}P_{0.33}$ layer; the dark contrast results from local cracks in the deposit. *(From Howard and Dobrott.[39])*

and origin of "hillocks" (localized surface protrusions) in these epitaxial layers. By carefully correlating the position of hillocks with (440) topographs of the GaAs substrate and the $GaAs_{0.67}P_{0.33}$ epitaxial layer, they were able to show that the hillocks probably originated at the substrate interface and had a composition of $GaAs_{0.85}P_{0.15}$. This striking series of topographs is shown in Figs. 8-19 to 8-21. Figure 8-22 is a photograph of the epitaxial layer. Subsequent angle lapping and staining of these hillocks confirmed the x-ray topographic analysis of the hillock phenomenon.

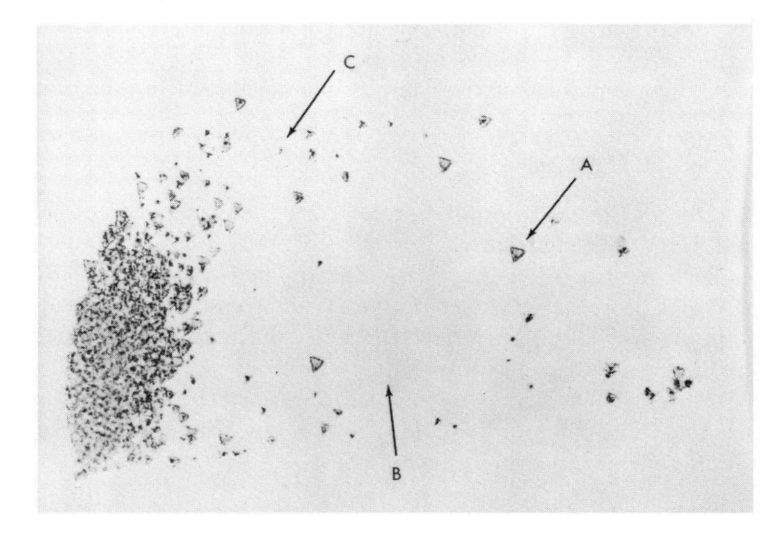

Fig. 8-21. The (440) topograph of the $GaAs_{0.85}P_{0.15}$ layer; only the hillocks are in diffracting position. (*From Howard and Dobrott.*[39])

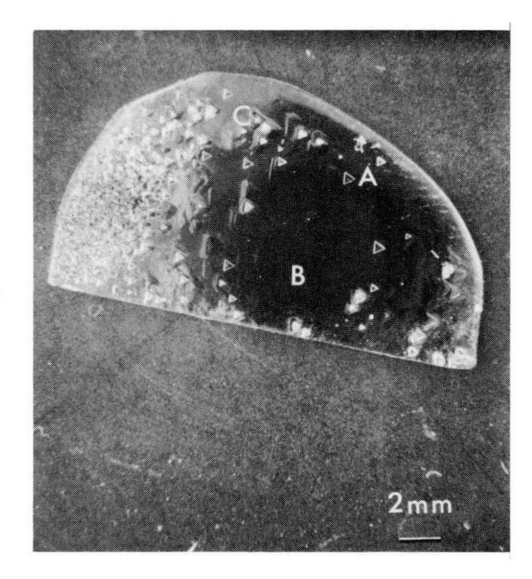

Fig. 8-22. Photograph of the epitaxial film showing numerous hillocks on a (111)-*B*GaAsP surface. (*From Howard and Dobrott.*[39])

The successful application of compositional x-ray topography depends on several factors: (1) the total epitaxial-layer thickness must be less than the depth of penetration of the x-radiation used to examine the film; (2) the separation of diffraction angles for the different compositions and substrate must be large enough not to exceed the angular-resolution capabilities of the equipment; and (3) the absorption coefficient must be constant for each compositional layer, or the topograph from the underlying diffracting layers will have little meaning.

The technique of compositional x-ray topography, when combined with the x-ray diffraction method, is a powerful nondestructive technique for the analysis of mixed III-V epitaxial films. Both planar and axial compositional inhomogeneities can be determined. The composition and origins of epitaxial imperfections such as hillocks can be investigated by this technique. Only compositional x-ray topography can give a picture of the composition of the entire area of a film.

8-15. ELECTRON MICROPROBE

The application of the electron microprobe to the determination of the composition of mixed III-V epitaxial alloy films is technically quite feasible. However, no work has been published on this particular application of the microprobe. As in most analytical procedures, the major problem would be the standardization. The logical approach would be to utilize the x-ray diffraction technique to analyze the epitaxial films and then use those films as standards. Small areas could be analyzed for the three elements, say Ga, In, and As, to examine stoichiometry and homogeneity. The microprobe could not provide the specificity or detail of the compositional x-ray topography technique.

8-16. DOPANT DISTRIBUTION

The incorporation of dopants in epitaxial films usually occurs from three sources: (1) outdiffusion from a heavily doped substrate into the epitaxial layer, (2) deliberate introduction of dopants into the film during growth, and (3) the adventitious incorporation of unwanted impurities into the film during growth. The last-mentioned source of impurities will be discussed along with the relevant analytical procedures in Sec. 8-38, Chemical Imperfections.

The characterization of an epitaxial film for its doping concentration and behavior is usually undertaken by radiochemical techniques and/or electrically. Here, as with bulk-semiconductor evaluation, the various analytical techniques are complementary and for maximum information should be used concurrently. Electrical measurements give only the net carrier concentration and cannot give a direct measurement on a particular dopant. On the other hand, radiochemical techniques give precise information on the dopant under investigation but cannot give any information about the electrical activity of that dopant or, even more important, the resultant net carrier concentration.

Epitaxial films are usually grown on heavily doped substrates (10^{18} to 10^{19} atoms/cm^3). This type of substrate is necessary for subsequent device fabrication. Frequently at the film-growth temperatures the dopant will form a mobile species or dissolve and diffuse into the epitaxial film, as shown[41] in Fig. 8-23. This type of outdiffusion or movement of the dopant from the substrate into the epitaxial film can completely negate the effect of a deliberately added dopant in an epitaxial layer and ruin the film for subsequent use in device fabrication.

8-17. RADIOACTIVE-TRACER STUDIES

The analysis for the distribution of a dopant in an epitaxial layer is analogous to the diffusion-analysis problem (Chap. 9). As with diffused regions, the volume of material available for analysis is small, and generally the concentrations are

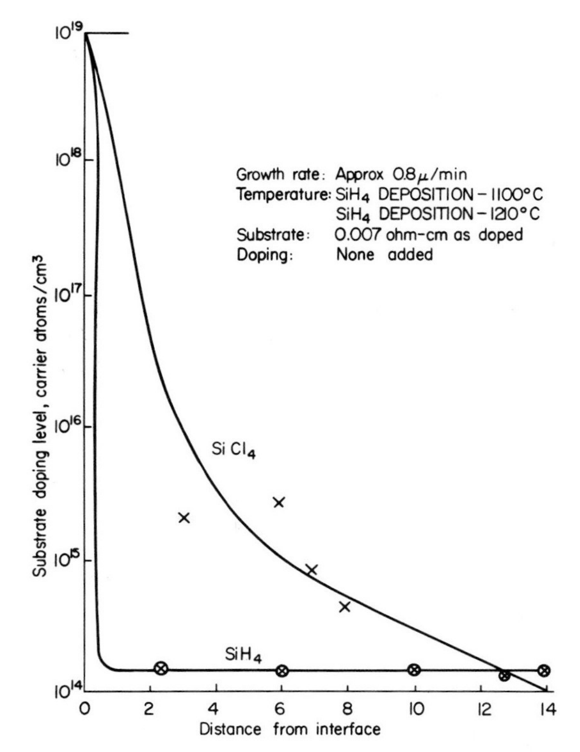

Fig. 8-23. Movement of the dopant from the substrate into the epitaxial film for silane (1100°C) versus silicon tetrachloride (1210°C) systems. *(From Bhola and Mayer.*[41]*)*

small. As a result, radiochemical techniques offer the most sensitive and straightforward approach to the problem. Both activation analysis and radiotracer techniques have been used in this type of work.

In the radiotracer technique, a single radioactive isotope is used to study the behavior of a dopant during the growth of the epitaxial films. The dopant under study can be in the substrate and outdiffusing into the new growth of epitaxial material. Conversely, the dopant can be studied during deposition from the gas phase into the growing epitaxial film. One of the most powerful aspects of this technique is the use of autoradiography to obtain a picture of the distribution of the dopant across the face of the slice.

Nakanuma[42] reported on the use of radiophosphorus-32 as a tracer to study the incorporation of phosphorus in epitaxially grown silicon. The phosphorus concentration in the epitaxial layer (from 10^{14} to 10^{18} atoms/cm^3) was shown to be a function of the phosphorus concentration in the vapor phase. Attempts to increase the dopant concentration markedly above 10^{18}, by increasing the PCl_3 concentration in the gas phase, resulted in a sharp deviation from the transfer function. Nakanuma felt that this may have been due to the effect of repulsive forces between phosphorus in the gas phase and that adsorbed on the growing surface, or the effect of having reached the solubility limit of phosphorus in silicon. No autoradiograms of the phosphorus-32 doped layer were reported.

Baker and Compton[43] used ^{131}I to measure the incorporation of iodine into germanium single-crystal films. The disproportionation of GeI_2 on a germanium substrate in a closed-tube system was used. Following film growth, samples were sectioned by hand lapping and the lappings counted. The results showed 10^{14} to 10^{15} atoms/cm^3 incorporated into the films. The authors felt that this was a moderately low concentration and could not find any correlation between the iodine concentration and the electrical effects. Baker and Compton also studied the incorporation of arsenic[44] using ^{76}As and gold[45] using ^{198}Au into germanium films. A one-to-one correlation between arsenic concentration and Hall coefficient electrical carrier concentration was observed. The gold-tracer work points out the value of this technique because of the difficulties in electrically evaluating gold-doped semiconductors. Maximum solubility data for gold in epitaxial germanium as a function of substrate orientation were also obtained from the tracer data.

Joyce[46] used radiotracers to study impurity redistribution from substrates into the epitaxial layers and into other layers downstream in a multislice flow system. Silicon slices labeled with either ^{32}P, ^{76}As, ^{122}Sb, or ^{72}Ga were placed in an epitaxial reactor with a number of unlabeled slices downstream. Following growth of the epitaxial layer each downstream slice was analyzed for the radioactive species that had been present in the tagged slice. Typical results are shown for arsenic in Fig. 8-24. The effect of sealing the backface of the doped slice and position of the slice relative to the doped slice is shown. It was found that oxide films provided better outdiffusion barriers, with resultant less downstream contamination, for the group V elements than for gallium. More significant was the observation that most of the downstream dopant transfer occurred during hydrogen pretreatment rather than during subsequent epitaxial growth.

Similar predeposition cleanup effects for tellurium-127 tagged gallium arsenide substrates have been observed in Texas Instruments laboratories.[47] In this work

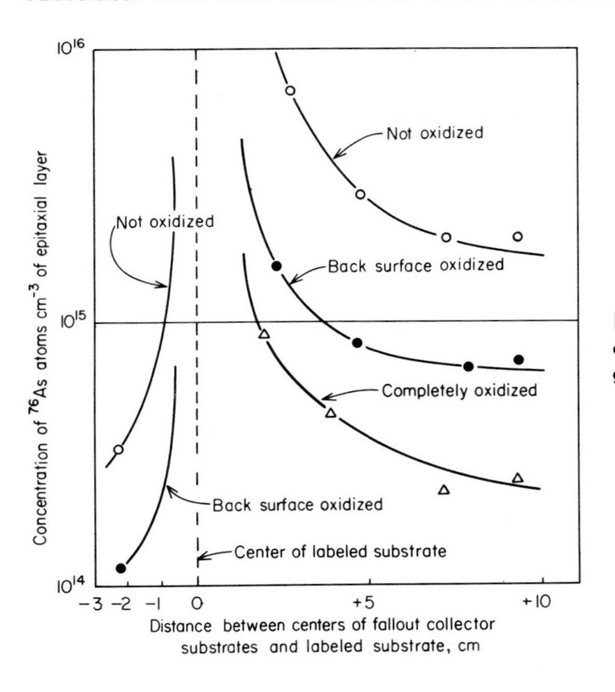

Fig. 8-24. Effects of oxide barriers on impurity transfer during epitaxial growth. (*From Joyce et al.*[46])

Fig. 8-25. Autoradiogram showing buildup of tellurium at the epitaxial-substrate interface.

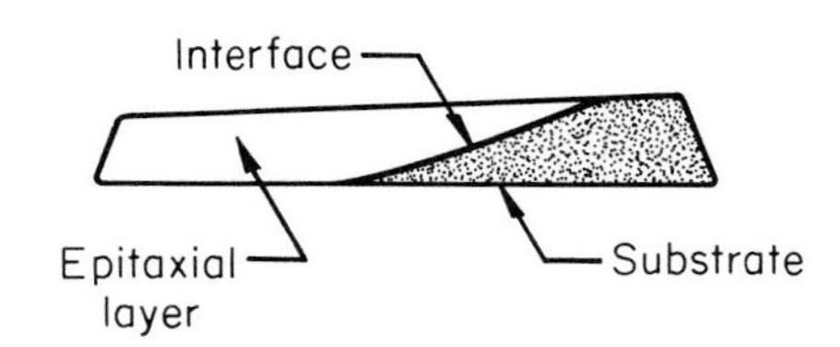

a slice of ^{127m}Te–doped gallium arsenide (1×10^{18} atoms/cm^3) was placed in an epitaxial reactor and cleaned by vapor etching with HCl at 700°C prior to high-purity GaAs epitaxial growth. A 50-μ epitaxial layer was then grown on the surface, and the sample was angle lapped. An autoradiogram of this undoped layer on the doped substrate is shown in Fig. 8-25. As can be seen from the dark band at the substrate-film interface, there was a pronounced increase in tellurium concentration in that area. Modifying the cleanup procedure eliminated this interfacial buildup and resulted in substantially improved microwave devices.

Radiotracer phosphorus-32 has been used in the formation of Ga(AsP) layers by diffusing phosphorus vapors into GaAs at 30 atm and 750 to 1125°C for times up to several days.[48] This procedure did not yield a single composition $GaP_xAs_{(1-x)}$, but rather a graded layer, and is, strictly speaking, not an epitaxial film. However, the technique is illustrative of the power of the radiotracer technique.

8-18. ACTIVATION ANALYSIS

Neutron activation analysis of impurities in epitaxial layers is readily applicable to silicon. The other semiconductors are very difficult to analyze by this technique because of the large amounts of activities produced by the matrix. However, while difficult, the problems are not insurmountable, and with ingenuity and perseverance the radiochemist can carry out such an analysis.

Abe and Sato[49] used neutron activation analysis to determine the distribution of arsenic atoms in epitaxial films deposited on heavily doped arsenic substrates. While outdiffusion into the film was observed, the profile of the distribution was found to be complex, and the tail region in the epitaxial film could not be fitted to the overall error-function distribution.

8-19. ELECTRICAL CHARACTERIZATION

The electrical characterization of a thin epitaxial layer is not as straightforward as one might imagine. As mentioned earlier, these films are generally deposited on very-low-resistivity substrates to aid in subsequent device fabrication. As a result, care must be taken in any electrical measurement to ensure that the electrical parameters being measured are those of the layer and are not being influenced by the substrate. This effect is illustrated in Fig. 8-26, where the current between two point contacts on an n-type epitaxial film on a heavily doped n^+ substrate flows through the substrate. In effect, the substrate electrically shorts out the epitaxial layer. This effect can of course be overcome by depositing the epitaxial film on a control slice of high resistivity or of opposite type. However, the assumption must be made that the control epitaxial layer grown on a different type of substrate will have the same characteristics. There is a sampling problem, and 100 percent inspection is not possible.

8-20. FOUR-POINT PROBE

The four-point-probe technique used for epitaxial-film resistivity measurements is the same as that described in Sec. 4-16 for bulk materials. The epitaxial layer must be electrically isolated from the substrate and exist as either an n-type film on a p-type substrate (n/p) or vice versa (p/n). Dobbs and Kovacs[50] described a four-point-probe procedure that can be used routinely in epitaxial production. The procedure involves placing the slice (1 in. diameter in this work) in a dark box and measuring the voltage V between two probes while applying a constant current I between two other probes. The volume resistivity ρ is then calculated,

$$\rho = 2\frac{V}{I}\frac{t}{86.9} \tag{8-9}$$

with the epitaxial-layer thickness t in mils. Dobbs and Kovacs devised a computing circuit to allow the input current to be set equal to $t/86.9$, leaving $\rho = V$.

Schumann and Hallenback[51] developed a modified four-point-probe technique whereby two probes were placed on top of the epitaxial layer and the other two

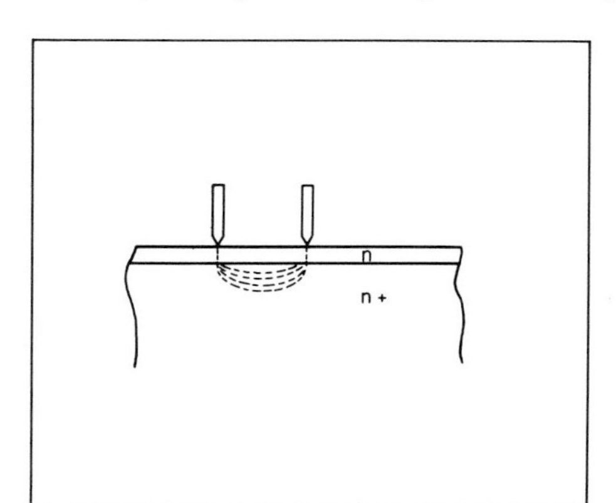

Fig. 8-26. Schematic showing how the heavily doped substrate will "short out" the point contacts on the epitaxial film.

on the bottom of the substrate slice (in contrast to the normal four probes in line configuration). This "over-under" configuration is then applicable to n/n^+ and p/p^+ epitaxial layers. The technique, as presented, was in its initial development stages, and the authors preferred the two-point-probe technique.

One advantage of the four-point-probe technique is its wide range of resistivity, from 0.001 to 1,000 ohm-cm before significant errors occur.

8-21. THREE-POINT PROBE

The three-point-probe technique for measuring the resistivity of epitaxial films has received far more attention than other electrical evaluation techniques. The advantage of this method is its direct application to n/n^+ and p/p^+ films that are used in device fabrication. Since the method depends on measuring the breakdown voltage through the film to the substrate, it cannot be used on n/p or p/n films. The technique is simple, nondestructive, and requires little equipment which makes it applicable to production testing.

Basically, the three-point-probe breakdown technique is a potentiometric measurement of the reverse breakdown voltage of a metal-to-semiconductor point-contact diode. The methodology[50,52] and theory[53,54] of three-point-probe evaluation of epitaxial films have been carefully studied. Figure 8-27 shows a typical experimental arrangement for an n on n^+ silicon sample. A 3.5-mil tungsten probe serves to form the metal-semiconductor point-contact diode. The other two probes are sharpened dumet wires, of which one carries current to complete the circuit and the second serves as the potentiometric probe. This high-impedance probe measures the potential drop across the depletion layer. Since the depletion region is of the order of microns, the proximity of the potential probe is of no consequence in present probe designs. As the potential on the reverse-biased contact is increased, the depletion region associated with that probe extends deeper into the epitaxial film. At some voltage, depending on the epitaxial-film resistivity, breakdown will occur. The high-impedance potential probe is generally monitored on an oscillo-

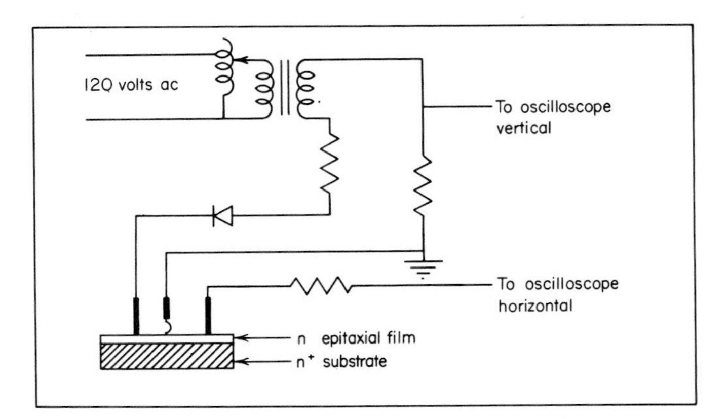

Fig. 8-27. Circuit diagram of the experimental arrangement for three-point-probe measurements on epitaxial films. (*After Dobbs and Kovacs.*[50])

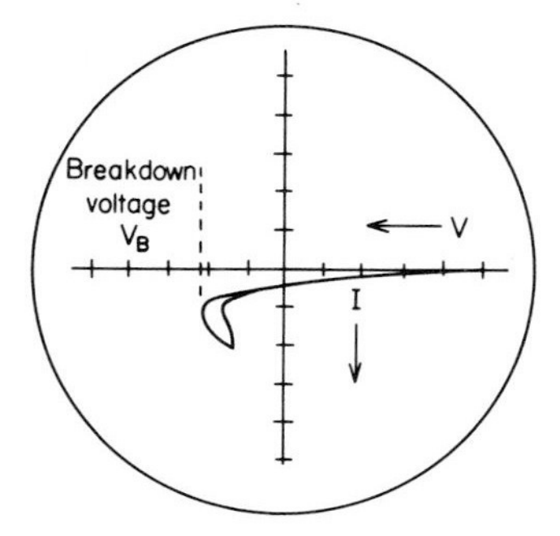

Fig. 8-28. Typical oscilloscope trace of voltage breakdown through an epitaxial film using a three-point probe. (*After Dobbs and Kovacs.*[50])

scope but can be fed to an X-Y recorder. In either case, the I-V trace will increase until breakdown and then sharply "snap back." A typical I-V trace is given in Fig. 8-28, showing the measured breakdown voltage. If the thickness of the epitaxial film is less than the depth of penetration of the depletion region, then the breakdown voltage is a function of both resistivity and thickness. A curve showing this relationship is given in Fig. 8-29. Since probe material, radius, pressure, and spacing are critical and difficult to control, the procedure must be empirically calibrated. The measurement range of the technique is generally limited to resistivities from 0.1 to 1.0 ohm-cm. Poorly defined breakdown voltages limit the lower end; and, while well-defined breakdown occurs at higher resistivities, it is usually controlled by factors other than resistivity.

Gardner and Schumann[54] applied a correction factor, to the three-point-probe measurements and got good agreement with the differential-capacitance

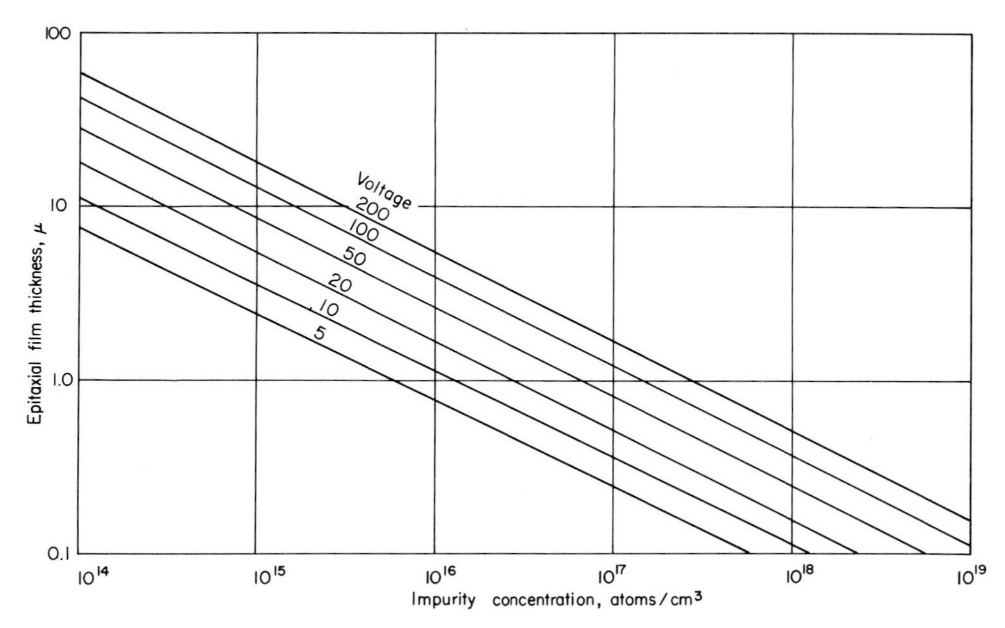

Fig. 8-29. Typical calibration curves for three-point-probe measurements. Note the effect of layer thickness on breakdown voltage at a given doping level.

technique (Sec. 8-24). They report a variation of ±5 percent over a 2-week period for measurements on a sample of n-type silicon.

8-22. TWO-POINT PROBE

The spreading-resistance probe, or two-point probe, is a modification of the four-point-probe technique and is usable where the four-point probe fails. Schumann and Hallenback[51] and Gardner et al.[54] developed and applied this technique to silicon and germanium epitaxial layers.

In this technique, two probes are placed on the n/n^+ or p/p^+ epitaxial surface as shown in Fig. 8-26, a small potential (~30 mv) is applied, and the resistance is measured. As long as the probe separation is greater than the layer thickness, the current will be restricted to the region below the contact on the layer and then spread widely in the substrate. This technique is sometimes termed *spreading-resistance probe* and obviously can be used only on the conventional n/n^+ or p/p^+ materials. The measured resistance is proportional to the resistivity of the epitaxial film. In order to calculate the film resistivity it is necessary to know the thickness of the epitaxial film accurately. This thickness is readily measured by any of the techniques described earlier (Secs. 8-6 to 8-10). Gardner[52] determined the precision of this technique to be ±15 percent on silicon, and Schumann[51] determined ±10 percent on germanium.

Schumann[51] made a comparison of the results of three-point- and two-point-probe techniques for n/n^+ silicon. A control slice of p-type material was processed along with the other material, and the resistivity of the n-type layer measured by using (presumably) the four-point probe. These results are shown in Table 8-2, and good agreement was obtained for the two-point and three-point-probe measurements. Poor agreement was obtained between the p-type control and both of the other two methods.

8-23. PULSED POINT CONTACT

Allen et al.[55] developed a pulsed-current technique which is really a semiautomatic, more rapid and more accurate three-point-probe technique. The experimental apparatus is shown in Fig. 8-30. A series of 500-μsec progressively increasing pulses drives the output of a constant-current generator, while a high-impedance voltage probe measures the potential drop across the depletion layer. In the conventional three-point-probe system, an oscilloscope I-V detector system is used to measure the breakdown voltage. In this system, a diode detector is used to charge a capacitor to the peak voltage, and the breakdown point is determined with a slope detector. This detector turns off the ramp generator when the rate of change of voltage, at the voltage probe, becomes zero or negative. The breakdown voltage is then read on a peak-reading voltmeter. The technique is empirical, and calibration curves must be determined experimentally.

In an evaluation of this pulsed-point-contact technique, a comparison of results on n/n^+ with the voltage-capacitance (V-C) diode technique (Sec. 8-24) showed good agreement (70 percent of values within ±20 percent with the V-C technique).

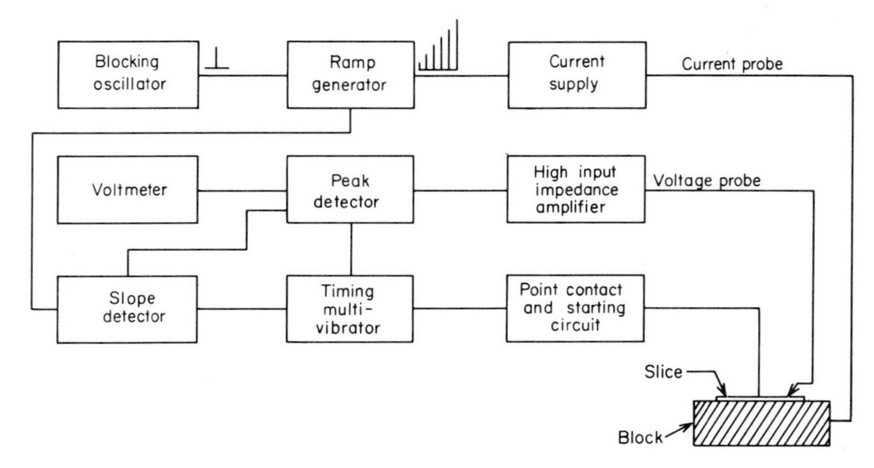

Fig. 8-30. Block diagram of semiautomatic measuring circuit for three-point-probe measurements. (*From Allen et al.*[55])

Table 8-2. Experimental Results Obtained on n/n⁺ Epitaxial Wafers with Two-point Probe Compared with Three-point Breakdown and p-type Control[51]

Layer thickness, μ	Layer resistivity, ohm-cm		
	p-type control	Three-point breakdown	Two-point probe
6.09	0.75	0.47	0.48
7.36	0.85	0.64	0.44
7.87	0.95	0.68	0.86
8.38	0.47	0.41	0.41
12.5	0.23	0.18	0.18
9.40	0.34	0.29	0.23
7.36	0.020	0.022	0.0093
9.90	6.0–8.2	....	3.3
8.88	0.39	0.35	0.36
10.2	0.10	0.09	0.11
9.90	0.10	0.09	0.15
10.7	0.12	....	0.12
10.7	4	....	8.8
10.9	0.75	0.53	0.52

On the other hand, resistivities measured on control slices made on the same run but deposited on p substrates did not agree with either technique (only 10 percent values within ±20 percent).

8-24. DIFFERENTIAL CAPACITANCE

The electrical-evaluation techniques described above give only the average resistivity of the epitaxial layer. The techniques cannot, directly, give information on the uniformity of the resistivity through the layer. Since there are several mechanisms whereby nonuniformity of dopant through the layer can occur, it is frequently necessary to characterize or profile the layer resistivity.

Thomas et al.[56] first described a differential-capacitance method for determining epitaxial dopant concentration profiles. The technique is sometimes referred to as the *capacitance-voltage method.* A diode is alloyed into the epitaxial layer, and the p-n junction has a built-in voltage; the carriers have diffused away from either side of the junction and formed a depletion region. Since the depletion region contains no free carriers, it behaves like an insulator with dielectric constant K. The two sides of the boundary layer are like parallel plates of a capacitor of area A, separation d, capacitance C, where

$$C = \frac{KA}{d} \tag{8-10}$$

This capacitance in a p-n junction is a function of the applied voltage (actually applied voltage + internal voltage), where

$$C = A\sqrt{\frac{qK\epsilon_0 N_D}{2(V_a - V_i)}} \tag{8-11}$$

where C = measured capacitance
A = junction area
q = electronic charge
K = dielectric constant
ϵ_0 = permittivity of free space
N_D = donor concentration in n-type epitaxial layer
V_i = built-in voltage (0.5 for silicon)
V_a = applied voltage

Substituting the required constants for silicon results in the equation

$$\frac{C}{A} = 2.91 \times 10^{-4}\left[\frac{N_D}{0.5 - V_a}\right]^{1/2} \quad \text{pf-cm}^{-2}$$

A plot of C/A versus $V_a + V_i$ results in a straight line on log-log paper, and any deviation from this straight-line relationship, with slope of $-\frac{1}{2}$, must be regarded as nonuniform dopant distribution. The experimental apparatus used by Kovacs and Epstein[57] is shown in Fig. 8-31. These authors give a detailed procedure for the application of this differential-capacitance technique to silicon. The technique has also been used for GaAs.[58,59]

Amron[60] developed a slide rule to aid in the calculation of the dopant concentration profiles of epitaxial films from the capacitance-voltage measurements. Amron[61] also carried out a detailed error analysis of this method of characterizing epitaxial films and showed that voltage errors as large as 10 percent were not significant but that errors in capacitance and diode diameter measurements could produce errors as high as several hundred percent.

8-25. MICROWAVE MEASUREMENTS

The application of microwave techniques to the determination of epitaxial-layer impurity profiles is highly specialized. The same information is probably more

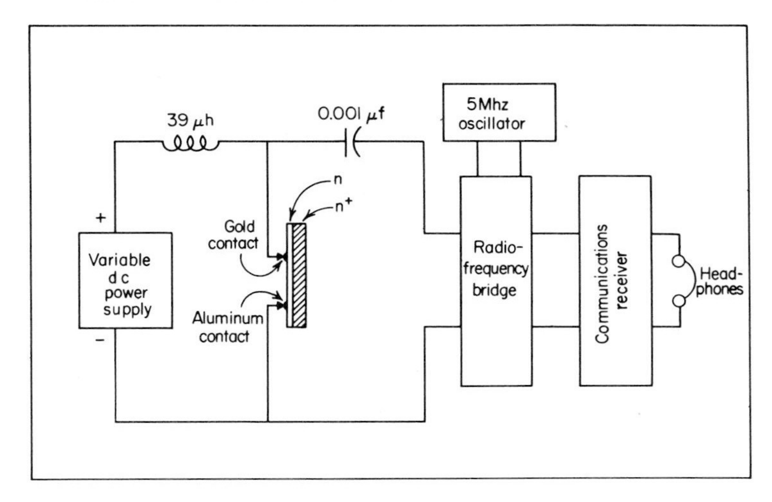

Fig. 8-31. Schematic diagram of the experimental arrangement used for measuring junction capacitance. (*After Kovacs and Epstein.*[57])

readily available by other techniques. Microwave diode measurements have been made on reverse-biased $p^+/n/n^+$ material[62] and a resistivity profile determined. Lindmayer and Kutsko[63] made microwave measurements at 25 and 100 Ghz but did not feel the technique would be applicable to epitaxial layers unless they were thicker than 50 μ and on low-resistivity substrates.

8-26. HALL AND VAN DER PAUW MEASUREMENTS

It is frequently necessary to determine the mobility, resistivity, and carrier concentration of an epitaxial film. These parameters are determined by standard ac or dc Hall coefficient measurements on shaped samples and Van der Pauw measurements on unshaped samples. The theoretical considerations and calculations are the same as those described in Sec. 4-17. However, the experimental arrangement for these electrical measurements on thin epitaxial films is different from that used on bulk samples.

For Hall and Van der Pauw measurements on epitaxial films it is necessary, as it was with the four-point probe, to deposit the epitaxial layer on a control substrate of either high resistivity or opposite type. Patrick[64] has described a procedure for measuring resistivity and mobility of silicon epitaxial layers on a control wafer of opposite type. In Patrick's procedure, a Hall spider is formed by vacuum evaporation of aluminum through a mask onto the surface of the epitaxial film. The sample is etched to remove the epitaxial silicon not covered by the aluminum, and this etching is continued until only the epitaxial film under the aluminum spider mask is left on the substrate. Concentrated HCl is used to remove the aluminum, and ohmic contact is made to the legs. The conventional Hall measurements are then performed on the sample. In the Texas Instruments laboratories, the preferred procedure is to cavitron a Hall bar from the epitaxial film, as shown in Fig. 8-32. Hall and Van der Pauw measurements performed on control slices suffer from

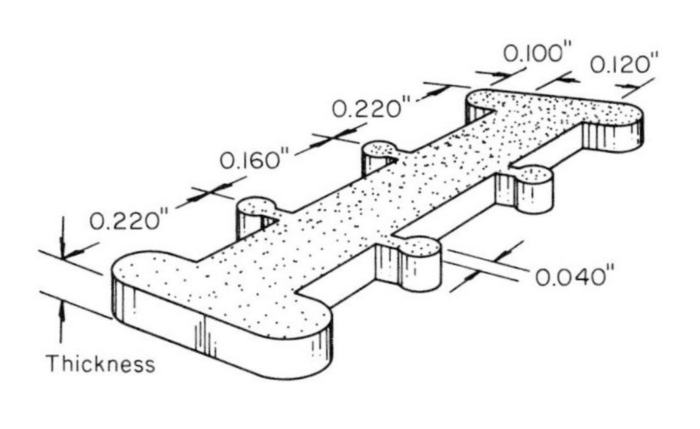

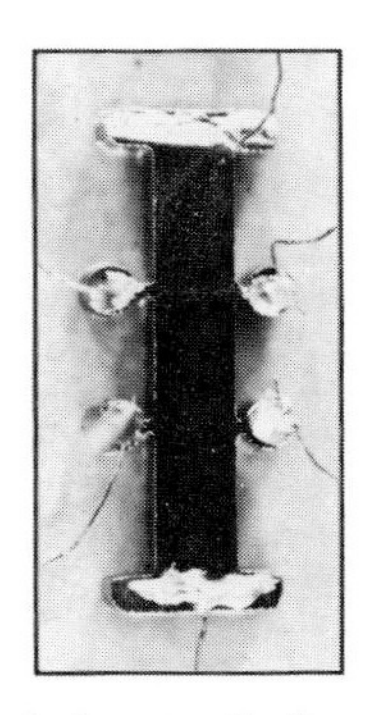

Fig. 8-32. Hall bar as cavitroned from epitaxial slice and a photograph showing the leads attached for Hall measurements.

the same criticisms directed at four-point-probe measurements. Patrick[64] has some very sound suggestions on how to maximize accuracy and reproducibility. Frank[65] demonstrated the influence of surface damage on Van der Pauw measurements on epitaxial germanium films.

8-27. INFRARED REFLECTIVITY

Rawlins[66] reported an interesting application of infrared reflectivity as a measure of the resistivity of epitaxial silicon layers. This technique makes use of the fact that the complex refractive index N of a material has two components:

$$N = n - ik \tag{8-12}$$

where n = refractive index
k = absorption index

For silicon epitaxial layers (0.1 to 100 ohm-cm), n will be constant and k extremely small as measured by conventional reflectivity techniques. By using attenuated total reflection techniques, it is possible to magnify k and to observe a variation of reflectivity with carrier concentration.

Experimentally, it is difficult to make the measurements, and the differences observed by Rawlins were not great. However, there exists a real need for this technique since it would be possible to measure the resistivity and thickness of an epitaxial film simultaneously. Considerably more work is required in this area to make the technique applicable to routine measurements.

8-28. PHYSICAL IMPERFECTIONS

The growth of a single-crystal, oriented epitaxial film on a substrate inevitably results in a film which has crystallographic imperfections. These imperfections are often visible to the unaided eye or can be seen microscopically. Other imperfections or defects must be delineated with etches prior to visual examination. Then there are those defects that can be detected only with the aid of such techniques as x-ray topography and electron microscopy.

The causes of these imperfections are many and varied. It is beyond the scope of this book to discuss in detail the preparation of epitaxial films. However, many of the causes of certain types of defects will be discussed with the techniques used to detect them. The most obvious contributing factor, and probably the most critical one, is the substrate surface which is the genesis of the epitaxial film, and this is the reason that Chap. 7 is devoted entirely to the characterization of semiconductor surfaces. One of the most powerful techniques for the examination of these substrate surfaces is x-ray topography, and this same technique is equally effective in the study of epitaxial films.

The crystallographic orientation of the substrate plays a significant role in the ultimate perfection of the epitaxial layer. There are two reasons for the effects of misorientation, and one is dependent on the other. Williams[67] demonstrated the effect of orientation on GaAs growth rate and electrical properties. Table 8-3 shows

Table 8-3. Effect of Growth Orientation on Growth Rate for Epitaxial GaAs[67]

Orientation	*Growth Rate (relative)*
<100>	1.0
<111>*A*	0.9
<110>	0.14
<111>*B*	0.10

this effect to be large, and differences in growth rate as high as a factor of 10 can be seen. Reisman and Berkenblit[68] used the {111} surfaces on germanium to study the effect of slight misorientation. There is threefold symmetry involved, and a misoriented (111) surface may be toward the (110) or (211) planes. These workers tilted the (111) from 3 to 5° off, toward either the (110) or the (211) planes, and observed pronounced differences in the quality of the epitaxial film which were directly attributable to the misoriented substrate.

One of the defects observed was what Lenie[69] called the "edge-ledge" defect, shown schematically in Fig. 8-33. This type of defect develops at the edge of an epitaxial film and is a crystallographic facet growing on the edge of the wafer. It appears on slightly misoriented substrates where the differences in growth rate of

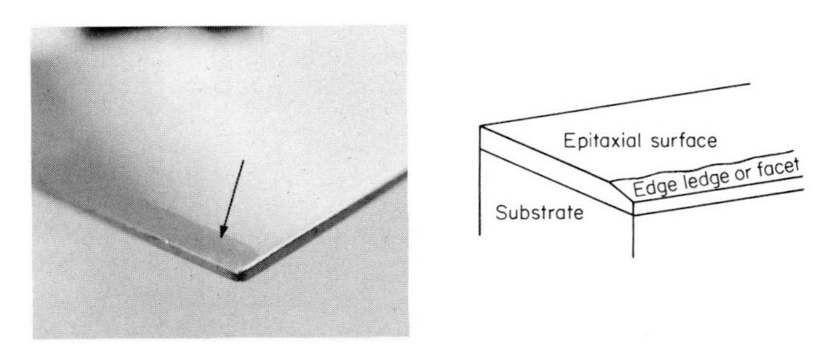

Fig. 8-33. Photograph and schematic of an edge-ledge or facet on the edge of an epitaxial gallium arsenide film.

various crystallographic planes can be magnified and result in growth on other than the desired plane.

The second effect which can appear as a result of this simultaneous nonuniform growth on several crystallographic planes is the anisotropic segregation of impurities in these areas. Anisotropic segregation in melt-grown bulk semiconductor crystals was discussed in Sec. 4-13, and Williams[67] demonstrated this same

Table 8-4. Ratio of Carrier Concentrations on Different Orientations as a Function of Dopant[67]

Dopant	n†		
	<111>B / <100>	<111>B / <111>A	<111>B / <110>
None	11.0 (6)	15 (1)	28 (1)
Zn	0.43 (7)	0.20 (1)	0.49 (2)
Te	7.4 (8)	~20 (3)	~15 (1)
Se	6.3 (3)	....	4.4 (1)
Sn	2.3 (2)		
S	1.4 (2)		

†Numbers in parentheses refer to the number of experimental runs.

crystallographic orientation dependence on final epitaxial-film carrier concentration for gallium arsenide. Table 8-4 shows Williams' results on a study of this orientation effect. Carrier concentration differences as large as 28 times were observed to result from different growth orientations. Mendelson[71] made a detailed study on the effect of orientation on the ultimate perfection of silicon epitaxial films.

8-29. VISUAL CHARACTERIZATION OF SURFACE DEFECTS

Many of the crystallographic defects which occur in epitaxial films can be characterized by visual examination using microscopy and etching techniques. Lenie[69] and Hallas et al.[70] have studied these defects and their causes. The nomenclature for some types of defects is peculiar to each laboratory, while other types such as stacking faults and pyramids are common to all workers in the field.

Lenie[69] recommended a visual inspection of the surface of the epitaxial film, against a dark background, under a 30-watt fluorescent light. This examination will show macro defects such as scratches, pits, orange peel, and pyramids. A more detailed examination should be carried out with an interference microscope. Under the interference microscope any projections from the surface will cause interference fringes. By using this technique it is possible to measure the size of the defect by counting the number of fringes. Surface defects such as spikes, pits, dimples, and surface scratches are readily measured by this technique.

8-30. STACKING FAULTS

The use of stacking faults as a measure of the thickness of epitaxial films was discussed in Sec. 8-8. These stacking faults also serve as a measure of the overall perfection of the film. It has been observed experimentally that the majority of these stacking faults originate at the substrate interface. Further, the number of defects depend on the physical perfection of the substrate surface and the amount of contamination of the surface. Since the epitaxial film nucleates at the surface, this behavior would be expected. When several slightly misoriented nucleation centers grow together, a stacking fault is produced and they propagate to the surface. They continue to grow in size as a direct function of film thickness. Newman,[8] in an excellent review of silicon and germanium epitaxy, reviewed stacking faults and their probable causes. Both oxygen and carbon contamination appear to be very suspect. Haneta[72] reported that contaminants such as water and oxygen in the carrier gas can reduce the number of stacking faults when introduced at the 10- to 1,000-ppm level. This was attributed to the etching characteristics of these two impurities under the conditions used to grow silicon epitaxial films. Nitrogen in this same concentration range was observed to increase the number of stacking faults. This was believed to be due to the formation of silicon nitrides in the films, which acted as nucleation centers for stacking faults.

These stacking faults can be observed under a phase-contrast microscope, but it is difficult to determine the density of faults by counting under these conditions. As a result, the preferred technique is the procedure recommended by Lenie,[69] where the wafer is etched with a chromic-hydrofluoric acid etch to delineate the imperfections. Some other etches sometimes used by other workers are shown in Table 8-5. These etch procedures tend to be destructive and therefore can be applied only to representative films from a larger production lot. The interpretation of the number, size, and shape of these stacking faults is best obtained by referring to the work of Booker,[73] Mendelson,[71] Newman,[8] and Batsford and Thomas.[74]

Table 8-5. Etches Used to Delineate Stacking Faults in Epitaxial Films

Material	Reference	Etch	Composition
Silicon	75, 69	Sirtl	50 g CrO_3 in 100 ml H_2O; add 75 ml HF. Etch 15–30 sec.
Silicon	76	Sailer	300 ml HNO_3, 600 ml HF, 2 ml Br_2, 24 g $Cu(NO_3)_2$. Use diluted 10:1 with H_2O. Etch 4 hr.
Indium arsenide	77	HNO_3–HF	3:1 HNO_3–HF (vol/vol)
Germanium		WAg	25 ml H_2O, 10 ml HF, 15 ml HNO_3, 0.2 g $AgNO_3$. Etch 15 sec in 5 sec increments.
Gallium arsenide	78	{100} GaAs	2 ml H_2O, 8 mg $AgNO_3$, 1 g CrO_3, and 1 ml HF, 65°C, 10 min
Gallium arsenide	78	As{111} faces	2 ml H_2O, 8 mg $AgNO_3$, and 1 ml HF, 65°C, 10 min

8-31. TRIPYRAMID DEFECTS

During the microscopic examination of an epitaxial film, a structure will appear that can best be described as a tripyramid or growth hillock. A photograph of such a tripyramid is shown in Fig. 8-34. These crystallographic imperfections are believed to be caused by the introduction of a few atomic layers of silicon carbide at the substrate interface.[79,80] Mendelson[81] has shown that while silicon carbide or other adsorbed impurities can initiate the growth of the tripyramid imperfections, these impurities are not unique or necessary, and he suggests a reentrant twin mechanism, shown in Fig. 8-35. The three-dimensional tripyramid-defect crystallographic model of Miller et al.[79] is shown in Fig. 8-36. The model proposed by Inoue[80] is different from that of Miller et al. However, both workers agree that the

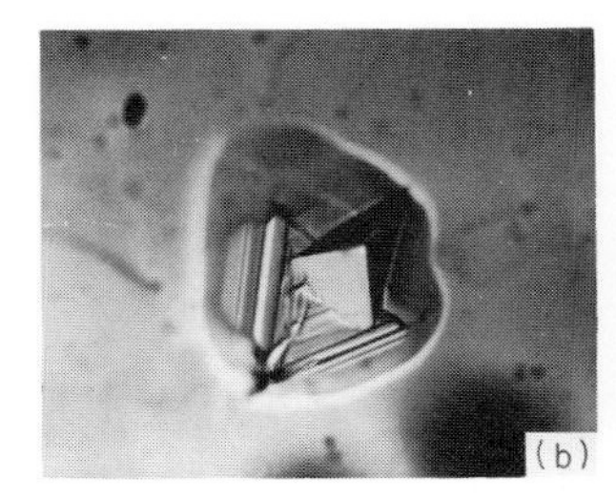

Fig. 8-34. Microphotographs of tripyramid growth on silicon (111). (*a*) **A typical tripyramid;** (*b*) **a monopyramid;** (*c*) **modified tripyramids.** (*Photographs courtesy of Inoue.*[80])

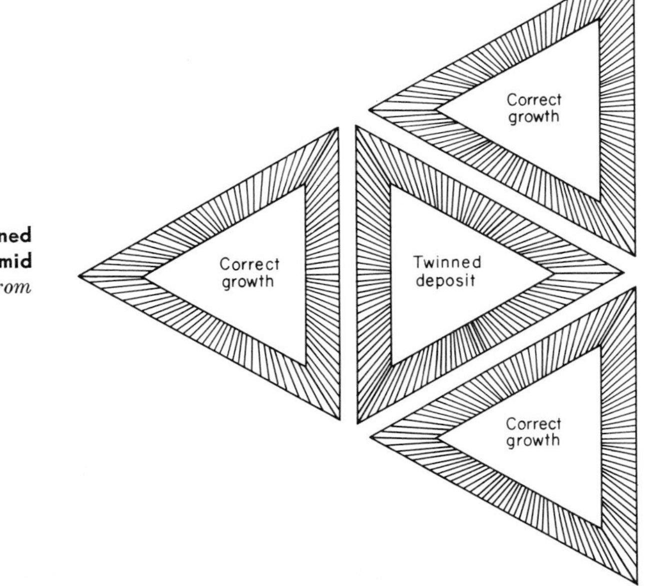

Fig. 8-35. Morphology of twinned nucleus for microtwin and tripyramid formation in epitaxial films. (*From Mendelson.*[81])

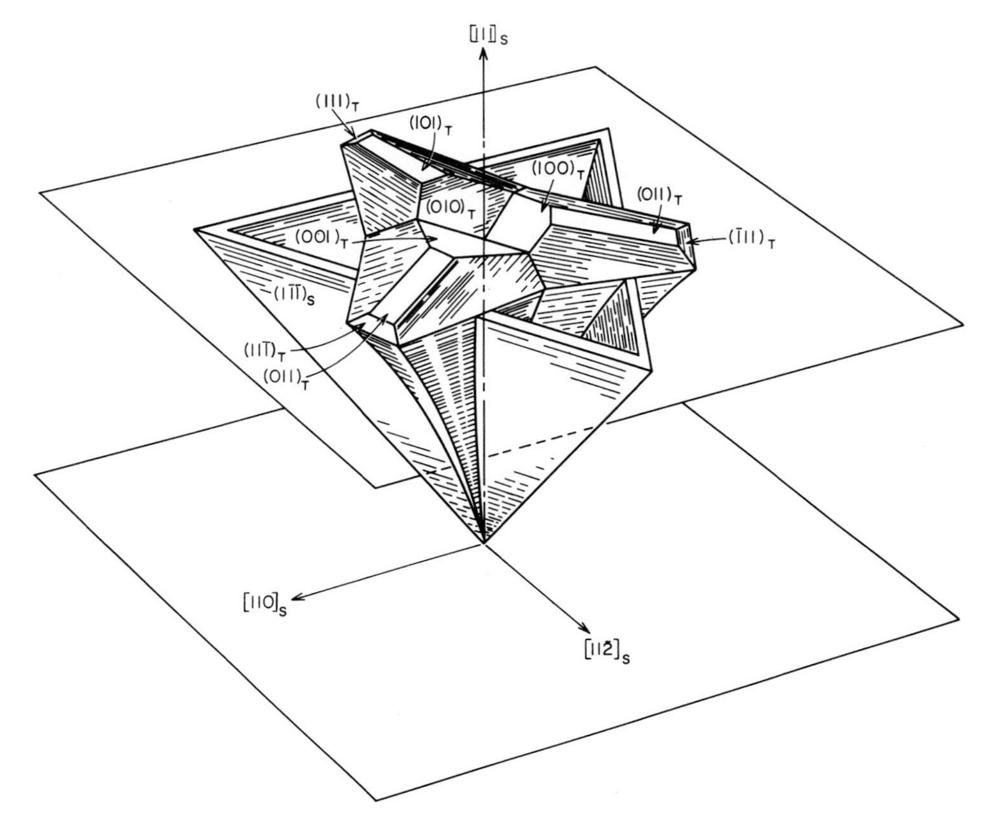

Fig. 8-36. Three-dimensional schematic showing the geometry of a tripyramid. (*From Miller et al.*[79])

cause of the defect is the presence of silicon carbide at the substrate-epitaxial interface.

While these tripyramids protrude from the surface of the epitaxial layer, Inoue[80] found that Sailer etch (Table 8-5) and iodine etch were very useful in delineating the defects.

8-32. ORANGE PEEL

Figure 8-37 shows schematically an "orange peel" surface on an epitaxial layer. This type of surface is a direct result of an "orange peel" substrate, which in turn is caused by improper mechanical and/or chemical polishing techniques prior to epitaxial-film growth.

8-33. MISCELLANEOUS PHYSICAL DEFECTS

Lenie[69] described a variety of other physical defects which are sometimes observed on or in epitaxial films. It is usually possible to determine, by microscopic examina-

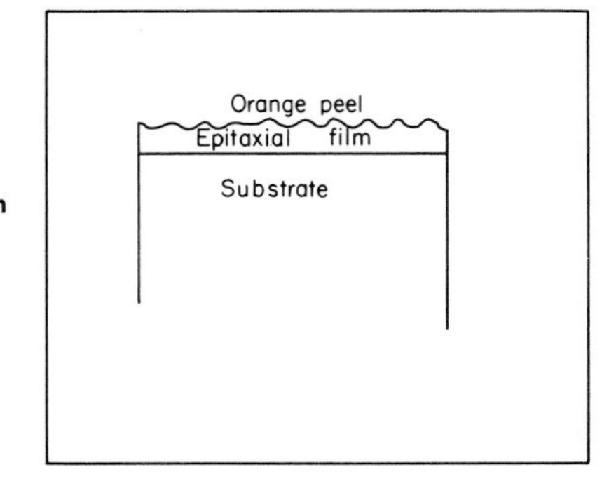

Fig. 8-37. Schematic showing "orange peel" surface on an epitaxial film.

tion, whether the defect was propagated up through the epitaxial film or resulted from mishandling of the slice after growth. Defects such as scratches, voids, spikes, crowns, and dimples are all readily distinguished on visual inspection of the epitaxial layer.

8-34. ELECTRON MICROSCOPY

Marcus[82] has reviewed electron-microscope and electron-diffraction techniques as applied to the characterization of thin films. By combining both these electron-beam techniques, it is possible to obtain information on surface morphology, crystal structure, and the defect structure of epitaxial films. The electron microscope enables these characterizations to be performed with resolutions from 5 to 20 Å. While scanning electron microscopy can provide pictures with excellent depth of field for surface-texture studies, it is capable at best of only 100 Å resolution. Similarly, x-ray topography yields good macro surveys of defects in epitaxial films, but it is not capable of providing information on a micro scale.

8-35. SURFACE MORPHOLOGY

The morphology or texture of grown epitaxial films is frequently controlled by the substrate surface. The same replicating techniques described in Sec. 7-13 are directly applicable to the examination of thin-film surfaces by electron microscopy. Carbon replicas of the surface are usually "shadowed" by evaporating a very thin layer of a heavy metal such as palladium or platinum at an angle onto the replica.

Reflection electron diffraction can also give valuable information on the surface morphology. The details of this technique are given in Sec. 7-20. If the surface of the epitaxial film is smooth and featureless, the diffraction pattern will show only strong Kikuchi lines (Fig. 8-38). If there are protrusions from the surface, a spotted pattern will result. These diffraction spots occur because the electron beam interacts with the projections instead of simply reflecting from a perfect surface. Figure 8-39 shows a typical spotted diffraction pattern from a semiconductor surface with surface projections. Charig et al.[83] used these techniques to study the initial mode of growth of silicon epitaxial films. Both techniques give only qualitative information on the crystallite size of the epitaxial film, and to obtain more complete information, it is necessary to use transmission electron microscopy.

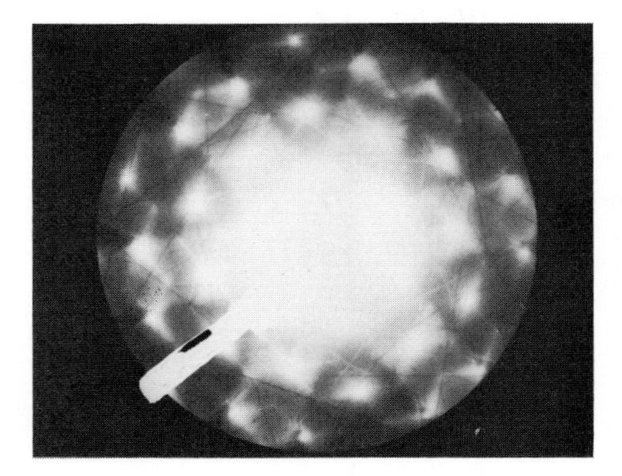

Fig. 8-38. Diffraction pattern obtained by electron reflection from a silicon epitaxial layer; showing Kikuchi lines which indicate high perfection.

Fig. 8-39. Reflected electron-diffraction spots from an epitaxial film, showing that other crystalline imperfections are present.

8-36. CRYSTAL STRUCTURE AND DEFECTS

Transmission electron microscopy is particularly well suited to the study of defects and microstructure of single-crystal epitaxial films. The films must be very thin (less than 1 μ) to allow the electron beam to be transmitted through the sample, and the removal of the substrate and ultimate thinning of the epitaxial film is an art. Finch and Queisser[84] prepared the thin foils of epitaxial silicon by chemically etching away the substrate. The silicon slice was wax mounted, with the epitaxial face down, onto a Teflon disk. The disk was then placed in a Teflon beaker tilted at 45° and rotated at 30 rpm. The rolling disk thus assured uniform etching in the 95:5 HNO_3–HF (vol/vol) etch solution. Etching was continued until the substrate was dissolved away and at least one hole was etched through the epitaxial layer. Transmission electron microscopy was carried out through the thin edges of these holes. Abrahams and Buiocchi[85] used similar techniques to study twins and stacking faults in epitaxial gallium arsenide. A transmission micrograph of a germanium epitaxial film deposited on calcium fluoride is shown in Fig. 8-40.

Transmission electron microscopy allows the observation of defects that cannot be resolved by any other technique. This aspect, coupled with surface replica studies and electron-diffraction work[86], makes electron microscopy one of the more valuable analytical tools in epitaxial-film-growth research.

Fig. 8-40. Transmission electron micrograph of a germanium epitaxial film which had been deposited on a calcium fluoride substrate.

8-37. DEFECTS BY X-RAY DIFFRACTION

The diffraction of x-rays by defects in single crystals (see Chap. 6) is frequently applied to the analysis of epitaxial layers for physical imperfections. Schwuttke,[87,88] using the technique of Lang,[89] applied x-ray diffraction microscopy to the study of silicon epitaxial films. The substrate was etched down as closely as possible to the interface without actually penetrating it. The thinned specimen was then examined by x-ray diffraction microscopy, as shown in Fig. 8-41. This method is based on the extinction contrast technique, where a defect-free specimen would yield a film of uniform contrast. A sample with defects which diffract the x-rays would of course result in a film showing the defects.

While this technique works extremely well for single-crystal materials, it is somewhat difficult to use for epitaxial films. The difficulty lies in the need to remove the substrate to obtain good x-ray transmission. While this can be done on small areas of a slice, it is virtually impossible to separate the epitaxial film from the substrate over the entire slice. The scanning-reflection x-ray topography method developed by Howard and Dobrott[39,40] is the best x-ray technique for analyzing epitaxial layers for physical or crystallographic defects. This technique was described in Sec. 8-14, where it was applied to compositional analysis. Since this is a reflection technique, an epitaxial film can be evaluated without contribution from the substrate by proper choice of the diffraction plane and energy of x-radiation. In this way the radiation will not penetrate the epitaxial layer, and the resulting reflection topograph will show only imperfections from the layer. Figure 8-42 shows the results of the application of this technique to the growth of gallium arsenide epitaxial films. The full potential of the reflection method becomes apparent when a topograph of the substrate prior to deposition can be directly

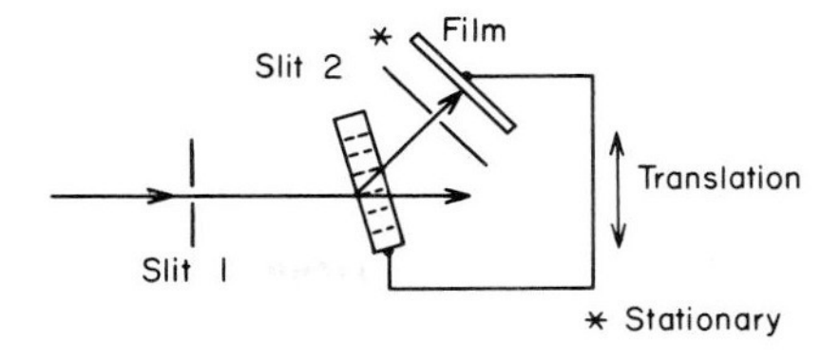

Fig. 8-41. Schematic illustrating transmission x-ray diffraction microscopy.

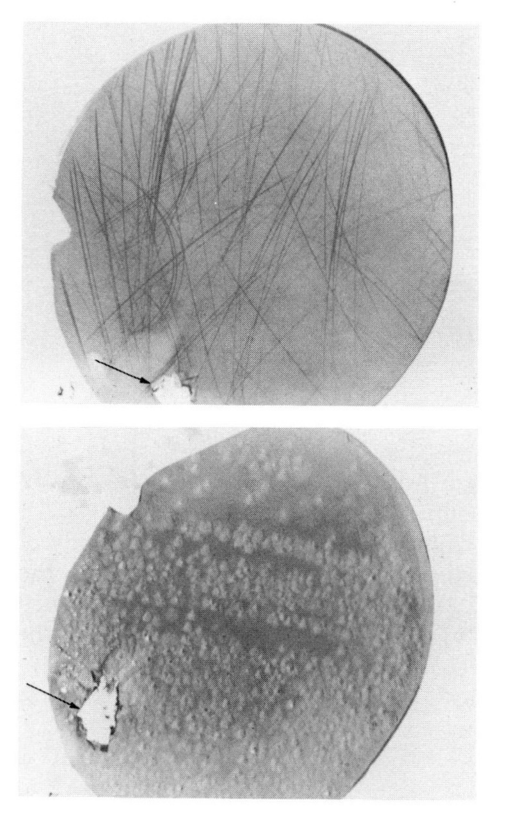

Fig. 8-42. (*Top*): **A (440) topograph reveals polishing damage in a GaAs substrate slice; a region of macromisorientation (arrow) is also observed.** (*Bottom*): **The (242) topograph of the subsequently deposited GaAs epitaxial film exhibits numerous surface protrusions (hillocks); the misoriented region is also present** (*arrow*). (*From Howard and Dobrott.*[90])

compared with a topograph of the epitaxial film deposited on that substrate. Meieran[91] applied this reflection topographic technique to the study of gallium arsenide epitaxy on germanium substrates.

8-38. CHEMICAL IMPERFECTIONS

The analysis of epitaxial layers for chemical impurities is severely hampered by the sample size and the difficulty of separating this minute sample from a heavily doped substrate. A typical silicon epitaxial layer, 10 μ in thickness and on a 1-in.-diameter slice, would be potentially a 9.3-mg sample. This total sample weight could never be realized since the only way to remove the epitaxial layer is by dissolution. This etching process must be closely controlled to prevent dissolution of the material at the film-substrate interface. Malkova et al.[92] circumvented this difficulty by depositing the germanium film on a glass substrate and then dissolving the germanium with aqua regia. The solution was evaporated on graphite powder, volatilizing the germanium as $GeCl_4$. The graphite powder, with residual impurities, was analyzed by emission spectrography. They report a sensitivity of 50 ng for In, Ga, Bi, Sb, and As in 2- to 40-mg germanium films. While this approach may be satisfactory to the analyst, it is totally unacceptable to the materials scientist. As discussed earlier (Sec. 8-19), there is considerable doubt that epitaxial films grown on control slices of opposite type are the same as those grown on heavily doped substrates for device fabrication. Further, epitaxial semiconductor films deposited on glass or other foreign substrates are not single-crystal, and impurities segregate and precipitate in polycrystalline materials. This approach cannot be recommended except in the most rudimentary studies.

Cheng and Goydish[93] determined Ga and In spectrophotometrically in germanium thin films with a sensitivity of 3 to 5 ng/cm^2 by using 1-2(2-pyridylazo)-2-naphthol (PAN). In general, it is apparent that the dissolution of part of an epitaxial layer followed by conventional chemical analysis can be used only for highly specialized analyses. In most cases the impurities in the acids and other reagents will be larger than those found in the epitaxial film.

8-39. MASS SPECTROSCOPY

Solids mass spectroscopy has been applied successfully to the analysis of surfaces and thin films. The major difficulty with this type of analysis for epitaxial films is ensuring that the excitation system samples only the film. Willardson[94] has reviewed the mass spectrographic analysis of thin films, and Table 8-6 shows a summary of the various excitation sources. As can be seen, spark sources are those most directly applicable to the analysis of epitaxial films. Ion-bombardment sources do not possess the sensitivity, and laser sources have great potential but are still under development.

Hickam and Sweeney,[95] using a very small probe as a counterelectrode, manually scanned the surface of metals and demonstrated that it was possible to sample volumes as small as 25 μ in diameter and 3 μ in depth. Because of the very small amounts of materials being sampled at each point, it was necessary to spark over large areas to achieve the necessary sensitivities. Hickam and Sandler[96] developed the rotating sample and stationary counterelectrode system shown schematically in Fig. 8-43. The volumes sampled by this system were 4000 Å in diameter and 2000 Å deep, with 1 to 2 μ spacings between sampled volumes. The surface was sparked at either 0.6 or 1.2 μ intervals. Roberts and Millett[97] used a similar technique, but, while the disk was being rotated, the counterelectrode was moved from the edge of the silicon epitaxial layer toward the center. The speed with which the counterelectrode was scanned toward the center was adjusted so that, while the mechanism resembled a phonograph record player, a constant linear speed was achieved for the spiral path from the edge of the rotating slice to the center. The volume removed with each spark was 50 to 80 μ in diameter and 20 μ deep. This slightly excessive depth can be corrected by using the spark conditions described by Hickam and Sweeney.[95] Roberts reported sensitivities in the order of 0.01 ppm, which is good for semiconductor epitaxial-film analysis.

The use of lasers as external excitation sources looks very attractive since it both minimizes contamination and allows the analysis of high-resistivity material, which is difficult to handle by conventional spark-source techniques. Honig[98] has used a pulsed 1-joule laser in the spark source of a mass spectrometer but obtained huge craters and excessive amounts of ion current. Board and Townsend[99] carried out an

Table 8-6. Sources of Ions for Analysis of Thin Films by Mass Spectrometry[94]

	Typical energy	Min. area analyzed, μ^2	Min. depth of analysis, μ	Max. penetration rate, μ/sec	Comments
Spark source:					
Conventional	20–100 kev 1 Mhz RF	10^3	3	10^6	Not applicable to most thin-film studies
Rotating probe	20–100 kev 1 Mhz RF	0.2	0.2	10^5	Detection limit about 0.1% unless scanning is used
Ion-bombardment source:					
Ion microprobe	1 ma 10 kev	10^5	10^{-2}	0.01	Sensitivity varies widely for different elements.
Ion microscope	10 μa 10 kev	1	10^{-2}	0.01	Qualitative analysis only
Low-voltage source	0.4 kev	Large	10^{-3}		Only useful for some organic films
Laser excited source:					
Ruby laser	4 pulses/min 1 joule (10^{-4} sec) pulses	10^3–10^4	10^2–10^3	10^7	Not applicable to most thin-film studies
He–Ne laser	10^3–10^4 pulses/sec 1–100 mjoule (2×10^{-7} sec) pulses	10^2	10^{-2}	10^5	Scanning required for detection limits below about 0.1%
CO_2 laser	1–100 watts continuous duty	10^3	10^{-2}	10^2	Minimum area analyzed limited by relatively long wavelength of light used

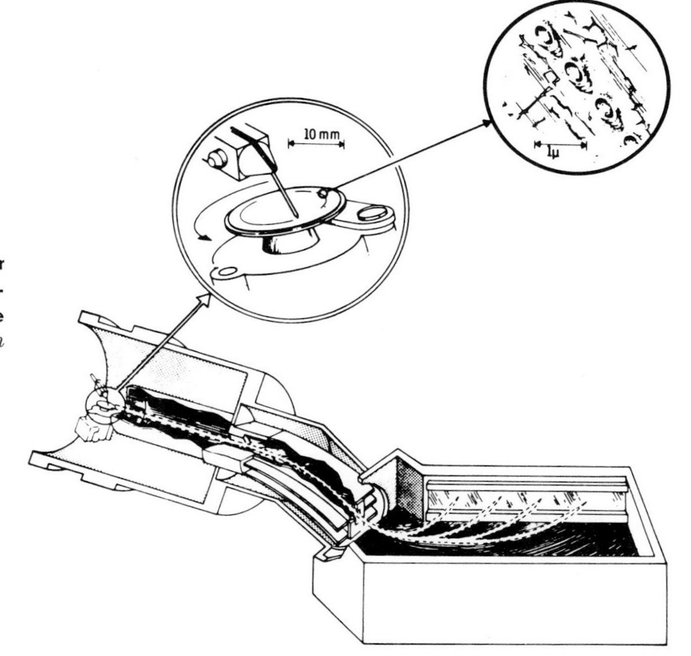

Fig. 8-43. Mass spectrometer of the Mattauch-Herzog design with a rotating electrode microprobe source. (*From Hickam and Sandler.*[96])

excellent study of the vaporization of thin metallic films with a focused laser beam. Their results should be directly applicable to laser excitation sources for spark-source mass spectrometers.

8-40. NEUTRON ACTIVATION ANALYSIS

Neutron activation analysis of epitaxial films has received very little attention. There are many sound reasons for this neglect, and one of the most serious arises from the heavily doped substrate used in device slice production. Since boron is not activated, heavily doped boron substrates can be used, but then one is back to the now familiar problem of using different substrates or control substrates. Conclusions drawn from epitaxial films grown on these control substrates are open to serious criticism if they are extrapolated to those produced on the production substrates.

Abe and Sato[49] used neutron activation analysis to study the outdiffusion of arsenic from a 0.0025 ohm-cm arsenic-doped <111> substrate. They obtained results that suggested that the diffusion constant in epitaxial films is a complicated function of position. There still remains considerable development work to be done because of the difficulties inherent in this particular application of activation analysis. The separation of the epitaxial film, by etching, from the heavily doped substrate without contamination of the etchant by the substrate is extremely tedious and very difficult.

If the deposition of a doped film on an undoped substrate or on a boron-doped substrate is acceptable, then the lateral distribution of dopant across the face of the slice can be studied by activating the slice and taking an autoradiogram of the epitaxial film. The amount of dopant in the epitaxial layer can be determined by gamma counting or gamma-ray spectroscopy, assuming no contribution from the substrate. If necessary, the epitaxial layer can be etched off and the substrate contribution measured in the same way.

REFERENCES

1. Royer, M. L.: *Bull. Soc. Franc. Mineral.*, **51**:7 (1928).
2. Dash, W. C.: "Growth and Perfection of Crystals," p. 361, John Wiley & Sons, Inc., New York, 1958.
3. Marinace, J. C.: *IBM J. Res. Develop.*, **4**:248 (1960).
4. Wajada, E. J., B. W. Kippenhan, and W. H. White: *IBM J. Res. Develop.*, **4**:288 (1960).
5. Wajada, E. S., and R. Glang: in R. O. Grubel (ed.), "Metallurgy of Elemental and Compound Semiconductors," p. 229, Interscience Publishers, Inc., New York, 1961.
6. May, J. E.: *J. Electrochem. Soc.*, **112**:710 (1965).
7. Handelman, E. T., and E. I. Povilonis: *J. Electrochem. Soc.*, **111**:201 (1964).
8. Newman, R. C.: *Microelectron. Rel.*, **3**:121 (1964).
9. Hale, A. P.: *Vacuum*, **13**:93 (1963).
10. Heavens, O. S.: *Proc. Phys. Soc. (London)*, **65B**:788 (1952).
11. Gasson, D. B.: Unpublished work (1962) reported by R. C. Newman, *Microelectron. Rel.*, **3**:121 (1964).
12. Runyan, W. R.: "Silicon Semiconductor Technology," McGraw-Hill Book Company, New York, 1965.
13. Mehal, E. W., and G. R. Cronin: *Electrochem. Tech.*, **4**:540 (1966).
14. Rubenstein, M., and E. Meyer: *J. Electrochem. Soc.*, **113**:365 (1966).
15. Conrad, R. W., P. L. Hoyt, and D. D. Martin: *J. Electrochem. Soc.*, **114**:164 (1967).
16. Bond, W. L., and F. M. Smits: *Bell System Tech. J.*, **35**:1209 (1956).
17. Anon.: *Semicond. Prod. Solid State Tech.*, **10**(3):72 (1967).
18. Dash, W. C., *J. Appl. Phys.*, **33**:2395 (1962).
19. Spitzer, W. G., and J. Tanenbaum: *J. Appl. Phys.*, **32**:744 (1961).
20. Albert, M. P., and J. F. Combs: *J. Electrochem. Soc.*, **109**:709 (1962).
21. Walsh, R. J.: *Semicond. Prod. Solid State Tech.*, **7**(8):23 (1964).
22. Spitzer, W. G., and H. Y. Fan: *Phys. Rev.*, **106**:822 (1957).
23. Groves, W. O.: *Semicond. Prod. Solid State Tech.*, **5**(12):25 (1962).
24. Cave, E. F., and B. R. Czorny: *RCA Rev.*, **24**:523 (1963).
25. Schumann, P. A., Jr., R. P. Phillips, and P. J. Olshefski: *J. Electrochem. Soc.*, **113**:368 (1966).
26. "1968 Book of ASTM Standards," ASTM, Philadelphia, 1968.
27. Hilton, A. R., and C. E. Jones: *J. Electrochem. Soc.*, **113**:472 (1966).
28. Jones, C. E., and A. R. Hilton (Texas Instruments Incorporated): Unpublished work, 1966.
29. Kruse, P. W., L. D. McGlauchin, and R. B. McQuistan: "Elements of Infrared Technology," p. 314, John Wiley & Sons, Inc., New York, 1962.
30. Rhodes, R. G.: "Imperfections and Active Centers in Semiconductors," p. 207, The Macmillan Company, New York, 1964.
31. Rubenstein, M.: *J. Electrochem. Soc.*, **112**:426 (1965).
32. Williams, E. W., R. H. Cox, R. D. Dobrott, and C. E. Jones: *Electrochem. Tech.*, **4**:479 (1966).
33. Tietjen, J. J., and J. A. Amick: *J. Electrochem. Soc.*, **113**:724 (1966).

34. Wendlandt, W. W., and H. G. Hecht: "Reflectance Spectroscopy," Interscience Publishers, Inc., New York, 1966.
35. Woolley, J. C., and K. W. Blazey: *Phys. Chem. Solids,* **25**:713 (1964).
36. Jones, C. E. : *Appl. Spectry.,* **20**:161 (1966).
37. Conrad, R. W., C. E. Jones, and E. W. Williams: *J. Electrochem. Soc.,* **113**:287 (1966).
38. Williams, E. W., and C. E. Jones: *Solid State Comm.,* **3**:195 (1965).
39. Howard, J. K., and R. D. Dobrott: *J. Electrochem. Soc.,* **113**:567 (1966).
40. Howard, J. K., and R. H. Cox, in A. R. Mallett, M. Fay, and W. M. Mueller (eds.), "Advances in X-ray Analysis," vol. 9, p. 35, Plenum Press, New York, 1966.
41. Bhola, S. R., and A. Mayer: *RCA Rev.,* **24**:511 (1963).
42. Nakanuma, S.: *J. Electrochem. Soc.,* **111**:1199 (1964).
43. Baker, W. E., and D. M. J. Compton: *IBM J. Res. Develop.,* **4**:269 (1960).
44. Baker, W. E., and D. M. J. Compton: *IBM J. Res. Develop.,* **4**:275 (1960).
45. Baker, W. E., and D. M. J. Compton: *IBM J. Res. Develop.,* **4**:296 (1960).
46. Joyce, B. A., J. C. Weaver, and D. J. Maule: *J. Electrochem. Soc.,* **112**:1100 (1965).
47. Larrabee, G. B., O. W. Wilson, and K. G. Heinen (Texas Instruments Incorporated): Unpublished work, 1966.
48. Goldstein, B., and C. Dobin: *Solid-State Electron.,* **5**:411 (1962).
49. Abe, T., and K. Sato: *Japan J. Appl. Phys.,* **4**:70 (1965).
50. Dobbs, P. J. H., and F. S. Kovacs: *Semicond. Prod. Solid State Tech.,* **7**(8):28 (1964).
51. Schumann, P. A., Jr., and J. F. Hallenback, Jr.: *J. Electrochem. Soc.,* **110**:538 (1963).
52. Gardner, E. E., J. F. Hallenback, Jr., and P. A. Schumann, Jr.: *Solid-State Electron.,* **6**:311 (1963).
53. Brownson, J.: *J. Electrochem. Soc.,* **111**:919 (1964).
54. Gardner, E. E., and P. A. Schumann, Jr.: *Solid-State Electron.,* **8**:165 (1965).
55. Allen, C. C., L. H. Clevenger, and D. C. Gupta: *J. Electrochem. Soc.,* **113**:508 (1966).
56. Thomas, C. O., D. Kahng, and R. C. Manz: *J. Electrochem. Soc.,* **109**:1055 (1962).
57. Kovacs, F. S., and A. S. Epstein: *Semicond. Prod. Solid State Tech.,* **7**(8):32 (1964).
58. Moest, R. R., and D. T. Lassota: *J. Electrochem. Soc.,* **114**:110 (1967).
59. Lawley, K. L.: *J. Electrochem. Soc.,* **113**:240 (1966).
60. Amron, I.: *Electrochem. Tech.,* **2**:327 (1964).
61. Amron, I.: *Electrochem. Tech.,* **5**:94 (1967).
62. Kressel, H., and M. A. Klein: *Solid-State Electron.,* **6**:309 (1963).
63. Lindmayer, J., and M. Kutsko: *Solid-State Electron.,* **6**:377 (1963).
64. Patrick, W. J.: *Solid-State Electron.,* **9**:203 (1966).
65. Frank, H.: *Solid-State Electron.,* **9**:609 (1966).
66. Rawlins, T. G. R.: *J. Electrochem. Soc.,* **111**:810 (1964).
67. Williams, F. V.: *J. Electrochem. Soc.,* **111**:886 (1964).
68. Reisman, A., and M. Berkenblit: *J. Electrochem. Soc.,* **112**:315 (1965).
69. Lenie, C. A.: *Semicond. Prod. Solid State Tech.,* **7**(9):41 (1964).
70. Hallas, C. E., and E. J. Patzner: *Semicond. Prod. Solid State Tech.,* **8**(11):20 (1965).
71. Mendelson, S.: in M. H. Francombe and H. Sato (eds.), "Single Crystal Films," p. 251, Pergamon Press, New York, 1964.
72. Haneta, Y.: *Japan J. Appl. Phys.,* **4**:69 (1965).
73. Booker, G. R.: *J. Appl. Phys.,* **37**:411 (1966).
74. Batsford, K. O., and D. J. D. Thomas: *Microelectron. Rel.,* **3**:159 (1964).
75. Sirtl, E., and A. Adler: *Z. Metallk.,* **52**:529 (1961).
76. Booker, G. R., and R. Stickler: *J. Appl. Phys.,* **33**:3281 (1962).
77. Cronin, G. R., R. W. Conrad, and S. R. Borrello: *J. Electrochem. Soc.,* **113**:1336 (1966).
78. Abrahams, M. S., and C. J. Buiocchi: *J. Appl. Phys.,* **36**:2855 (1965).

79. Miller, D. P., S. B. Watelski, and C. R. Moore: *J. Appl. Phys.*, **34**:2813 (1963).
80. Inoue, M.: *J. Electrochem. Soc.*, **112**:189 (1965).
81. Mendelson, S.: *J. Appl. Phys.*, **38**:1573 (1967).
82. Marcus, R. B.: in B. Schwartz and N. Schwartz (eds.), "Measurement Techniques for Thin Films," p. 1, Electrochemical Society, New York, 1967.
83. Charig, J. M., B. A. Joyce, D. J. Stirland, and R. W. Bicknell: *Phil. Mag.*, **7**:1847 (1962).
84. Finch, R. H., and H. J. Queisser: *J. Appl. Phys.*, **34**:406 (1963).
85. Abrahams, M. S., and C. J. Buiocchi: *Phys. Chem. Solids*, **28**:927 (1967).
86. Booker, G. R.: *Phil. Mag.*, (8) **11**:1007 (1965).
87. Schwuttke, G. H.: *J. Appl. Phys.*, **34**:3127 (1963).
88. Schwuttke, G. H.: *J. Electrochem. Soc.*, **109**:27 (1962).
89. Lang, A. R.: *J. Appl. Phys.*, **30**:1748 (1959).
90. Howard, J. K., and R. D. Dobrott: *Appl. Phys. Letters*, **7**:101 (1965).
91. Meieran, E. S.: *J. Electrochem. Soc.*, **114**:292 (1967).
92. Malkova, O. P., A. N. Zhukova, and N. K. Rudnevski: *Zh. Anal. Khim.*, **19**:312 (1964).
93. Cheng, K. L., and B. L. Goydish: *Anal. Chim. Acta*, **34**:154 (1966).
94. Willardson, R. K.: in B. Schwartz and N. Schwartz (eds.), "Measurement Techniques for Thin Films," p. 58, Electrochemical Society, New York, 1967.
95. Hickam, W. M., and G. G. Sweeney: in "Eleventh Annual Conference on Mass Spectrometry and Allied Topics, San Francisco, Calif., 1963," p. 319, ASTM Committee E-14, 1963.
96. Hickam, W. M., and Y. L. Sandler: in J. I. Bregman and A. Dravnieks (eds.), "Surface Effects in Detection," p. 189, Spartan Books, Washington, 1965.
97. Roberts, J. A., and E. J. Millett: in "Proceedings of the Sixth Annual MS7 Mass Spectrometer Users Conference," p. 39, AEI Ltd., Manchester, 1966.
98. Honig, R. E.: in "Twelfth Annual Conference on Mass Spectrometry and Allied Topics, Montreal, Canada, 1964," p. 233, ASTM Committee E-14, 1964.
99. Board, K., and W. G. Townsend: *Microelectron. Rel.*, **5**:251 (1966).

9

Diffusion

9-1. INTRODUCTION

In the semiconductor industry, the controlled diffusion of impurities into the bulk semiconductor is one of the basic and probably most important device-fabrication processes in use today. The process of diffusion, when stated in its simplest form, is the redistribution of matter in such a way that there will be a decrease in the concentration gradient. During a diffusion into a semiconductor, a high concentration of impurity will be presented to the semiconductor surface and conditions adjusted to allow this impurity to redistribute into the semiconductor. The important questions that the device engineer or solid-state physicist will ask are: How fast does this impurity diffuse, how does this rate of diffusion change with temperature, what is the shape of the diffusion profile, and is the impurity uniformly distributed across the face of the diffusion front?

The analytical chemist finds himself in the position of having to carry out at least the analyses of the diffused semiconductor. More often he is presented with a new semiconductor material, say indium arsenide, and asked to look at the diffusion of some impurity, say cadmium, into this semiconductor. It is essential that the chemist understand the basic concepts of diffusion before he can undertake such a program.

9-2. SOLID-STATE DIFFUSION

Generally the semiconductor material and the impurity will be heated together in a controlled atmosphere, either in an evacuated quartz ampule or in a flowing-gas open-tube system, for a given length of time. The impurity, which is at a high concentration at the surface, will redistribute itself into the bulk of the semiconductor in a manner that is described by the diffusivity or diffusion coefficient D. In the simplest case, the diffusion coefficient for an impurity in any one semiconductor material is a constant. (The units of D are always square centimeters per second.) This process and D are described in Fick's first law of diffusion, where the *amount* of material diffusing through a plane per unit time is described by J, the flux.

This flux across a plane can be taken to be proportional to the concentration at that point in the semiconductor. This flux of impurity, J, is equal to the diffusivity of the impurity times the concentration gradient. The term is negative because the concentration is decreasing into the crystal.

$$J = -D \frac{\partial N}{\partial x} \tag{9-1}$$

or Flux = (diffusivity) $\times$ (concentration gradient)

where J = flux or diffusion current
D = diffusivity or diffusion coefficient
N = concentration of diffusing dopant
x = distance

For diffusions into semiconductor materials, the concentration of diffusant is generally low (as compared with metallurgical metal-metal diffusions) and the depth of the diffusion small. Under these and other ideal conditions, it can be assumed that D is constant, and Fick's second law of diffusion can be derived. This second law simply examines the diffusion of the impurity into the semiconductor under other than steady-state conditions. That is, the concentration is changing with time. In Fig. 9-1, the rate at which an impurity is doping a given column of a semiconductor can be examined by looking at the flux of impurity entering one plane (x_1) and leaving a closely adjacent plane (x_2). The amount of impurity doping this small volume will then be the difference in fluxes (i.e., flux 1 minus flux 2).

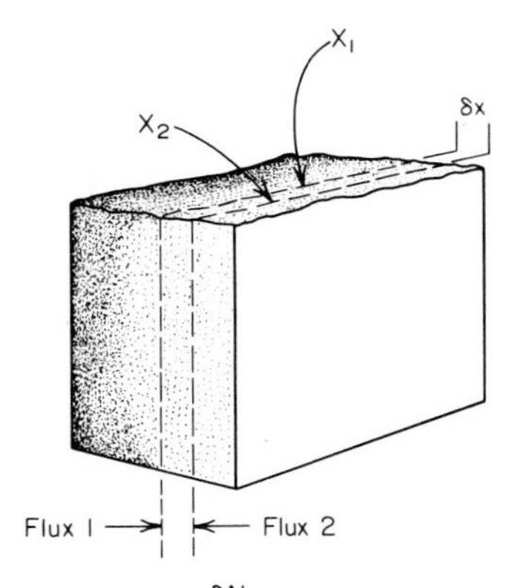

Flux (1) = $J = -D \frac{\delta N}{\delta X}$

Through plane X_1

Flux (2) = $J + \frac{\delta J}{\delta X} = -D \frac{\delta N}{\delta X} + \left[-\frac{\delta}{\delta X} \left(D \frac{\delta N}{\delta X} \right) \right]$

Through plane X_2

Flux (2) − Flux (1)

$$\frac{\delta J}{\delta X} = -\frac{\delta}{\delta X} \left(D \frac{\delta N}{\delta X} \right)$$

Fig. 9-1. Schematic illustrating the derivation of Fick's second law of diffusion

Since $\partial J/\partial x$ is the same as the negative rate of concentration change $-\partial N/\partial x$, we arrive at Fick's second law:

$$\frac{\partial N}{\partial t} = \frac{\partial}{\partial x} D \frac{\partial N}{\partial x} \tag{9-2}$$

$$\frac{\partial N}{\partial t} = D \frac{\partial^2 n}{\partial x^2} \tag{9-3}$$

This partial differential equation then describes the spatial distribution of the diffusing species with a diffusion coefficient D as a function of time. All diffusion studies are solutions of this differential equation with different boundary or experimental conditions. The reader is referred to Crank[1,†] for a detailed mathematical study of the many different solutions of this partial differential equation. Runyan,[2] Boltaks,[3] and Shewmon[4] also have excellent descriptions of many of these solutions with specific examples.

The first two solutions of Fick's second law of diffusion that will be discussed here are the so-called "infinite-source" and "limited-source" diffusions. These two types of diffusion are probably the most frequently considered in diffusion studies but are not all-inclusive. The reader is cautioned to examine carefully the experimental evidence and to watch for diffusions with surface-rate limitations, grain boundary diffusions, field-aided diffusions, or other complicating factors.

9-3. LIMITED-SOURCE DIFFUSION

Since the basic diffusion process is a redistribution of an impurity toward decreasing concentration, it is important to know how much impurity is available for redistribution. If only a small and limited amount of impurity is presented to the semiconductor surface for diffusion, then the solution of the partial differential equation (9-3) must include this quantity of material Q. This particular diffusion problem has the simplest mathematical solution and is shown as

$$N_{(x,t)} = \frac{Q}{\sqrt{\pi D t}} e^{-x^2/4Dt} \tag{9-4}$$

Equation (9-4) describes the diffusion of a limited amount of impurity Q, deposited at time $t = 0$ and in the plane $x = 0$. Figure 9-2 shows the effect of diffusion time t on the shape of the diffusion profile. This profile is a gaussian distribution of the limited source Q. Since there is a fixed or limited quantity of impurity available for diffusion, the total area under the curve stays constant, the surface concentration N_0 decreases with increased diffusion time, and the depth of penetration of the impurity into the semiconductor increases with the square root of time.

An experimental evaluation of the diffusion profile of this limited-source diffusion can be obtained by taking the natural logarithm of Eq. (9-4).

$$\ln N_{(x,t)} = -\frac{x^2}{4Dt} + \ln \frac{Q}{\sqrt{\pi D t}} \tag{9-5}$$

†Superscript numbers indicate References listed at the end of the chapter.

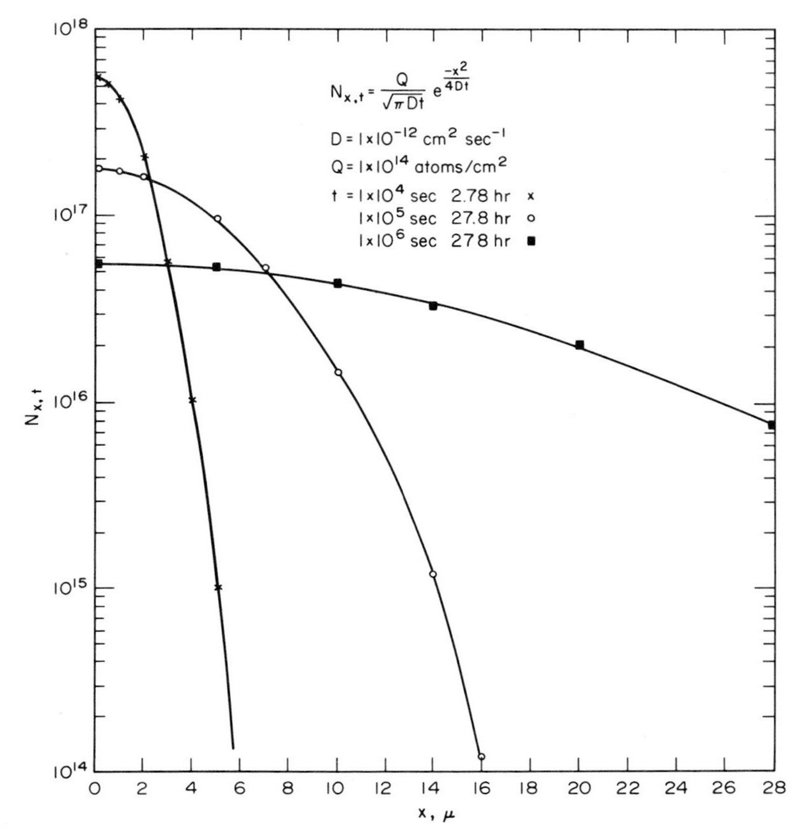

Fig. 9-2. Diffusion from a limited source into a semi-infinite body, illustrating the effect of diffusion time.

Thus a plot of ln N versus x^2 will yield a straight line with slope $-(4Dt)^{-1}$. Since the time of diffusion, t, is always experimentally known, it is possible to calculate D from this slope.

9-4. INFINITE-SOURCE DIFFUSION

It is often possible to arrange the experimental conditions so that there is an infinite amount of impurity available for diffusion into the semiconductor. Thus the surface concentration N_0 will remain constant throughout the entire diffusion. This infinite source can be simply the vapor of an impurity sealed in an evacuated ampule with the semiconductor, or in an open-tube flow system it can be a doping gas flowing over the surface. As diffusion proceeds, the amount of impurity diffusing into the bulk is replenished from the vapor, N_0 remains constant, and its maximum value is controlled only by the ultimate solubility of that impurity in the semiconductor at the temperature of the diffusion.

The mathematical solution to the partial differential equation describing Fick's second law of diffusion for a constant surface concentration N_0 is

$$N_{(x,t)} = N_0\left(1 - \operatorname{erf}\frac{x}{2\sqrt{Dt}}\right) \tag{9-6}$$

The error function erf which appears in mathematical solutions of this type of problem is a numerical evaluation of a very difficult integral and is available in tabular form.[5] Noting that erf $0 = 0$ and erf $\infty = 1$, it can be seen that at the surface ($x = 0$) for any diffusion time t, $N = N_0$ and the surface concentration remains constant. Figure 9-3 shows the effect of time on diffusion from an infinite source. As can be seen, the surface concentration remains constant while the depth of penetration increases with diffusion time. By comparing this figure with Fig. 9-2, the differences between the two types of diffusion become more obvious.

9-5. VARIATION OF D WITH CONCENTRATION

All the mathematical solutions of the differential equation describing Fick's diffusion law assume a constant diffusion coefficient D at any given temperature.

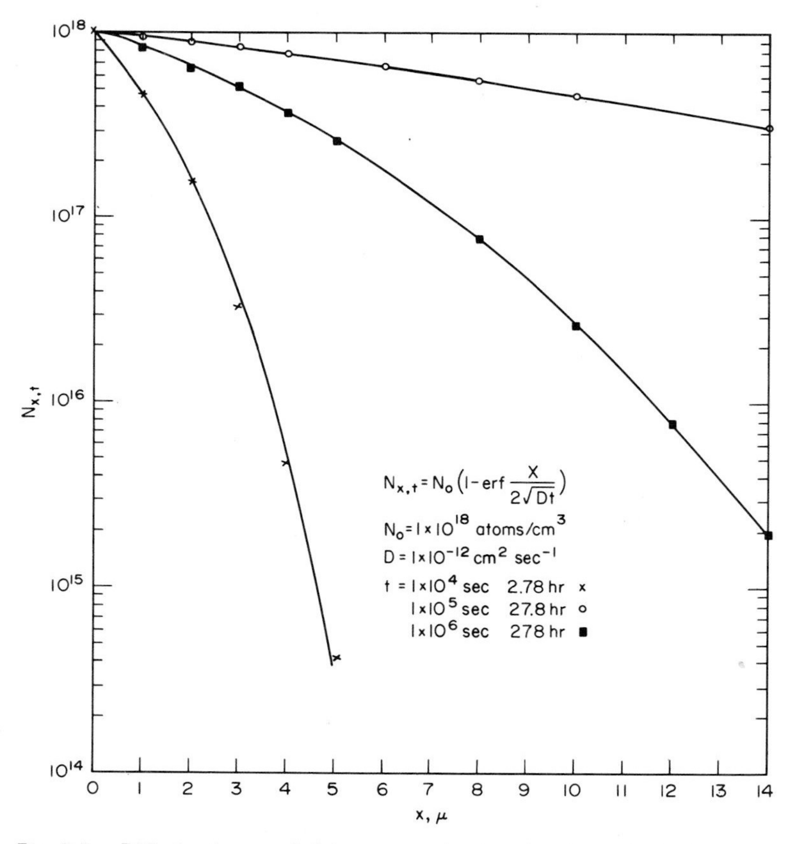

Fig. 9-3. Diffusion from an infinite source, showing the effect of diffusion time.

Generally a constant D is observed for any given set of boundary conditions for the diffusion of impurities in silicon and germanium. Diffusion profiles of impurities into the III-V intermetallic compounds, notably GaAs, have been consistently difficult to interpret. It has been found that the diffusion coefficient varies with concentration in the crystal, which results in a diffusion coefficient for *each* impurity concentration throughout the profile. The mathematical analysis of this type of data is complex. An example of a concentration dependent D is shown in Fig. 9-4 for the diffusion of zinc into GaAs at 900°C.[6-8] As can be seen, the diffusion coefficient is strongly influenced by the zinc concentration. This behavior results in a diffusion profile that cannot be fitted by either an error function or a gaussian distribution. Typical diffusion profiles of this type are shown in Fig. 9-5. Notice the flat tops on the profiles followed by very rapid decreases in zinc concentration. Weisburg and Blanc[9] reported some success in fitting diffusion profiles of this type to a modified error function distribution where D was concentration-dependent. In this modification D was proportional to the product of the surface diffusion coefficient D_{sur} and the square of the ratio of concentration to surface concentration so that

$$D = D_{sur}\left(\frac{N}{N_0}\right)^2 \tag{9-7}$$

$$N_x = N_0\left[1 - \operatorname{erf}\frac{x}{2\sqrt{D_{sur}(N/N_0)^2t}}\right] \tag{9-8}$$

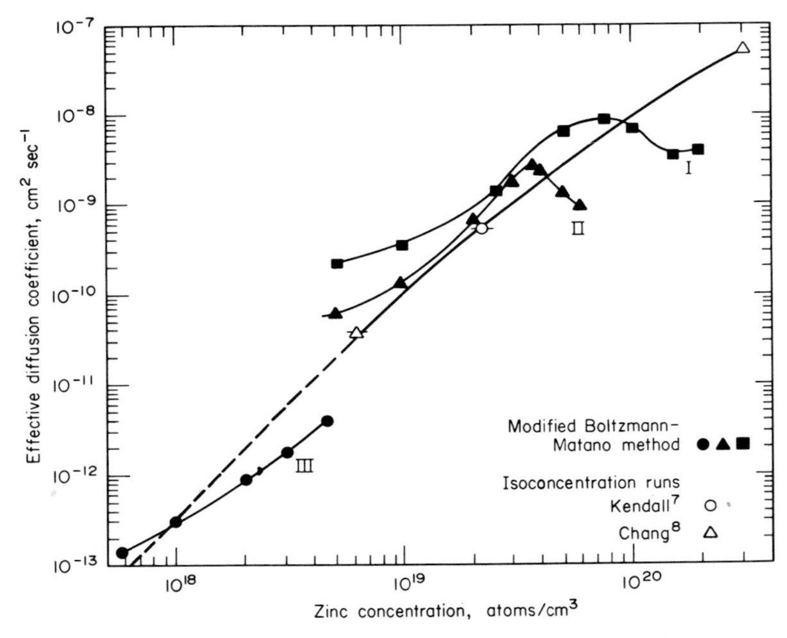

Fig. 9-4. Analysis of concentration-dependent diffusion profiles using the Boltzmann-Matano method. (*From Kendall.*[6])

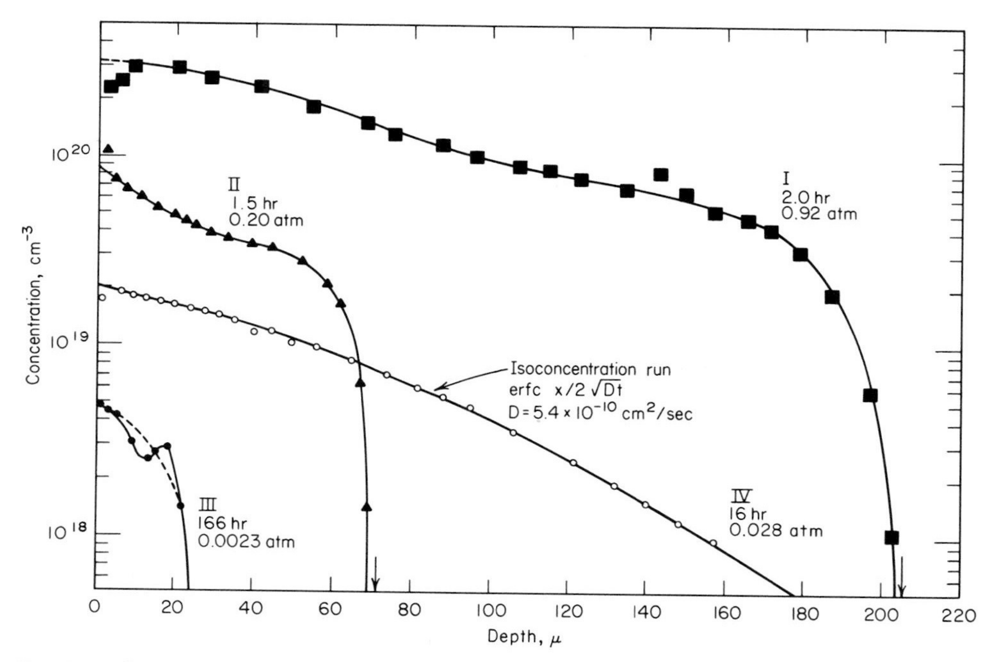

Fig. 9-5. Diffusion profiles of zinc in gallium arsenide at 900°C showing the effects of diffusion time and zinc partial pressure. (*From Kendall.*[6])

This approach has received little attention for the analysis of concentration-dependent diffusion profiles. Matano's[10] method of applying Boltzmann's[11] solution for concentration-variable diffusion coefficients has been widely used. Boltzmann assumed that at a given temperature the diffusion coefficient was a function of a single variable λ which was equal to $x/t^{1/2}$. The solution to the differential equation can be shown[1,3] to simplify to

$$D_c = -\frac{1}{2}\frac{\partial\lambda}{\partial c}\int_0^c \lambda\, dc \tag{9-9}$$

The application of this equation to data from a diffusion profile will be discussed in Sec. 9-25.

9-6. VARIATION OF D WITH TEMPERATURE

In all previous discussions of diffusions of impurities it has been assumed that D is constant. The effect of temperature on the diffusivity has been found empirically to be

$$D = D_0 \exp\left(-\frac{Q}{RT}\right) \tag{9-10}$$

where R is the gas constant, T is the absolute temperature, Q is the activation energy, and D_0 is sometimes called the frequency factor but is better known to the

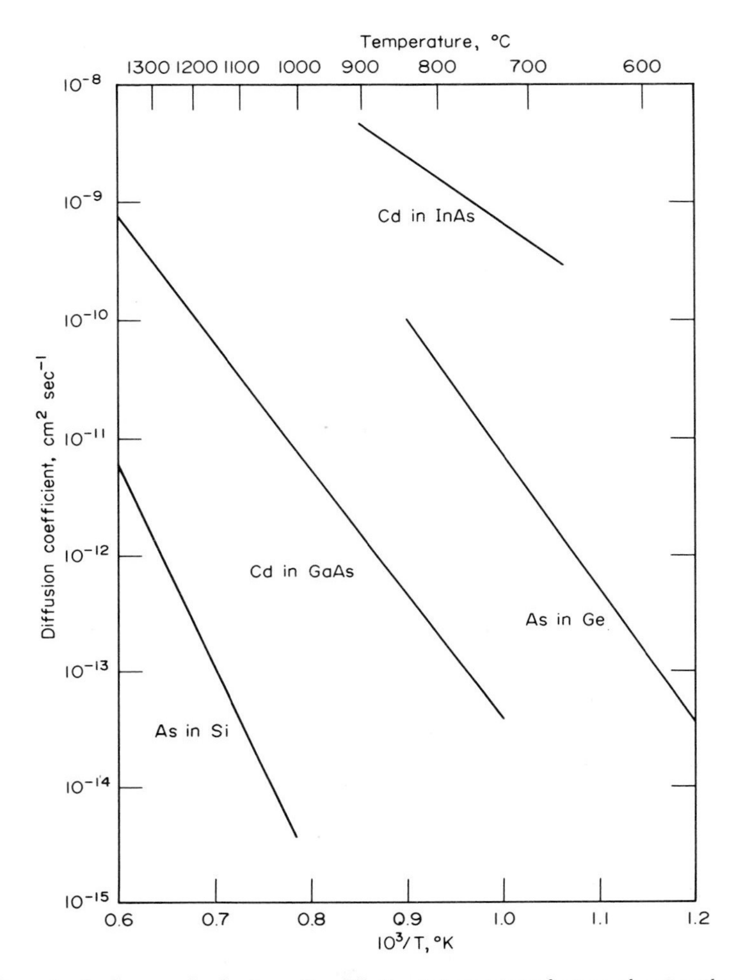

Fig. 9-6. Some typical examples for impurity diffusion into semiconductors, showing the variation of the diffusion coefficient as a function of temperature.

chemist as ΔS. This is the standard Arrhenius equation and can be solved for D_0 and Q by experimentally determining values for D at various temperatures and plotting $\ln D$ versus $1/T$. The slope of the straight line is $-Q/R$, and the intercept at $1/T = 0$ is $\ln D_0$. Figure 9-6 shows the variation of D with temperature for several typical impurities in semiconductors.

Since the diffusion coefficient varies exponentially with temperature, the need becomes apparent for the accurate experimental control of temperature. The diffusion furnaces used in the semiconductor industry are controlled by proportional controllers and have long, flat temperature zones. As a result, temperature profiles such as that shown in Fig. 9-7 are accepted as typical, and variations of less than ± 1°C over the length of the zone and ± 1°C in the zone as a function of time are necessary. Diffusion-furnace manufacturers[12] at the time of this writing are

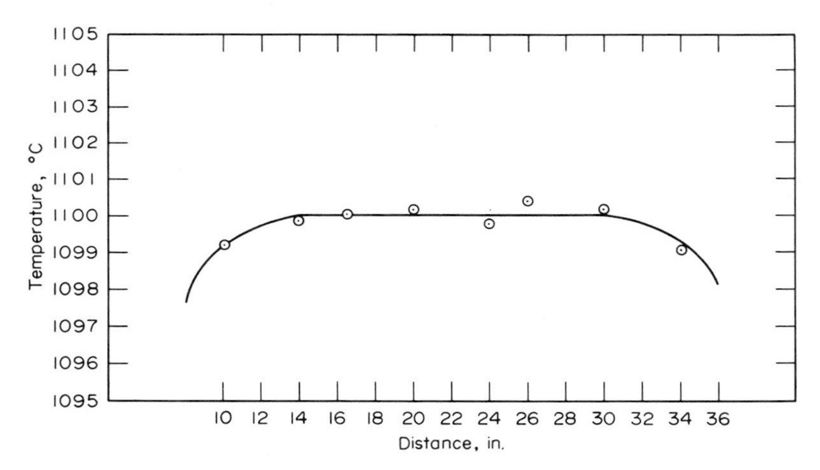

Fig. 9-7. Typical temperature profile of a diffusion furnace used in the semiconductor industry.

claiming variations of ± 0.1 to $0.5°C$ at operating temperatures of 1000 to 1300°C. The experimentalist is cautioned not to overlook this important parameter in planning his experiment. A variation of only a few degrees from end to end of a sealed quartz ampule in the diffusion of impurities, in gallium arsenide, for example, will have a pronounced effect on the experimental results.

9-7. VARIATION OF D WITH TIME

Experimentally in the study of diffusions it is necessary to introduce the sample and the diffusant into a diffusion furnace and allow sufficient time for the sample to come up to temperature. Since the diffusion coefficient varies exponentially with temperature, the sample is being diffused with an impurity with a varying diffusion coefficient until constant temperature is achieved (see Fig. 9-6). A similar problem is observed at the end of the diffusion, when the sample is cooled. As a result, the time of diffusion at constant temperature should be very long when compared with the temperature rise and fall at the beginning and end of the diffusion run. The worst possible procedure would be to introduce the sample or quartz ampule into a cold furnace, allow the furnace to heat up, and, at the end of the diffusion run, shut off the furnace and allow the sample to cool with the furnace.

Mathematically any variation of D with time (experimental or real time-dependent effects) can be handled by Fick's second law of diffusion when a new time diffusion coefficient R is introduced and D is integrated over the diffusion time t:

$$R = \int_0^t D(t)\, \partial t \tag{9-11}$$

This equation can be used to correct for the diffusion that occurs during the heating and cooling cycle. Shewmon[4] has considered this experimental problem and has shown that the amount of diffusion is negligible until the sample has reached 80 percent of the temperature at which the diffusion will be carried out. However, the time required to reach thermal equilibrium over the last 20 percent of the temperature rise can have a significant effect on the value of the diffusion coefficient derived from the experiment. In the case where a large ampule is introduced into a small or underpowered furnace, the time required to reach the diffusion temperature will be large and will have a significant effect that must be considered.

9-8. EXPERIMENTAL METHODS OF EVALUATION

A careful and complete evaluation of the diffusion of an impurity into a semiconductor involves many different characterization techniques. All techniques have been found to be complementary, and none are mutually exclusive. The method most successfully used has been the radioactive-tracer technique. In this method the impurity under study is tagged with a radioactive tracer, for example, ^{65}Zn in zinc or ^{64}Cu in copper, and the exact profile of the impurity distribution determined by lapping or etching techniques. The amount of radioactivity in the removed lap is determined by counting techniques, and the concentration of impurity in each lap is calculated from the known specific activity.† This technique will be discussed in detail in Secs. 9-13 through 9-25. Other evaluation techniques include angle lap and stain, interferometry, and many electrical techniques.

It is both important and convenient to use the electrical effects of the impurity in the semiconductor. If the impurity under study is of opposite conductivity type to that of the crystal, then an electrical junction (p-n junction) will occur when the number of donors equals the number of acceptors, $N_D = N_A$. For diffusions in semiconductors, the carrier concentration of the host crystal is generally known or can be determined from Hall measurements (Sec. 4-17). Therefore, at the p-n junction the concentration of diffusing impurity (N_x) is equal to the carrier concentration (n_0) of the host crystal. The distance between the p-n junction and the surface (x) can be determined by any of the techniques described below (Secs. 9-9 and 9-10). These two values n_0 and x along with the diffusion time t are sufficient to solve any of the equations [for example, (9-4) and (9-6)] for the diffusion coefficient.

9-9. ANGLE LAP AND STAIN

The distance of the p-n junction from the surface can be obtained by angle lapping the diffused area and chemically staining the lapped junction. Table 9-1 lists some of the stains used in the semiconductor industry to stain and delineate p-n junctions. The use of these stains is somewhat of an art[17] but is generally quite simple and reproducible unless areas of high resistivity are encountered.

The diffused slice of material is mounted on a lapping jig, as shown in Fig. 8-5. The angle of the jig must be precisely known, and in the example shown in Fig. 9-8, the angle is 10.0°. The choice of this angle determines the amount that the p-n

†Specific activity is the unit used in radiochemistry to describe the amount of inactive isotope associated with the radioactive isotope, e.g., atoms/cpm, μg/cpm, etc.

Table 9-1. Stains Used for p-n Junction Delineation

Semiconductor	Etch	Reference
Silicon	0.1% HNO_3 in HF	16
Gallium arsenide	10% (vol) HNO_3	14
Indium arsenide	$1HF:3HNO_3:2H_2O$	
Germanium	$15HF:25HNO_3:15CH_3COOH$ with 0.1 part bromine	15

junction will be spread out. At 10° the y component in Fig. 9-8 is 5.75 times larger than the depth x. An angle of 3° makes it 19.1 times larger. The spread p-n junction y is measured by using a microscope with a calibrated eyepiece. The actual depth x can then be calculated from the known angle θ of the lapping jig by $x = y \sin \theta$.

Other workers[13,14] have used optical interferometry to measure the stained depth of the angle-lapped p-n junction. This technique was covered in Sec. 8-7. Optical interferometry is generally accepted to be accurate to ± one fringe, 0.2945 μ for sodium light. This accuracy is adequate for normal diffusion studies in semiconductors where it is possible to work with diffusion depths of 3 μ or greater. Since the depth of diffusion varies with the square root of the diffusion time, the experimental conditions can be optimized to obtain the desired diffusion depth.

Junction delineation or staining is frequently accomplished by electrochemical techniques.[18] When a small ac bias is applied across either a silicon or a germanium p-n junction in a dilute electrolyte, the n-type region will be preferentially etched. After treatment the p-type region will look rough and textured or have a dark

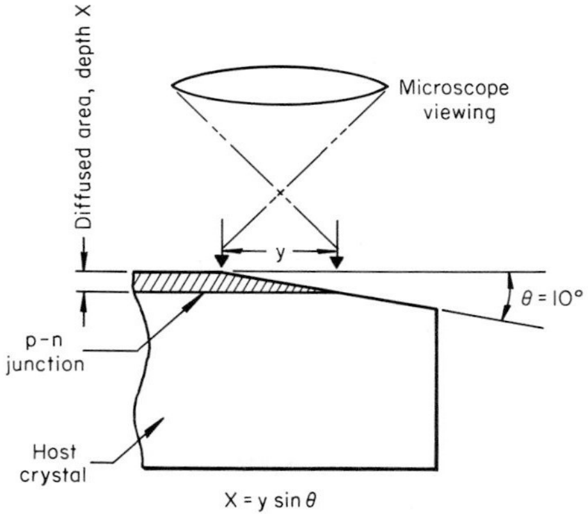

Fig. 9-8. Schematic illustrating the angle-lap technique for measuring the depth of diffusion.

surface deposit, while the n-type region will be smooth and shiny. If the electrolyte contains some metal which can be deposited,[19,20] then p-n junction decoration can be used. In a reverse-biased p-n junction, this results in plating on the p-type region because it is negative with respect to the n-type region. Plating at the p-n junction interface can be enhanced by making contact to the n-type region and applying a negative bias with respect to an external electrode. A dilute copper nitrate or sulfate solution is the most frequently used electrolyte.

9-10. ELECTRICAL PROBE AND RESISTIVITY

Since the purpose of the diffusion of an impurity into a semiconductor is to change the conductivity, or resistivity, the electrical properties of the diffused material are frequently used to aid in determining diffusion properties. In a manner analogous to the angle-lap-and-stain technique, the electrical properties can be utilized by an angle-lap-and-probe method. The angle lapping of the diffused p-n junction is performed in exactly the same manner as described in Sec. 9-9.

If the diffusion depth is sufficiently deep and the angle of the lap fixture can be chosen to obtain a large spread of the p-n junction, then an electrical probe can be used to locate the junction. This is sometimes accomplished by probing along the lapped junction with a small wire probe and watching for the type change. Because of the difficulties in ensuring ohmic contact, a wire probe can present problems unless the material is of very low resistivity.

9-11. SHEET RESISTIVITY BY FOUR-POINT PROBE

Another more useful electrical technique is to measure the sheet resistance of the diffused region by using a four-point probe[21-23] (Sec. 4-16). This technique is, of course, useful only when a p-n junction is formed so that the diffused region will be electrically isolated from the bulk of the slice. The procedure for characterization then consists in measuring the resistivity, removing a thin layer, and remeasuring the resistivity. The difference in resistivity is related to the electrical properties of the thin layer that was removed. This lapping procedure is continued with a four-point-probe resistivity measurement after each lap. It can be shown[3,21] that the reciprocal of the difference in resistivity is equal to the conductivity σ for the thin layer removed and that

$$\sigma = ne\mu\,\Delta l \tag{9-12}$$

where $\sigma = \dfrac{1}{4.5}\left(\dfrac{I_2}{V_2} - \dfrac{I_1}{V_1}\right)$

n = net carrier concentration

$e = 1.60 \times 10^{-19}$ coul

μ = mobility

Δl = thickness of layer removed

Thus from the measured conductivity, and by knowing the thickness of diffused material removed, it is possible to calculate the net carrier concentration. The four-point-probe resistivity measurements must be very accurately measured since it is the *difference* between the two measurements that is used to calculate the diffusing impurity concentration.

As pointed out earlier, these electrical techniques are very important and, when coupled with radioactive-tracer measurements on the removed laps, supply a wealth of information. However, electrical measurements alone can be misleading if the electrical properties of the impurity are not well known or if other electrically active impurities are present. The latter problem has plagued the semiconductor industry for silicon,[24] germanium,[25] and gallium arsenide.[26] It is known as "thermal conversion"; and when it occurs, the bulk semiconductor undergoes a complete type change (say n to p) during the diffusion cycle. Very fast-diffusing impurities such as copper have been shown to cause this conversion. When thermal conversion occurs, all electrical measurements are meaningless and the sample is lost. If radiotracers are used, then the sample can still be profiled and a diffusion coefficient determined.

9-12. PROBLEMS ENCOUNTERED WITH OTHER ANALYTICAL TECHNIQUES

Not all analytical laboratories have radioactive-tracer facilities. Other analytical chemical methods can be used but only with considerable difficulty. The main reason for these difficulties is the low level of impurity concentration diffused into the semiconductor coupled with shallow diffusion depths. For example, in Fig. 9-9, a typical diffusion of phosphorus into 10^{14} atoms/cm^3 boron-doped silicon, the total number of phosphorus atoms in the entire diffusion is only 1×10^{15} atoms or 0.05 μg phosphorus. Table 9-2 shows the levels of phosphorus that would have to be determined if 10 laps of 1 μ were made on a 1-cm^2 diffused slice of silicon. Almost any analytical technique would be hard pressed to analyze for these levels of phosphorus. The reagent blanks for any analytical method would make the validity of final results questionable. Nevertheless, emission spectroscopy has been used in a study of the diffusion of magnesium into gallium arsenide[27] (radiotracer techniques could not be used here because of the low specific activity and prohibitively high price of magnesium-28). In this work, the diffused slice was incrementally etch-

Table 9-2. Amount of Phosphorus in Incremental Laps for Typical Diffusion Profile into Silicon (1 cm^2 Surface Area)

Depth, μ	Total P, atoms	Avg P conc, atoms/cm^3	Weight P in lap, μg
0–1	4.5×10^{14}	4.5×10^{18}	2.2×10^{-2}
1–2	3.0×10^{14}	3.0×10^{18}	1.5×10^{-2}
2–3	1.5×10^{14}	1.5×10^{18}	7.5×10^{-3}
3–4	7.0×10^{13}	7.0×10^{17}	3.5×10^{-3}
4–5	2.0×10^{13}	2.0×10^{17}	1.0×10^{-3}
5–6	6.0×10^{12}	6.0×10^{16}	3.0×10^{-4}
6–7	1.5×10^{12}	1.5×10^{16}	7.5×10^{-5}
7–8	5.0×10^{11}	5.0×10^{15}	2.5×10^{-5}
8–9	1.2×10^{11}	1.2×10^{15}	6.0×10^{-6}

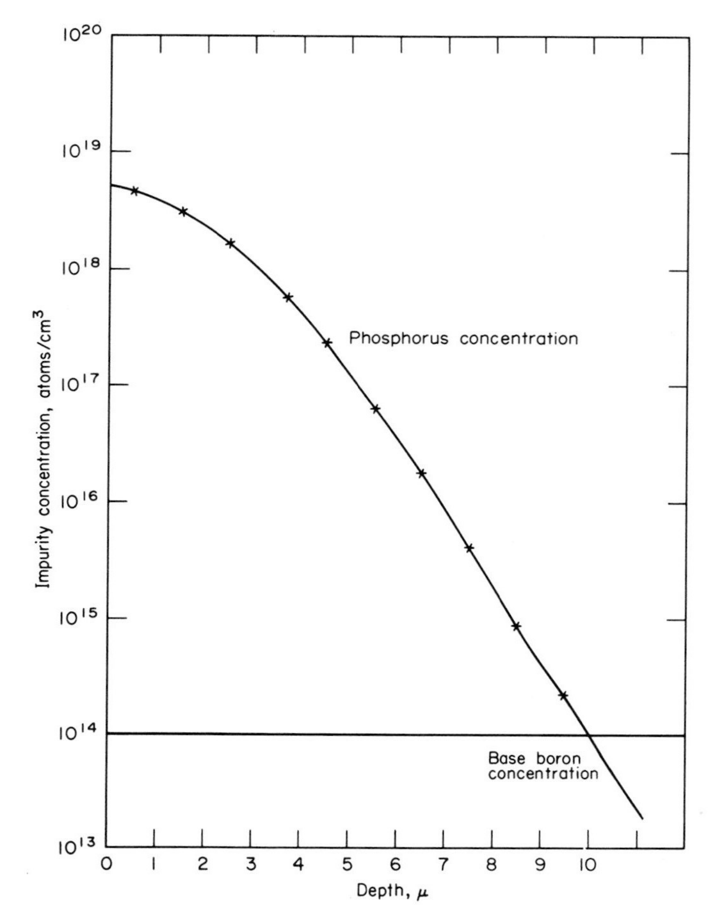

Fig. 9-9. Diffusion profile for phosphorus diffused into silicon with a boron concentration of 1×10^{14} atoms/cm^3.

lapped, and the magnesium concentration in the etch was determined by emission spectroscopy. The sheet resistance measurements carried out at the same time indicated a one-to-one correlation between chemical and electrical carrier concentrations. It is only under almost ideal conditions that this approach can be fruitful.

9-13. RADIOCHEMICAL DIFFUSION TECHNIQUES

Radiochemical techniques are ideally suited to the study of the diffusion of impurities into semiconductors. Reagent blanks and sample contamination are virtually nonexistent. It is also possible to choose the sensitivity one needs in any particular problem by adjusting the specific activity. In semiconductor diffusion studies, it is convenient to use the units atoms/cpm for specific activity. It is necessary for the radiochemist to know the lowest and highest concentration that will be encountered in any diffusion problem and then to adjust the specific activity by adding the required amount of radioactive isotope to the inactive species. It is important to keep in mind that unless the radioactive isotope is supplied "carrier-free,"† there will be some inactive material associated with the radioactive species. In the example given in Fig. 9-9 and Table 9-2 for the diffusion of phosphorus into silicon, a specific activity of around 1×10^9 atoms/cpm would give 20 cpm in lap 10 and 5.4×10^5 cpm in lap 1. The presently available beta-counting equipment could handle these levels of activity without difficulty.

It is also possible, and quite often preferable, to use activation analysis when studying the diffusion of impurities into a semiconductor, such as silicon, which will not produce activities that interfere in subsequent lapping and counting operations. The only silicon isotope that will activate is silicon-30, and it is only 3.12 percent of all the stable silicon isotopes with a relatively low capture cross section of 0.11 barn. The product of the neutron irradiation ^{30}Si (n,γ) ^{31}Si has a half-life of 2.65 hr. It will quickly decay to a level that will not interfere in the counting of the activated impurity under study. Once the neutron activation is completed, there are no longer any contamination problems since the impurity under study is radioactive and the probability of contaminating with the same radioactive species is remote. This technique of activation analysis is particularly attractive to the materials scientist because all sample preparation can be carried out away from the radiochemistry laboratory by using standard diffusion procedures with inactive materials. It is then possible for the analytical chemist to analyze and determine diffusion profiles and diffusion coefficients from samples prepared in a production area, or in a solid-state physicist's laboratory. If the alternative technique using a radioactive tagged impurity in the diffusion study is chosen, then it is necessary that the diffusion and sample handling be performed under the direction of the radiochemist or radiation safety officer and preferably in a controlled area such as the radiochemistry laboratory.

9-14. EXPERIMENTAL TECHNIQUES

There are five basic steps to be carried out in the radiotracer study of the diffusion of an impurity into a semiconductor: (1) semiconductor sample preparation, (2) the diffusion process, (3) incremental lapping of the diffused sample, (4) radiochemical assay of each lap, and (5) reduction of the raw data and the determination of D, the diffusion coefficient.

9-15. SEMICONDUCTOR SAMPLE PREPARATION

The importance of the preliminary sample preparation for diffusion cannot be overemphasized. The results of the entire diffusion depend on these initial steps; and if any errors or omissions are made, then the resulting diffusion coefficient and diffusion profile have little meaning. If electrical measurements are to be made concurrently, then every effort must be made to prevent the introduction of other

†Carrier-free radioisotopes have no added inactive isotopic carrier and have the highest attainable specific activity.

impurities that will diffuse into the semiconductor along with the impurity under study. In Table 9-2, it can be seen that during the last five laps of the diffusion profile only 0.004 μg of electrically active impurity is present. An equal amount of electrically active impurity of opposite type would drastically change the position of the electrical p-n junction. It would be very difficult to correlate radiotracer measurements with electrical measurements, and misleading conclusions about the electrical activity of the impurity of interest would be possible. Contamination of the semiconductor slice by contact with etch and wash solutions is one of the most probable causes[28,29] (Sec. 7-15). The importance of the use of very pure distilled or deionized water is of course obvious. Since many impurities are adsorbed by a Freundlich type of physical adsorption mechanism, it is imperative that flowing rinses be employed. Batch rinsing techniques are not effective unless they are replaced after each rinse and a large number are used in succession.

In principle, any size of sample can be used to study diffusions in semiconductors. In practice the sample can be too small or too large. If the diffused portion of the semiconductor is to be incrementally lapped at 1 μ intervals, then a surface area of 1 cm^2 will yield a volume lap of 10^{-4} cm^3. Generally, the amount of material removed is determined by weighing the sample both before and after the lap. A 10^{-4} cm^3 volume lap weighs 232 μg for silicon and 577 μg for indium antimonide. Weight changes of this size can be readily determined by using a microbalance. If it becomes necessary to use a smaller sample, then the weight of each lap is proportionately smaller, and the accuracy of the depth removed is also smaller. It is often necessary to use less than 1 μ laps if the diffused layer is very thin. In this case a sample size of much less than a square centimeter does not leave much weight in each lap. Too large a sample (for example, a 3.8 cm slice or 11.2 cm^2) presents difficulties in lapping, particularly mechanical lapping, leading to a nonplanar front. Problems encountered in lapping will be discussed later.

The semiconductor samples used for diffusion will generally be sawed slices between 500 and 1,250 μ in thickness. All sawed semiconductor surfaces are abrasive-damaged as deep as 80 μ;[30,31] this must be removed prior to diffusion (Sec. 7-1). Removal of work damage is accomplished by polishing with successively finer polishing compounds on a flat glass plate. This polishing also flattens the face of the slice so that subsequent lapping of 1 μ increments into the diffused layer is planar. It has been found that in polishing, the crystal should be moved over the abrasive in a figure eight rather than a circular motion. If the crystal is rotated frequently in this fashion, the edges will not round and the slice will be polished quite flat.

Following the finest polishing grit, it is always necessary to etch the semiconductor slice. This etch removes the final small amount of surface mechanical damage left by the polishing compound. The etch also removes the surface impurities and should leave a highly polished and flat, clean surface ready for diffusion.

The flatness of the surface is usually determined by placing an optical flat on the polished face and looking at the optical pattern or fringes at the interface. Figure 9-10 shows one of the fringe patterns observed. If the slice is not flat, the fringes are curved. If the flat portion is not large enough for profiling, then it is necessary to relap and again etch the slice.

The etches recommended for polishing are shown in Table 9-3. If the wrong etch

Fig. 9-10. Typical fringes observed when a flat semiconductor is viewed through an optical flat. (*Photograph courtesy of Crane Packing Co.*)

is chosen, the semiconductor slice will be unevenly etched, or etch pits will develop, or oxides and other surface films will cover the face of the crystal.

Semiconductor diffusions are almost always carried out in quartz tubes. In the production areas, these are usually open tubes with doped gases (for example, PCl_3, BBr_3) flowing over the slices. However, in research studies of impurity diffusions, particularly in radioactive studies, the diffusions are carried out in sealed quartz ampules. Figure 9-11 shows the shape of the ampules used. The bulb on the end in used to hold the dopant and aids in keeping it from contacting the semiconductor slice during diffusion.

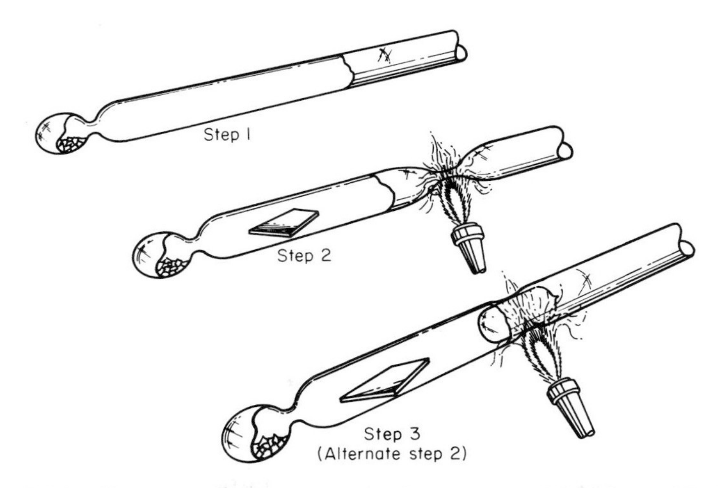

Fig. 9-11. Quartz ampules in preparation for vacuum sealoff prior to diffusion.

The quartz ampule is cleaned prior to introducing the dopant or the semiconductor slice by soaking for 1 to 2 hr in 1:10 HF–H_2O, rinsing well with deionized water, and drying by pumping a good vacuum on the tube. The outside of the tube is usually fired, during the pumping, with a torch to ensure that all moisture has been removed from the walls of the tube.

When the dopant and the semiconductor slice have been introduced into the tube, the upper end is necked down, as shown in step 2 of Fig. 9-11, by using a hydrogen-

Table 9-3. Etches Recommended for Polishing Semiconductor Faces to Ensure a Good Flat for Profiling

Semiconductor	Etch	Composition
Silicon	Planar	15 parts HNO_3, 5 parts CH_3COOH, 2 parts HF
Germanium	CP-4A	3 parts HF, 5 parts HNO_3, 3 parts CH_3COOH
Gallium arsenide	. . . . . .	20 parts H_2SO_4, 75 parts H_2O, 5 parts 30% H_2O_2
Indium arsenide	. . . . . .	99.6 parts CH_3OH,0.4 parts Br_2
Indium antimonide	CP-4A	3 parts HF, 5 parts HNO_3, 3 parts CH_3COOH
Gallium antimonide	. . . . . .	1 part HF, 9 parts HNO_3

oxygen torch. If the tube is large, it is often easier to introduce a sealed quartz bulb, as shown in step 3. The tube is then pumped down to less than 1 μ of pressure and the tube sealed off by heating the necked-down portion in step 2, or around the sealed bulb of step 3. The sealed ampule is ready for insertion into the furnace for diffusion.

9-16. DIFFUSION PROCESS

In sealed-tube diffusions, it is important that there be no temperature gradients from one end of the tube to the other, or the dopant, or even the arsenic from a III-V compound semiconductor such as gallium arsenide, will move to the cool end of the ampule and negate the entire experiment. The ampule is introduced into the furnace with the dopant bulb going in last at the beginning of the diffusion and coming out first at the end of the diffusion. This ensures that a minimum of the dopant will condense on the semiconductor slice since the slice will always be warmer than the dopant bulb.

At the end of the diffusion time, the quartz ampule is removed and usually air-quenched. When the ampule is broken open, the sample is etched in some etch that will remove any excess or condensed radioactive impurity on the surface (e.g., for zinc-65 in GaAs, use warm 12 *M* HCl). This etching is repeated until no more radioactivity can be detected in the etch solution. This etching step is important because the only radioactive impurity of interest is that which has

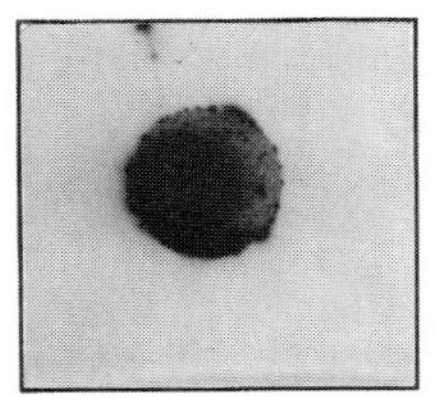

^{198}Au in Si

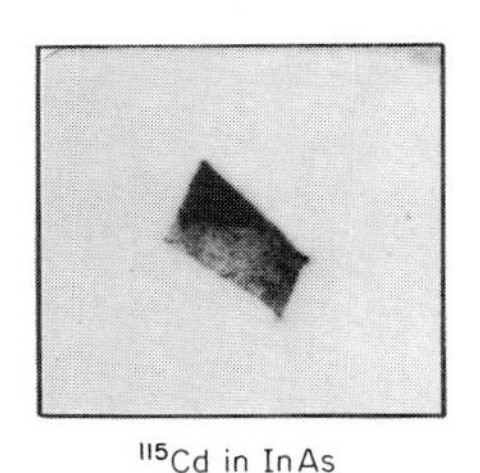

^{115}Cd in InAs

^{35}S in GaAs

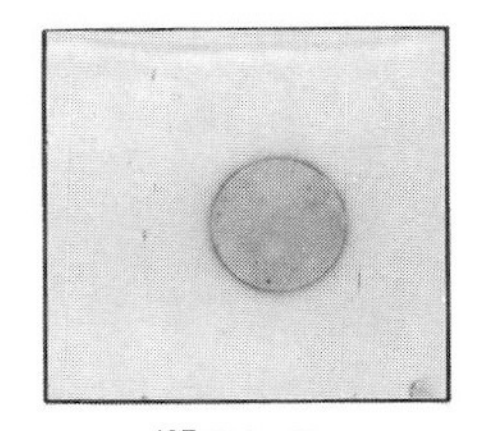

^{197}Pt in Si

Fig. 9-12. Series of autoradiograms taken after diffusion, showing the types of nonuniform impurity distribution that can be encountered.

dissolved and diffused into the semiconductor. If this etching is not carried out, artifically high surface concentrations N_0 will be obtained and will complicate data interpretation.

9-17. AUTORADIOGRAPHY

Autoradiograms are always run on the diffused sample after surface cleaning to look for nonuniform diffusion. Figure 9-12 shows a series of autoradiograms that were taken prior to lapping the diffused area. As can be seen, failure to take autoradiograms will always leave doubt about the uniformity of the radioactive impurity over the surface. This is particularly true in dealing with a new impurity whose diffusion characteristics are not well known. The autoradiogram will also aid in choosing a uniform area to be cut out of the crystal slice for use in profiling. In any case it is necessary to remove the edges of the slice either by scribing or lapping to remove the *edge effect,* which sometimes occurs during profiling. The radioactive impurity diffuses in from the edge as well as the two faces of the crystal, and this edge diffusion must be removed before starting to lap into the crystal face. Failure to remove these edges completely will cause a tailing in the concentration profiling, as shown in Fig. 9-13.

9-18. DETERMINATION OF IMPURITY-CONCENTRATION DIFFUSION PROFILE

The technique most frequently used to determine the concentration-versus-depth profile involves successive removal of layers. In practice, if the radioactive impurity under study is a gamma emitter, the layers that have been lapped away can be counted directly by using a sodium iodide detector. An impurity that is a pure beta emitter must be considered on its own merits, depending on the energy of the beta. For example, in a study of the diffusion of phosphorus in silicon, the only radioactive isotope available is ^{32}P, which has a beta maximum energy of 1.74 Mev. A beta with

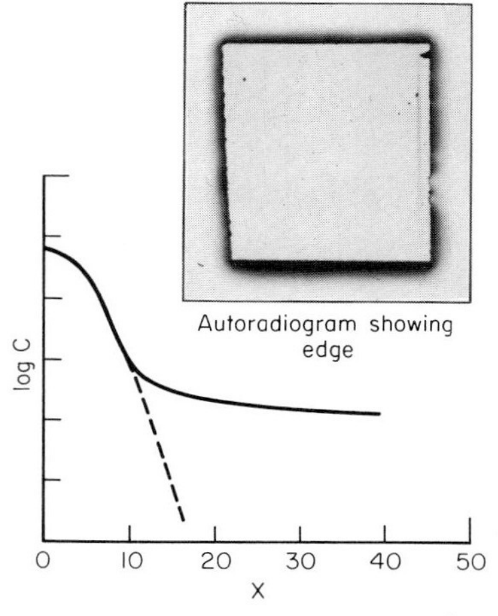

Fig. 9-13. Autoradiogram showing the heavy concentration at the edges of a crystal after lapping through the diffused face. The effect of these edges on the diffusion profile is to cause a long tail, as shown on the graph.

this energy is comparatively easy to count and will readily penetrate the small amounts of lapping compound used to remove the layer. As long as the standards are prepared in exactly the same manner, good counting accuracy is obtained. Isotopes such as ^{35}S with E_{max} 0.167 Mev, ^{14}C with E_{max} 0.155 Mev, and ^{45}Ca with E_{max} 0.254 can be handled in the same manner but with some sacrifice in accuracy due to absorption of the weak beta in the lappings.

9-19. MECHANICAL OR HAND LAPPING

A number of mechanical devices have been developed and used to lap[32-35] away successive layers of the diffused slice. These mechanical lapping machines all move the crystal slice in the same direction against the lapping pad and lapping compound. This causes one side of the crystal to be preferentially worn away, and successive laps will not be planar with the face of the crystal slice.

The hand lapping technique used in the Texas Instruments laboratories has proved to be one of the most reliable abrasive methods found. In this technique, illustrated in Fig. 9-14, high-density alumina lapping disks that have been ground optically flat are used both to lap on and to count the lappings removed. In this technique, a small amount of lapping compound, e.g., Linde AA, 0.05 μ slurry, is placed on the optically flat and highly polished lapping disk. The previously prepared diffused slice, with edges removed, is placed face down in the lapping slurry and lapped in a figure-eight motion using the finger (protected with a finger cot). This figure-eight motion is moved around the disk, and the crystal itself is rotated frequently. This lapping is continued until sufficient material has been removed; usually about 1 min is sufficient. The crystal is lifted from the lapping plate and carefully wiped clean with small pieces of tissue moistened with alcohol. These tissue wipes and the finger cot are carefully placed on the lappings, and the disk is carefully wrapped in thin plastic film and gamma counted.

The results of this lapping technique are recorded on the "diffusion profile" worksheet shown in Fig. 9-15. The data from this worksheet can be hand calculated or transferred to IBM data cards for computer reduction (Sec. 9-24).

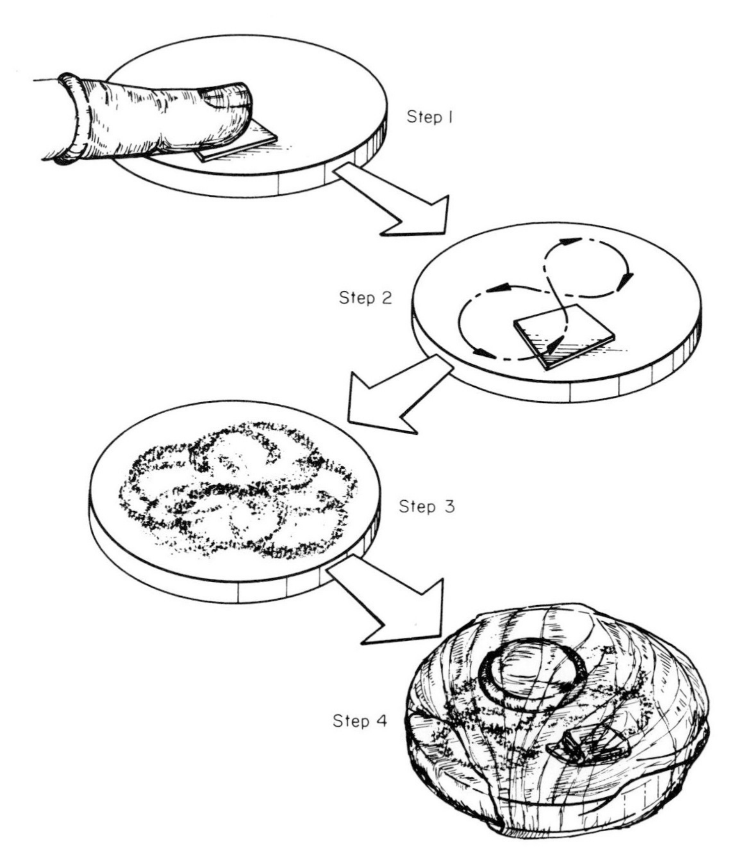

Fig. 9-14. The mechanical lapping technique using optically flat alumina disks.

9-20. ETCH LAPPING

A second lapping technique that works better and is easier to use than abrasive lapping is to etch away successive layers of the diffused slice. This technique can be applied only if an etch is available that will remove the entire crystal face in a planar manner. Very few etches will dissolve a surface in this manner, and most will generally etch more at the edges than in the center of the slice. Also, unless the crystal is of fairly high perfection, etch pits will develop and invalidate the experiment, although proper choice of an etch can minimize this.

In this technique, the back of the diffused slice, with edges removed, is masked with some etch-resistant material. Often Apiezon wax or other wax is melted onto the warmed slice to seal the back. Polypropylene tape has been found to seal effectively against many types of etches. Some etches used for this type of incremental lapping of diffused slices are shown in Table 9-4.

Table 9-4. Etches Used for Etch Lapping of Semiconductors in Diffusion Studies

Semiconductor	Etch composition by volume	Etch rate, μ/min
Silicon	15 parts HNO_3, 5 parts CH_3COOH, 2 parts HF (3-day life)	8
Germanium	5 parts silicon etch, 2 parts CH_3COOH (3-day life)	3.5–4.0
GaAs	20 parts H_2SO_4, 75 parts H_2O, 5 parts 30% H_2O_2 (mix H_2SO_4, H_2O; chill; then add H_2O_2)	2.1
InAs	CH_3OH:0.1–1% Br_2 (vol)	0.5–2.5

9-21. ELECTROCHEMICAL LAPPING

Silicon can be anodically oxidized in certain electrolytes[36,37] to yield a very thin uniform silicon dioxide layer. The forming rate of this oxide is approximately 4Å /volt. The upper voltage limit is around 500 volts, which means that the largest layer that can be formed or, conversely, can be removed from the silicon is in the order of 2000 Å. However, for quite shallow diffusions, laps on the order of 1000 to 2000 Å (0.1 to 0.2 μ) are preferable, and this technique is to be preferred over earlier discussed methods. Unfortunately, n-type silicon is quite difficult to anodize because conduction in the electrode is by the limited supply of minority carriers, holes. Schmidt[36] discusses these problems in his paper (see also Sec. 10-4). Following the anodic formation of the silicon dioxide, the film is removed with hydrofluoric acid and the solution counted. The procedure is repeated for successive laps until no further radioactivity is detected. The thickness of SiO_2 formed can be determined by ellipsometry (Sec. 10-15).

9-22. RADIOCHEMICAL COUNTING TECHNIQUES

The quantitative evaluation of the amount of radioactive material in the bulk semiconductor or incremental lap must rest upon the ingenuity of the radiochemist in devising effective counting systems. Counting gamma emitters really presents few problems as long as sufficient care is taken in preparing the standards. Such factors as duplication of geometric size and proximity of sample and standard are important. Problems with signal to background or interference from other radioactive species present can often be circumvented by simply moving the zero base line of the counting system. For example, sodium-24 can be readily counted in the presence of silicon-31 and other activities normally encountered in neutron activation of ultrapure silicon such as copper-64 and gold-198. This is accomplished by moving the zero base line up to 1.30 Mev, which excludes all gamma radiation with energy less than 1.30 Mev but includes both sodium photopeaks, 1.36 and 2.75 Mev, where most of the sodium-24 activity is located. The background is always considerably lower when the base line is raised, which of course shows up as better signal to noise (generally evaluated as S^2/N) and, usually, lower detection limits.

Counting through single-channel pulse-height analyzers can be equally effective.

The semiconductor samples and the incremental laps which are normally counted are comparatively thin. As a result, gamma counting virtually eliminates the sample self-absorption problem which is always of concern in beta counting.

In neutron activation analysis of the diffused semiconductor slice it is indeed very infrequent that the only activity produced is the one of interest. Some commonly encountered isotopes include gold-198, copper-64, and sodium-24. If the sample were deliberately doped before diffusion, then dopants such as arsenic-76, antimony-124, and gallium-72 would complicate straightforward gamma counting. When other extraneous activities are present, it is necessary to use gamma-ray spectroscopy to assay the sample. The interpretation of a complex gamma spectrum to obtain quantitative results is difficult. A computer program developed by Helmer et al.[38] has proved to be very useful in the reduction of complex gamma spectra obtained during neutron activation analysis of silicon samples. This program is a linear least-squares analysis and, on the assumption that all components of the spectra are known, minimizes the sum of the residuals in the gamma spectra (Sec. 5-5).

Often the diffusant under study does not have a radioactive isotope that is amenable to gamma counting. Either it is a pure beta emitter, e.g., phosphorus-32, sulfur-35, or nickel-63, or the gamma yield from the decay scheme is very low, e.g., cadmium-115^m. Each beta-counting problem must be handled separately on its own merits.

The direct measurement of the beta activity from the sample would at first appear to be a questionable technique. There are three complicating factors. The first two are basic to beta counting and assume that the impurity is uniformly distributed into the bulk (which is, of course, not true in a diffused sample). Firstly, unlike alpha and gamma emission from radioactive nuclei, the energy distribution or spectrum for beta emitters is a continuum ranging from zero energy up to a maximum value which is characteristic of that nuclear species. Secondly, beta particles are so light that they are easily deflected as they traverse a thickness of the sample. This absorption is represented by the exponential

$$I = I_0 e^{-\mu x} \tag{9-13}$$

where I = intensity of radiation transmitted
I_0 = intensity with no absorption
x = thickness traversed
μ = linear absorption coefficient

Each beta particle will have a discrete energy from the continuum and will have its own linear absorption coefficient. It will be absorbed differently, and the amount of absorption will depend upon the depth x in the sample at which it originated. All diffused samples will have the radioactive beta emitter distributed into the sample according to a gaussian or erfc distribution. Both of the above effects will be combined with the effect due to the diffused-impurity distribution profile, and the direct beta counting of the sample will be complicated.

If the depth of diffusion of the impurity is shallow, and this is most often the case with semiconductor diffusions, then it is possible to count the sample directly. In this method, the sample is counted after each successive lap, and the difference between two consecutive laps is assumed to have been the activity in the earlier lap. This technique works well, provided the difference in counts between the two laps is greater than the expected two-sigma variation of the total number of accumulated counts in each lap. The validity of this approach was verified in the Texas Instruments laboratories when it was shown that the distribution profiles of sulfur diffused into gallium arsenide obtained by beta counting the sample as described above and by precipitation and counting the sulfur-35 in the etch lap were the same.

The real criterion that determines whether this beta-counting technique can be used for any given beta emitter is the depth of the diffusion. A good rule of thumb that can be used is one-half the range of the beta in the semiconductor material. This rough measure should not be applied rigorously, but it does serve as a useful guide.

9-23. ANALYSIS OF EXPERIMENTAL RESULTS

At the end of the experimental portion of the study of a diffusion of an impurity into a semiconductor, the investigator finds himself faced with a mass of data that must be reduced to some meaningful form. Usually, a number of diffusion experiments have been run since it requires very little more effort to diffuse and lap four samples than one because of delays in counting and weighing; one operation can be performed while the other is waiting.

Where there are large amounts of data to be processed, the use of a computer will give faster, more accurate results and cost less than hand calculation by a technician. The amount of computer time used is minimal, probably in the order of 1 sec, with the rest of the time used for reading in the data cards and printing out the results. A typical total time for four samples would be in the order of 30 to 40 sec and would cost from $2 to $3, depending on the type of computer used.

The computer programs used in the diffusion studies in Texas Instruments laboratories are designed to be used sequentially. The first program, designated L002, is a general data-calculation program that reduces the experimental data to a form that can be analyzed by the chemist for subsequent data processing. It also punches out the results on IBM cards for use in subsequent computer programs at the same time that it prints the results.

9-24. DATA REDUCTION

The experimental data from the "diffusion profile" sheet (Fig. 9-15) are transferred to an IBM data sheet and punched onto IBM punched cards. These cards are then arranged or stacked in order and placed with the program deck in the computer. The flowchart for this first computer program L002 is shown in Fig. 9-16. The form of the output from the computer is shown in Fig. 9-17, and, as can be seen, there are several ways of manipulating the data to aid in the subsequent analysis. The program estimates the surface concentration C_0 by a least-squares fitting of the data points and extrapolation back to zero depth. The concentration for each incremental lap is calculated, and the depth for that concentration is calculated

DIFFUSION PROFILE TI–4506

SAMPLE NUMBER	DATE	PAGE

WEIGHT OF SOURCE	ACT. COEFFICIENT	WEIGHT	THICKNESS	IMG = μ

CONDITIONS:

LAP NO.	TIME	COUNTS	C.P.M.	BKGR COUNT	CPM-BKGR	WEIGHT	WEIGHT CHANGE	μ	CPM/mg	ATOMS CM³	DECAY FACTOR

Fig. 9-15. Worksheet used for diffusion studies.

as the average depth for that lap. For example, if lap 3 had removed 1.00 μ from a depth of 3.00 to 2.00 μ, the concentration for that lap would be calculated at 2.50 μ. The datum of most interest is, of course, C(XAV), the concentration at X(AVE), the average depth for each lap. The counting statistics are evaluated as FY, which states that there are 90 chances out of 100 that the error of the concentration, C(X), corrected for background is less than X.X percent. The remaining columns of data are used to decide what type of distribution or diffusion profile is being analyzed; they are discussed in Sec. 9-25.

9-25. DATA ANALYSIS

The data from the first computer program, described in Sec. 9-24, must be evaluated to determine whether the distribution of diffusing impurity is gaussian or error function and whether the diffusion coefficient is concentration-dependent. It is necessary to determine whether the diffusion profile is made up of more than one diffusing species, say a fast and a slow component, that might cause overlapping error-function profiles. The computer output, shown in Fig. 9-17, has been designed to aid in this evaluation by providing calculated values of certain diffusion results.

If, in the experiment, the amount of dopant or radioactive impurity available for diffusion into the semiconductor were small, or somehow limited, then a gaussian distribution (Sec. 9-3) should be suspected. When this is the case, a plot of the log of the concentration CXAV against the square of the distance into the crystal, X(AVE)**2, should give a straight line. Equations (9-14) and (9-15) are the exponential and natural logarithms of Eq. (9-4) and show why a straight-line relationship applies. The slope of the straight line is $(-0.4343/4Dt)$, and the intercept is $\log Q/\sqrt{\pi Dt}$.

$$\log N_{(x,t)} = \log \frac{Q}{\sqrt{\pi Dt}} - \frac{x^2}{4Dt} \log e \tag{9-14}$$

$$\log N_{(x,t)} = \log \frac{Q}{\sqrt{\pi Dt}} - \frac{0.4343}{4Dt} x^2 \tag{9-15}$$

Figure 9-18 shows an example of the results obtained for a gaussian and an error-function distribution for such a plot. Both diffusions have the same diffusion

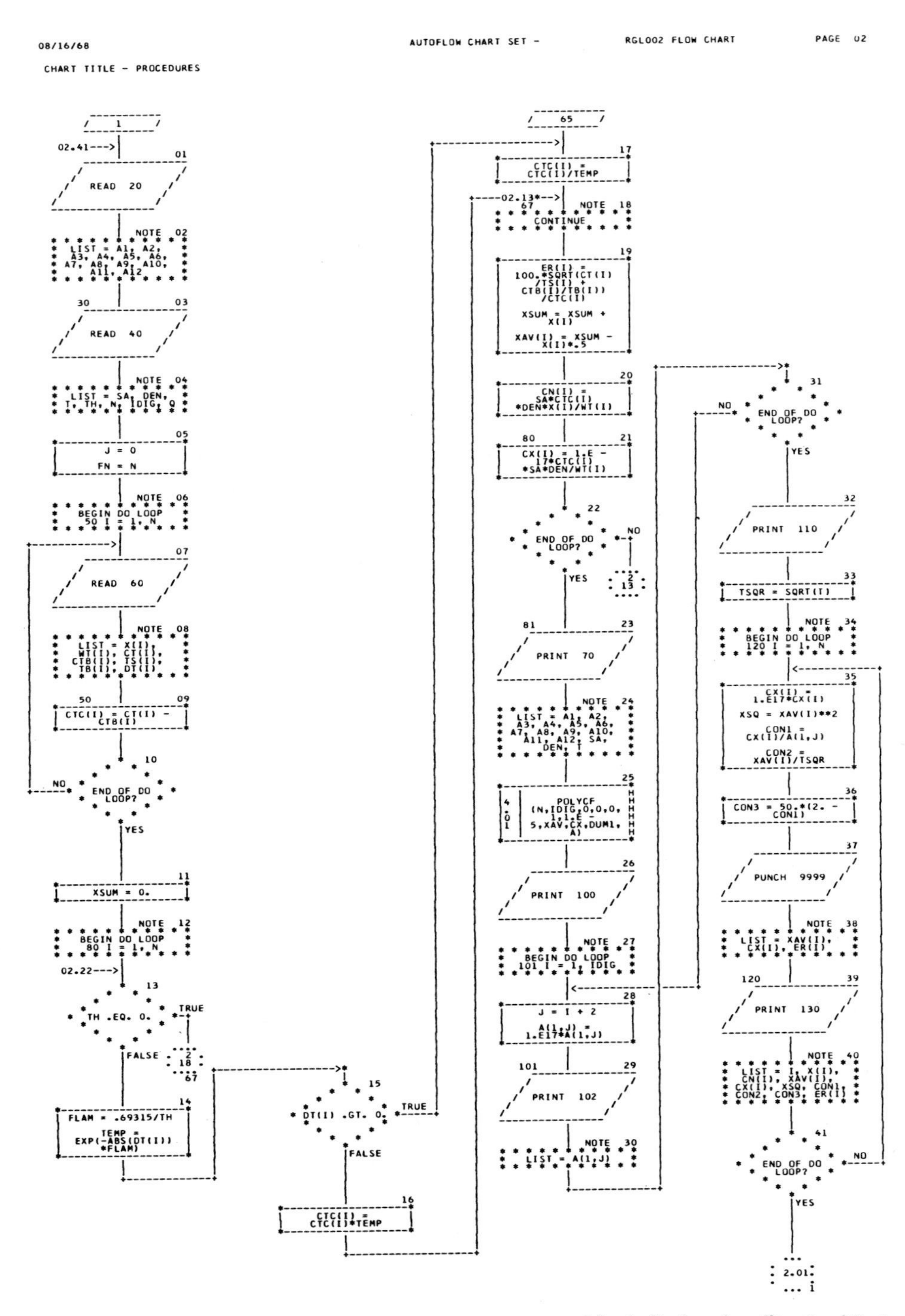

Fig. 9-16. Flow diagram for the L002 computer program used for bulk data handling in diffusion studies using radiotracers.

```
L002,        DIFFUSION PROFILE CALCULATION

             TIN 113 GAAS 6 HOUR DIFFUSION

SA(AT/CPM) =  1.8699996E 11  DEN(MG/CM3) =  5.3200000E 03       T(SEC) =  2.1600000E 04

 CO(AT/CM3)

6.5123344E 07
4.2288184E 02
3.1639723E 16
```

LAP	X	C(X)	X(AVE)	C(XAV)	X(AVE)**2	C(XAV)/CO	XAV/SQ(T)	50(2-C/CO)	FY
1	5.5000E-04	1.777E 13	2.7500E-04	3.231E 16	7.562E-08	1.021E 00	1.871E-06	4.894E 01	5.32
2	9.2300E-04	2.173E 13	1.0115E-03	2.354E 16	1.023E-06	7.440E-01	6.882E-06	6.280E 01	4.47
3	1.2760E-03	2.822E 13	2.1110E-03	2.212E 16	4.456E-06	6.990E-01	1.436E-05	6.505E 01	3.58
4	1.2560E-03	2.474E 13	3.3770E-03	1.970E 16	1.140E-05	6.225E-01	2.298E-05	6.887E 01	4.00
5	1.4760E-03	1.863E 13	4.7430E-03	1.262E 16	2.250E-05	3.990E-01	3.227E-05	8.005E 01	5.10
6	1.6350E-03	1.648E 13	6.2985E-03	1.008E 16	3.967E-05	3.186E-01	4.286E-05	8.407E 01	5.68
7	1.8920E-03	1.030E 13	8.0620E-03	5.442E 15	6.500E-05	1.720E-01	5.485E-05	9.140E 01	8.68
8	2.1090E-03	7.526E 12	1.0062E-02	3.569E 15	1.013E-04	1.128E-01	6.847E-05	9.436E 01	11.62

```
IHC217I

TRACEBACK FOLLOWS-   ROUTINE     ISN   REG.  14

                     IBCOM             A2044B74

                     MAIN              000044E8

ENTRY POINT=  40042C80
```

Fig. 9-17. Computer output from the L002 computer program.

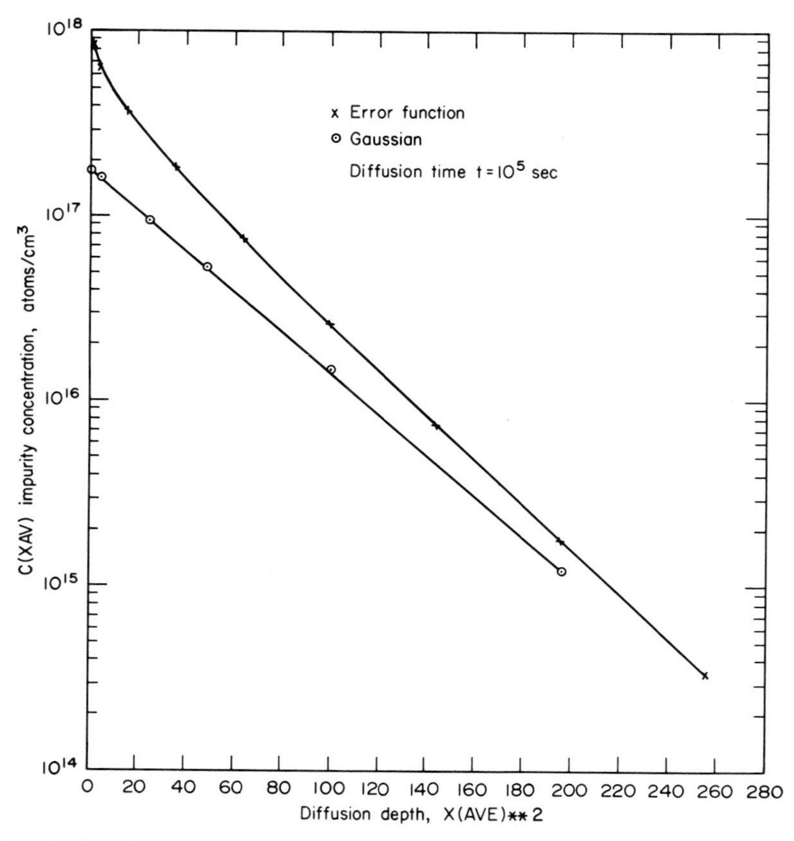

Fig. 9-18. A plot of average concentration versus the square of the diffusion depth, showing the difference between an error function and a gaussian distribution.

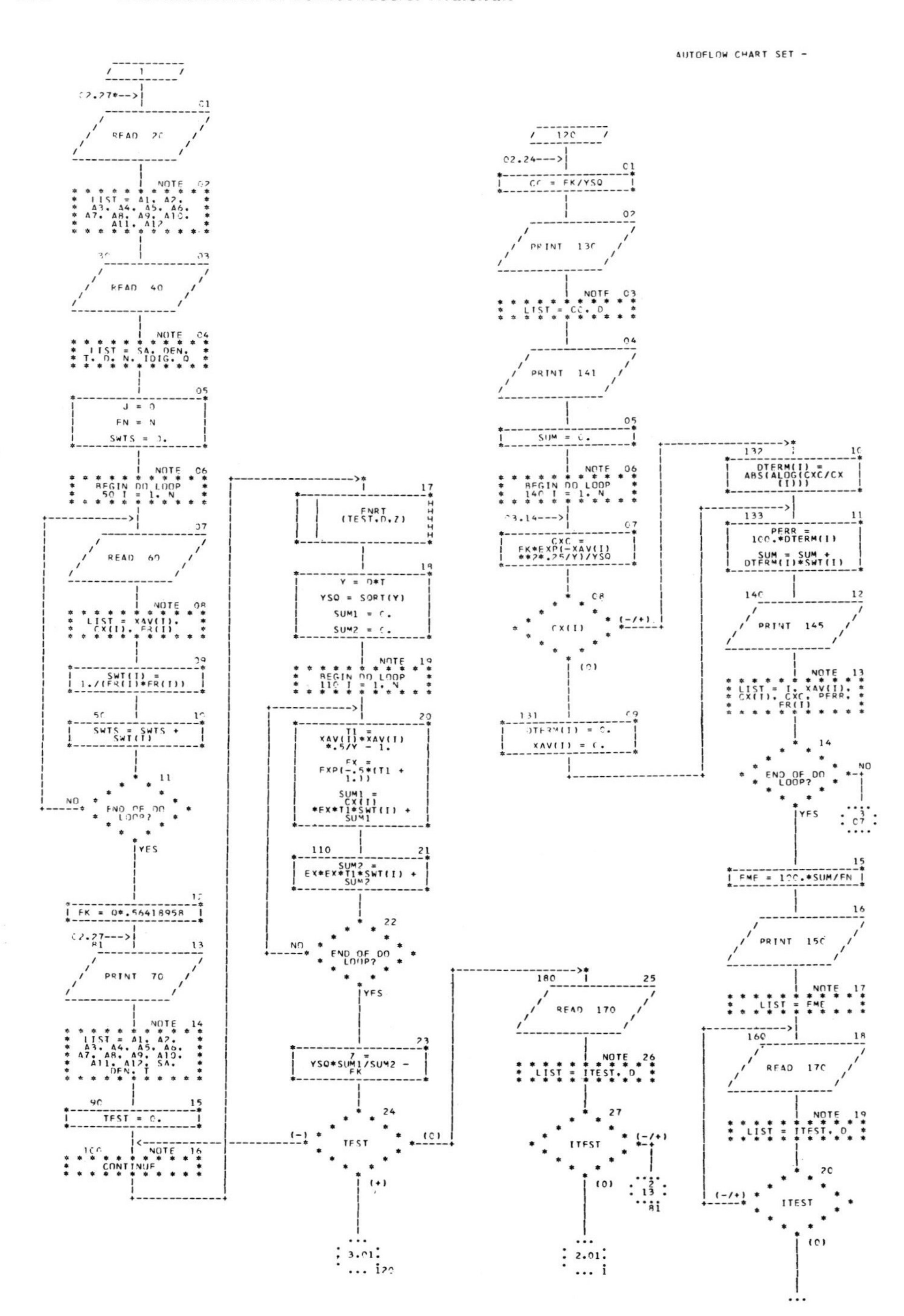

Fig. 9-19. Flow diagram for the L022 computer program used for fitting diffusion data from a limited-source diffusion to a gaussian distribution and determining the diffusion coefficient.

```
     L022          DIFFUSICN IN GALLIUM ARSENIDE, LIMITED SOURCE

                              PROGRAM  TEST  GAUSSIAN

  SA(AT/CPM) =  1.000000CE 10  DEN(MG/CM3) =  2.3000000E 03       T(SEC) =  1.0000000E 04

  CO(AT/CM3) =  5.6477924E 17   D(CM2/SEC) =  9.9791300E-13
```

LAP NO.	X(AVE) CM	C(X) ATOMS/CM3	C(X) CALC	ERR (PC)	FY (PC)
1	5.000000E-05	5.320000E 17	5.304917E 17	0.284	1.00
2	1.000000E-04	4.390000E 17	4.396206E 17	0.141	1.00
3	2.000000E-04	2.070000E 17	2.073366E 17	0.162	1.00
4	3.000000E-04	5.640000E 16	5.924784E 16	4.926	1.00
5	4.000000E-04	1.030000E 16	1.025812E 16	0.407	1.00
6	5.000000E-04	1.090000E 15	1.076122E 15	1.281	1.00

```
    M.E.(PER CENT) =  1.2004177E 00
```

Fig. 9-20. Computer output for the best computer fit to the experimental data along with the calculated diffusion coefficient.

coefficient D and were diffused for the same time t. Only the gaussian distribution yields a straight line.

When a straight-line relationship is observed, then the concentration and depth data on the punched cards from the L002 computer program are assembled for the L022 or gaussian computer program. The flow diagram for this program is shown in Fig. 9-19. In the program the data are fitted to Eq. (9-15) after weighting each concentration point by the factor FY, which in turn is based on the counting statistics for that point. A least-squares subroutine is used to analyze the data. The output from the computer program is shown in Fig. 9-20 and includes the best value for D and calculated values of concentration for that D at each depth. The difference between the computer-calculated concentrations for that D and the experimental concentration at each depth is a measure of the accuracy of the curve fitting and the resulting D value.

If the plot of log concentration versus the square of the distance into the crystal is not a straight line, a limited-source diffusion or gaussian distribution can be eliminated, and error-function distributions must be considered. To further confirm this decision and also to determine whether the diffusion coefficient is concentration-dependent (Sec. 9-5), Hall's[39] method of analysis is applied. From the infinite-source solution of Fick's second law of diffusion [Eq. (9-6)], it can be shown that

$$\frac{N}{N_0} = 1 - \operatorname{erf} \frac{x}{2\sqrt{Dt}} \tag{9-16}$$

$$50\left(2 - \frac{N}{N_0}\right) = 50\left(1 + \operatorname{erf} \frac{x}{2\sqrt{Dt}}\right) \tag{9-17}$$

and if D is not concentration-dependent (i.e., mathematically is a constant), then this is a linear equation. When plotted on probability paper where $P = 50(2 - N/N_0)$ is plotted versus $\lambda = x/\sqrt{t}$, a straight line should be obtained with slope $1/\sqrt{D}$. The two values P and λ are available from the output of the first computer program L002 (Sec. 9-24, Fig. 9-17). If the diffusion coefficient is not concentration-dependent, the straight line should intersect at the origin at $P = 50$. Figure 9-21

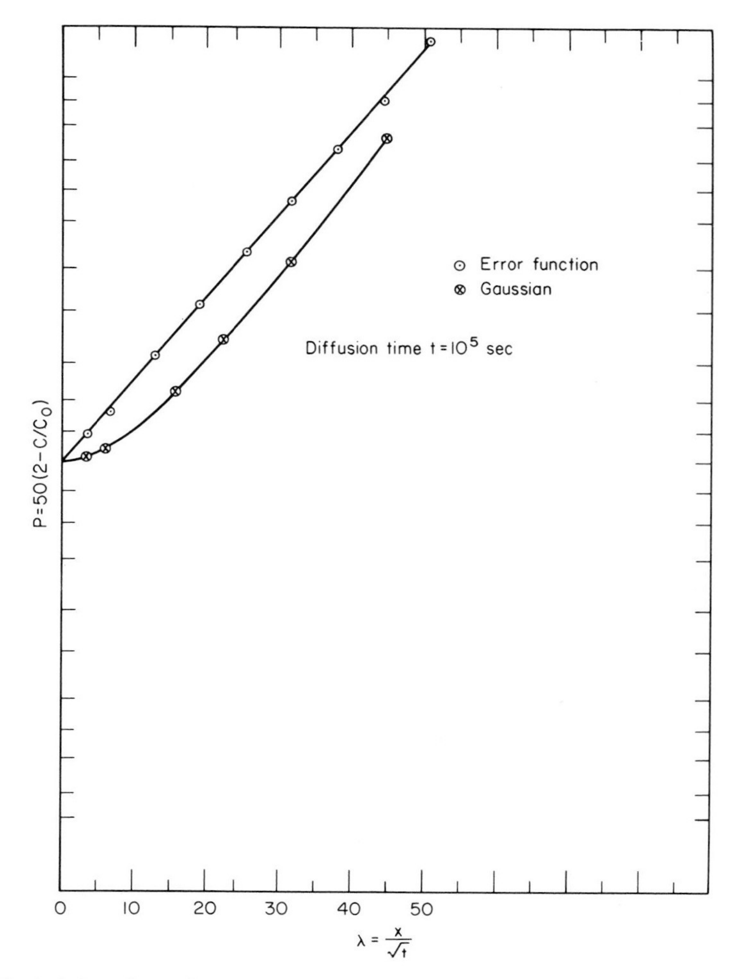

Fig. 9-21. Probability plot of P versus λ to determine whether the diffusion under study is concentration-dependent.

shows the results obtained by plotting a gaussian and an error-function distribution in this manner. As can be seen, the error-function distribution gives a good straight-line fit which intercepts at $P = 50$.

The D value for the error-function diffusion under study can be obtained from the slope of this straight line. However, a computer program L021 has been developed in the Texas Instruments laboratories to carry out an error-function distribution analysis of a set of diffusion data. This program is analogous to that shown in Fig. 9-19 except that the data are fitted to an error function [Eq. (9-6)] after all the concentration data points are weighted by the counting statistics FY. This FY value ensures that any data that were collected near the end of the diffusion tail, where very little radioactivity remains, will not adversely affect the least-squares fitting

```
L021,        DIFFUSICN IN GALLIUM ARSENIDE

                         PRCGRAM TEST   ERROR FUNCTION

SA(AT/CPM) =  1.0000J0CE 1C   DEN(MG/CM3) =  2.3000000E 03        T(SEC) =  1.0000000E 04

CO(AT/CM3) =  1.0009C79E 18    D(CM2/SEC) =  9.9881908E-13
```

LAP NO.	X(AVE) CM	C(X) ATCMS/CM3	C(X) CALC	ERR (PC)	FY (PC)
1	1.0C0CCCE-C4	4.790C00E 17	4.796756E 17	0.141	1.00
2	2.000000E-04	1.57CCCOE 17	1.571966E 17	0.125	1.00
3	3.C000COE-04	3.39CCCCE 16	3.382024E 16	0.236	1.00
4	4.C00000E-04	4.68CCCOE 15	4.657590E 15	0.480	1.00
5	5.C00000E-04	4.C700CCE 14	4.041121E 14	0.712	1.00
6	6.000000E-C4	2.210000E 13	2.186475E 13	1.070	1.00

```
M.E.(PER CENT) = . 4.6065205E-01
```

Fig. 9-22. Computer output for the analysis of experimental data based on an error-function distribution.

program. The form of the output of the program is shown in Fig. 9-22. In the examples used in Figs. 9-20 and 9-22, it can be seen that both have the same diffusion coefficient (1×10^{-12} cm^2/sec) and diffusion time (1×10^5 sec). The difference in surface concentration N_0 and the profile graphically illustrates the sharp differences resulting from different diffusion processes.

If a straight-line relationship was not obtained in either the log versus x^2 (gaussian) or the P versus λ (probability) plots, it must be assumed that a more complicated relationship exists. Generally, the P versus λ plot will show a curved line, which indicates a concentration-dependent D. In this case it is necessary to carry out a Boltzmann-Matano analysis of the data.

A computer program developed by Hartley and Hubbard[40] is used in the Texas Instruments laboratories to carry out this analysis. In this computer program, Eq. (9-9) is solved for D as a function of distance x into the crystal. This is accomplished by using the experimental data, concentration and distance x to construct a concentration-penetration curve using either a first-, second-, or third-degree least-squares polynomial on a plot of Z versus x. The value of Z is determined by normalizing the concentrations and utilizing an error-function relationship. The Matano interface must be found to complete the mathematical solution. The Matano interface is that distance x into the crystal chosen so that area A and area B in Fig. 9-23 are equal. The distance axis is shifted so that $x_m = 0$. Then at each

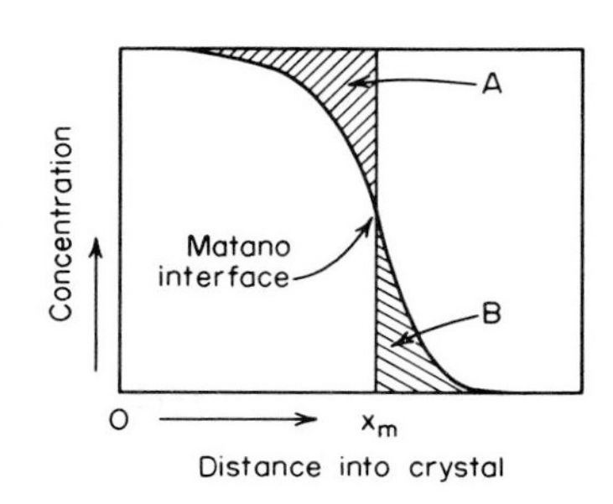

Fig. 9-23. Diffusion profile showing how the Matano interface x_m is determined.

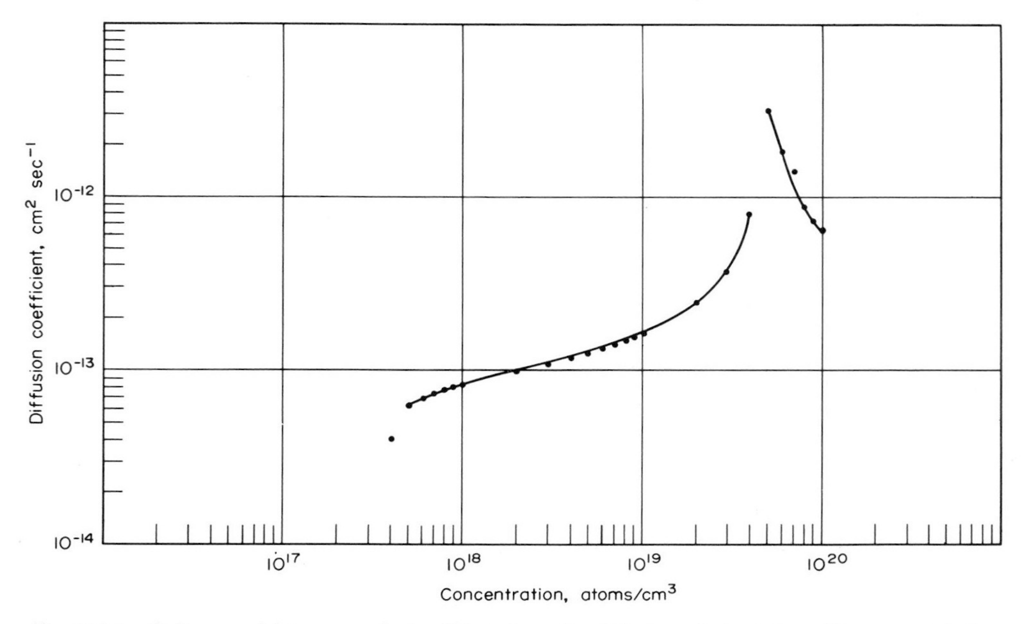

Fig. 9-24. Boltzmann-Matano analysis of data from the diffusion of zinc into gallium arsenide for 100 hr at 700°C.

distance x, which corresponds to a concentration, the slope of the curve is calculated. The slope of the curve is, of course, $\partial x/\partial c$. The diffusion coefficient at that point is the product of $1/2t$, $\partial x/\partial c$, and the area A^* [Eq. (9-18)]. The area A^* is the area in that segment of the diffusion out to the point x.

$$D = -\frac{1}{2t}\frac{\partial x}{\partial c}\int x\,\partial c \tag{9-18}$$

where t = diffusion time, sec
$\partial x/\partial c$ = slope, cm^4/wt
A^* = area = $\int x\,\partial c$, wt/cm^2

The output of reduced data from the Boltzmann-Matano computer program is of course in a different form from that of the previously discussed computer programs. In this case a diffusion coefficient is calculated for each concentration. The data are usually plotted as shown in Fig. 9-24 with D_c versus N. The data shown here are for a diffusion of zinc, using ^{65}Zn, into gallium arsenide.

There are, of course, other solutions to Fick's second law of diffusion, depending on the mathematical boundary conditions. The analyst is cautioned to examine the data carefully in order to ensure that the correct mathematical solution is used to match the experimental diffusion conditions.

REFERENCES

1. Crank, J.: "The Mathematics of Diffusion," Oxford University Press, Fair Lawn, N.J., 1956.

2. Runyan, W. R.: "Silicon Semiconductor Technology," McGraw-Hill Book Company, New York, 1965.
3. Boltaks, B. I.: "Diffusion in Semiconductors," Academic Press Inc., New York, 1963.
4. Shewmon, P. G.: "Diffusion in Solids," McGraw-Hill Book Company, New York, 1963.
5. Mathematical Tables Project: "Table of Probability Functions," vol. 1, Federal Works Agency, Works Projects Administration, New York, 1941.
6. Kendall, D. L.: in R. K. Willardson and A. C. Beer (eds.), "Semiconductors and Semimetals," vol. 4, p. 163, Academic Press Inc., New York, 1968.
7. Kendall, D. L., and M. E. Jones: *Solid State Device Research Conference*, Stanford, 1961.
8. Chang, L. L., and G. L. Pearson: *J. Appl. Phys.*, **35**:1960 (1964).
9. Weisburg, L. R., and J. Blanc: *Phys. Rev.*, **131**:1548 (1961).
10. Matano, C.: *Japan. J. Phys.*, **8**:109 (1933).
11. Boltzmann, L.: *Ann. Physik*, **53**:959 (1894).
12. Donovan, R. P., and R. M. Burger: *Semicond. Prod. Solid State Tech.*, **10**(10):40 (1967).
13. Bond, W. L., and F. M. Smits: *Bell System Tech J.*, **35**:1209 (1956).
14. Yeh, T. H., and A. E. Blakeslee: *J. Electrochem. Soc.*, **110**:1018 (1963).
15. Bertolotti, M., V. Grasso, T. Papa, and D. Sette: *Alta Freq.*, **32**(2):13 (1963).
16. Fuller, C. S., and J. A. Ditzenberger: *J. Appl. Phys.*, **27**:544 (1956).
17. Robbins, H.: *J. Electrochem. Soc.*, **109**:63 (1962).
18. Pankove, J. I.: in P. J. Holmes (ed.), "The Electrochemistry of Semiconductors," p. 290, Academic Press Inc., New York, 1962.
19. Iles, P. A., and P. J. Coppen: *J. Appl. Phys.*, **29**:1514 (1958).
20. Turner, D. R.: *J. Electrochem. Soc.*, **106**:701 (1959).
21. Valdes, L.: *Proc. IRE*, **42**:420 (1954).
22. Smits, F. M.: *Bell System Tech. J.*, **37**:711 (1958).
23. Logan, M. A.: *Bell System Tech. J.*, **40**:885 (1961).
24. Fuller, C. S., and J. A. Ditzenberger: *J. Appl. Phys.*, **28**:40 (1957).
25. Fuller, C. S., J. A. Ditzenberger, and K. B. Wolfstern: *Phys. Rev.*, **93**:1182 (1954).
26. Cunnell, F. A., J. T. Edmond, and W. R. Harding: *SERL Tech.*, **9**:104 (1959).
27. Massengale, J. (Texas Instruments Incorporated): Unpublished work, 1961.
28. Larrabee, G. B.: *J. Electrochem. Soc.*, **108**:1130 (1961).
29. Holmes, P. J.: in P. J. Holmes (ed.), "The Electrochemistry of Semiconductors," p. 331, Academic Press Inc., New York, 1962.
30. Buck, T. M.: in H. C. Gatos (ed.), "The Surface Chemistry of Metals and Semiconductors," p. 107, John Wiley & Sons, Inc., New York, 1960.
31. Jones, C. E., and A. R. Hilton: *J. Electrochem. Soc.*, **112**:908 (1965).
32. Leblans, L. M. L. J., and M. L. Verheijke: *Philips Tech. Rev.*, **25**:191 (1963/64).
33. Slifkin, L. M., and W. M. Portnoy: *Rev. Sci. Instrum.*, **25**:865 (1954).
34. DeBruin, H. J., and R. L. Clark: *Rev. Sci. Instrum.*, **35**:227 (1964).
35. Pawel, R. E., and T. Lundy: *J. Appl. Phys.*, **35**:435 (1964).
36. Schmidt, P. F., and W. J. Michel: *J. Electrochem. Soc.*, **104**:230 (1957).
37. Duffek, E. F., E. A. Benjamini, and C. Mylroie: *Electrochem. Tech.*, **4**:75 (1965).
38. Helmer, R. G., R. L. Heath, R. L. Metcalf, and D. D. Cazier: IDO 17015, 1964.
39. Hall, L. D.: *J. Chem. Phys.*, **21**:87 (1953).
40. Hartley, C. S., and K. Hubbard: *Rep.* ADT-TDR-62-858, Wright-Patterson Air Force Base, Ohio, November, 1962.

10

Characterization of Thin Films

10-1. INTRODUCTION

The thin films discussed in this chapter are the insulating films which have revolutionized semiconductor-device technology. These films are the backbone of the "planar" process, which in turn is the technology which has made integrated circuits and large-scale integration possible. For all intents and purposes the film under consideration is silicon dioxide on silicon. However, thin-film technology is moving rapidly, and silicon dioxide now can be used on germanium, gallium arsenide, or any other semiconductor. Other insulating films such as silicon nitride, aluminum oxide, and titanium dioxide are used in some device areas and will be discussed in lesser detail. However, these last-mentioned insulating films are growing in importance and will present many challenging characterization problems.

In order to appreciate the importance of these insulating films and the role of the analyst in characterizing them for both chemical and physical imperfections, several examples of how these films are used on devices will be given. The process taken as a whole may look very complicated to the reader, but in fact each step is quite simple. However, there are many steps in the fabrication of a device or an integrated circuit. Basically, the process involves growing a thin oxide film, cutting holes in the oxide, and diffusing through these holes. New oxide is grown, new holes are cut, and new areas diffused. This process is continued until the integrated circuit or discrete device is completed. Figure 10-1 shows a typical process starting with a p-type silicon slice that will have an n-type epitaxial layer deposited on it. A 1000 to 2000 Å silicon oxide layer is grown on the surface, and holes or channels are cut in the oxide to allow an impurity to be diffused into selected areas. After epitaxial film growth, a p-type impurity such as boron is diffused, creating isolated islands of n-type material. These islands are isolated physically and electrically and are used to build transistors, diodes, resistors, and capacitors. After the p isolation diffusion, more oxide is grown, more holes cut, and another diffusion carried out. This process is continued until, as shown in the last step of Fig. 10-1, all the devices are complete. At this point, still more oxide is grown to provide surface passivation and protection. The oxide also supports the metallized leads which are used to

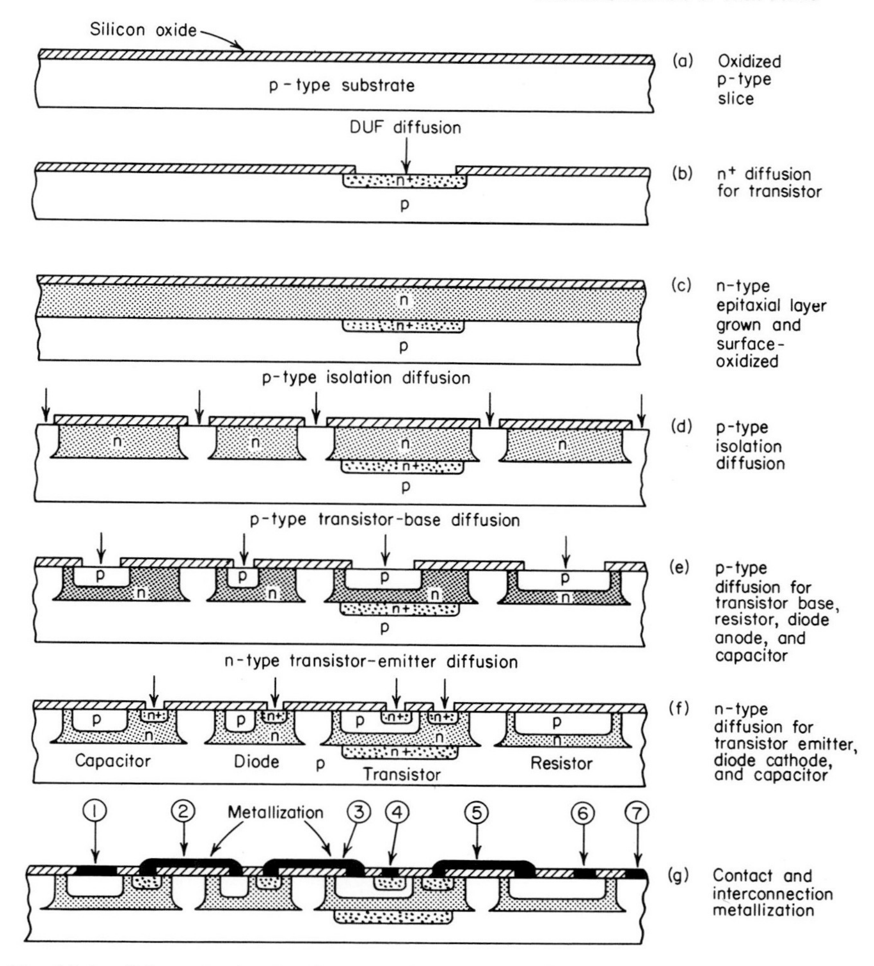

Fig. 10-1. Schematic showing how a wafer is processed in integrated-circuit production.

interconnect the circuit. Electrical contact is made through small holes that are etched into this last oxide.

From the above it can be seen that the oxide film must be of high physical perfection in order to provide a good diffusion barrier for each step of the fabrication process. It must be a highly uniform amorphous film to allow etching of very small (1 mil) holes and channels. The film must be both physically perfect and chemically pure to provide good surface passivation and electrical isolation. In the case of the capacitor, and more particularly in the discrete field-effect transistor, the chemical purity of the oxide is of prime importance because the oxide becomes an active part of the device. In these latter devices a potential is applied across the oxide from a top metal electrode to an underlying semiconductor. Any mobile impurities in the oxide will of course move and cause device instability.

The importance of the chemical and physical integrity of these insulating films

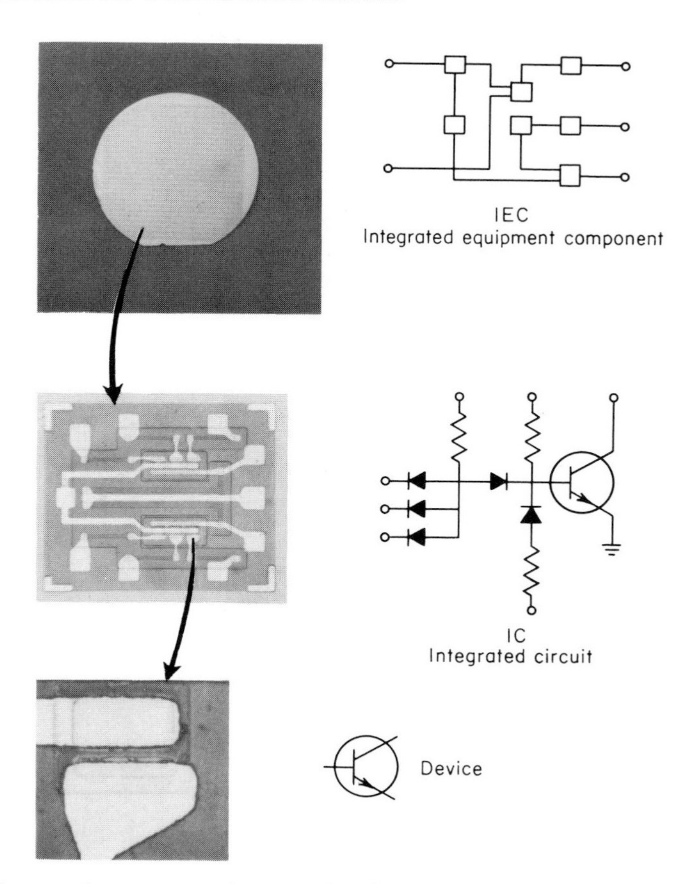

Fig. 10-2. Photographs comparing the sizes of a slice as used for an IEC device, one integrated circuit on that slice, and one transistor in that integrated circuit.

over the entire slice is shown in Fig. 10-2. The top picture shows a typical 2-in. silicon slice which can have as many as 300 to 400 integrated circuits. These circuits can be directly interconnected on the slice, and this is the so-called "LSI" or "large scale integration" device. The circuit can be scribed apart as shown in the second picture, and each integrated circuit can have 80 to 100 discrete devices on a 40 × 40 mil chip. The last picture shows a transistor from the above integrated circuit which illustrates the minute sizes of the individual elements of the transistor. The oxide or insulating film provides the groundwork for the formation of each part of each device on each integrated circuit over the entire slice. The chemist must analyze these insulating films for both chemical and physical imperfections and transmit the results to the engineer or scientist in a form that will result in improved films.

10-2. FILM FORMATION

Since silicon is the most widely used semiconductor, it is perhaps not surprising that the silicon dioxide films on silicon have received the most attention. All silicon

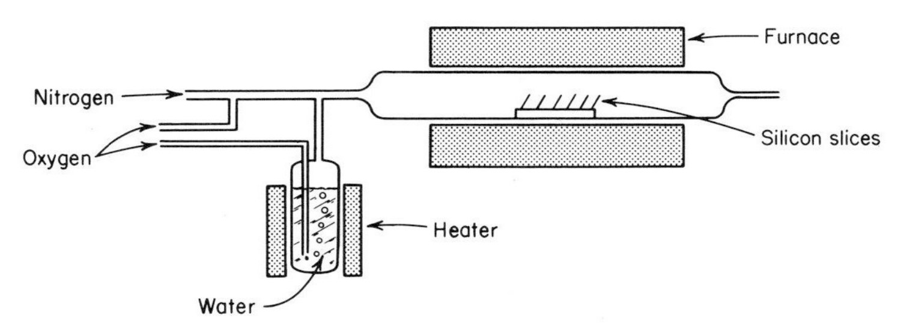

Fig. 10-3. Schematic showing a thermal oxidation system for silicon.

has a native oxide of 10 to 50 Å in thickness on its surface. In device fabrication it is only necessary to continue this natural growth of oxide to the desired thickness (1000 to 10,000 Å). This is most readily accomplished by thermal oxidation.

Other semiconductors such as germanium or the III-V intermetallics do not have native insulating oxides that are usable for these device applications. As a result, it has been necessary to develop methods of depositing SiO_2 or other insulating films on these semiconductors to aid in device fabrication. Chemical deposition from a flowing-gas system has become the preferred technique.

10-3. THERMAL GROWTH

The thermal growth of silicon dioxide on silicon can be carried out in a closed pressure system or a flowing-gas system. In the closed-tube or high-pressure bomb system, the silicon slices are sealed in a bomb made of some inert material with sufficient high-purity water to achieve the desired pressures. The growth of silicon dioxide films in high-pressure steam is linear with time and directly proportional to the pressure. Typical conditions are 59 atm pressure at 650°C for 1 to 2 hr.

The open-tube techniques are the only oxide growth techniques now in use for high-volume routine production. Experimentally, the growth procedure is relatively simple. The silicon slices are stacked vertically in quartz holders and inserted in a tube furnace, as shown in Fig. 10-3. The film growth is then carried out by passing either dry oxygen, wet oxygen, or steam through the tube furnace. Typical oxidation conditions and oxide growth rates are shown in Fig. 10-4.

The final properties of the oxide films are related to the growth conditions since different growth mechanisms are operative in the dry oxygen and steam techniques. Considerable work has been reported on the growth rate and growth mechanisms for these thermally grown oxide films.[1-7,†] In general, it is agreed that a mobile oxygen species diffuses through the oxide film to the silicon-oxide interface, where it reacts to form new SiO_2. In this way the film continues to grow by moving away from the silicon with the new oxide at the interface and the first grown oxide on the outer surface. There is strong evidence that in wet oxygen or steam the diffusing species is undissociated water and in dry oxygen it is the oxide ion (O_2^-).

Deal[1] compared the properties of the oxides formed by the three thermal proc-

†Superscript numbers indicate References listed at the end of the chapter.

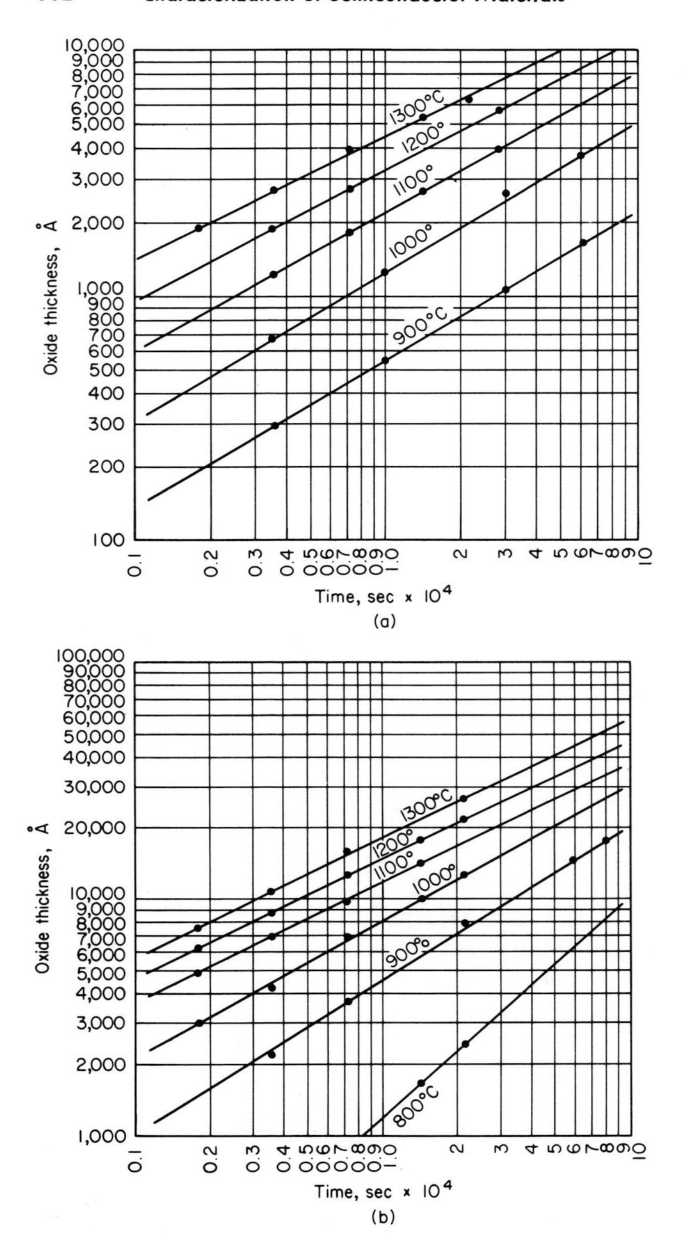

Fig. 10-4. Oxide thickness as a function of growth time and temperature for (111) oriented silicon surfaces. (*a*) **dry oxygen;** (*b*) **steam.** (*From Evitts, Cooper, and Flaschen.*[2])

esses. Table 10-1 lists some of the properties. Other properties such as refractive index, infrared absorption, and etch rate would also be affected.

10-4. ANODIC GROWTH

While the thermal growth of oxide is the principal production method used in all silicon-device fabrication, certain specialized silicon devices utilize electrolytically

Table 10-1. Effect of Mode of Oxide Film Growth on Density and Dielectric Strength[1]

Growth mode	Temperature, °C	Density, g/cm^3	Dielectric strength, volts/μ
Dry oxygen......	1000	2.27	565
	1200	2.15	520
Wet oxygen......	1000	2.18	530
	1200	2.21	540
Steam...........	1000	2.08	490
	1200	2.05	485

grown oxide. Schmidt and Michel[8] published the first comprehensive study on the anodic formation of oxide films on silicon. The anodization was carried out in *N*-methylacetamide made 0.04 *N* in potassium nitrate. As in the case of all anodically grown oxides,[9] the ultimate thickness was shown to be linearly related to the forming voltage with a thickness increment of 3.8 Å/volt. The maximum thickness that could be grown was limited by the voltage at which breakdown occurred and was 560 volts or about 2100 Å. Duffek et al.[10] reported the use of ethylene glycol solutions as electrolytes for the anodic growth of silicon dioxide.

The mechanism of anodic growth of silicon dioxide is interesting because the oxide is continually turning inside out with the mobile species silicon diffusing through to the outer oxide surface to react and form new SiO_2. The electrode reactions have been studied[11] and the excess oxygen content of the oxide shown to be a function of the water content of the electrolyte. This is important to the materials scientist because it affects the ultimate properties of the film.

Doped oxides have been grown anodically,[12,13] and Schmidt and Wonsidler[14] have reported a technique for anodically converting silicon nitride films to silicon dioxide. Since silicon dioxide is readily soluble in dilute HF, this provides a method for opening windows in silicon nitride films for device fabrication.

10-5. CHEMICAL DEPOSITION

As mentioned earlier, the growth of thick (>2000 Å) films of oxide on silicon is easily accomplished by thermal oxidation. Other semiconductors, such as germanium and the III-V intermetallics, that need a thin dielectric film for device fabrication must utilize chemically deposited films. The deposition of silicon dioxide has received considerable attention[15] because it has been well characterized on silicon. Subsequent developments have seen the application of silicon nitride and aluminum oxide films. The method used to deposit these films chemically must be known by the analyst since it will ultimately affect both the chemical and the physical imperfections in the film. The general technique of chemical deposition has the advantages that relatively low substrate temperatures can be employed (400 to 800°C) and, since all reactants are transported in the vapor form, films of higher purity are easily attained. The films are deposited on the heated substrates

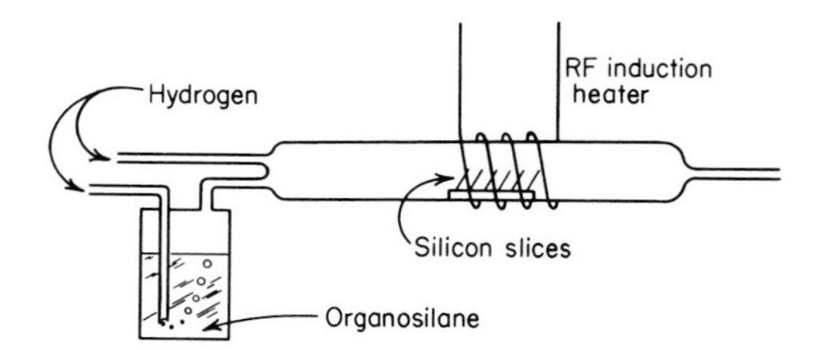

Fig. 10-5. Schematic showing apparatus used for pyrolytic deposition of silicon dioxide. Only the silicon slices are heated by the induction heater.

as amorphous films by using a gas-flow-reaction apparatus much like that shown in Fig. 10-5. Carrier gases and reactants flow down the tube to the heated substrate and react heterogeneously to deposit a dense amorphous adherent dielectric film.

10-6. SILICON DIOXIDE FILMS

Silicon dioxide can be deposited either by hydrolysis of a silicon halide or by pyrolysis of an alkoxysilane. The hydrolysis of a silicon halide is particularly well suited to incorporation into epitaxial systems where silicon halides are already being transported. The general chemical reaction involved is

$$SiX_4(g) + 2H_2O(g) \rightarrow SiO_2(s) + 4HX(g) \qquad (10\text{-}1)$$

The basic problem is to mix the easily hydrolyzed silicon halide with water vapor and at the same time allow the reaction to proceed only at the heated substrate. This problem has been circumvented by substituting a carbon dioxide–hydrogen mixture from which water is produced in situ at temperatures in excess of 400°C by the following reaction:

$$H_2(g) + CO_2(g) \rightarrow H_2O(g) + CO(g) \qquad (10\text{-}2)$$

As a result, water vapor is produced only at the heated substrate in the reaction tube, and SiO_2 is deposited only in that area. Steinmaier and Bloem[16] first reported the CO_2–H_2 process, and it has been widely used for deposition on silicon[17] and germanium.[18]

The deposition rates for the CO_2–H_2 process are about the same as the thermal-growth processes. The oxides usually contain from 0.5% chlorine to 2% bromine, depending on the deposition conditions. Rand[19] developed a nitric oxide process where the NO reacts with the hydrogen, producing water in the reaction zone (800 to 1200°C), which in turn reacts with the silicon tetrahalide.

$$2NO(g) + 2H_2(g) \rightarrow N_2(g) + 2H_2O(g) \qquad (10\text{-}3)$$

$$2H_2O(g) + SiX_4(g) \rightarrow SiO_2(s) + 4HX(g) \qquad (10\text{-}4)$$

The deposition rates are three to five times faster than the CO_2–H_2 or thermal process. Chu and Gruber[20] used HF as the transport agent in a closed-tube system to deposit SiO_2 on silicon, germanium, and gallium arsenide. In this reaction, shown in Eq. (10-5),

$$SiF_4(g) + 2H_2O(g) \rightleftarrows 4HF(g) + SiO_2(s) \qquad (10\text{-}5)$$

heat forces the reaction to the right, causing SiO_2 deposition, while low temperatures favor the formation of SiF_4, thereby enabling transport to the reaction region. These films contain about 1% fluorine and have a somewhat higher dissolution rate than other oxide films.

Silica films are also deposited by the pyrolysis of alkoxysilanes in a gas flow system. The organosilanes that have received the most attention are tetraethoxysilane, $Si(OC_2H_5)_4$, and ethyltriethoxysilane, $(C_2H_5)Si(OC_2H_5)_3$. Klerer's[21] excellent work on the mechanism of deposition of SiO_2 by pyrolytic decomposition points out that there are two types of film which can be deposited. In the 600 to 850°C range, radicals are formed such as R_3Si and $R_3Si–O$ which bond to the surface either at metal or oxygen atom sites. The organic part of this radical thermally cracks, and the process continues. Above 850°C the films were cloudy and probably contained carbon. No film deposition was obtained in the 250 to 500°C range. Film deposition was obtained on 95 to 250° substrates placed downstream from the 725°C cracking zone. These films were shown to be organic polymers of silicon. These organosilanes can be cracked and SiO_2 deposited at lower substrate temperatures by glow discharge[22] or RF discharge[23] in a reaction chamber. Chu[15] and Klerer[21] have compared the properties of the various types of oxide films. These comparisons are summarized in Table 10-2.

Table 10-2. Properties of Silica Films Grown by Various Techniques

Reaction	Substrate temperature, °C	Density, g/cm³	Refractive index	Etch rate in 1.8 *M* HF, Å/sec
Steam oxidation. .	. . . . . .	2.32	1.475	3.7
Thermal (dry O_2) .	. . . . . .	2.23	1.450	3.3
Anodic.	25	1.80	1.362	360
$SiCl_4(H_2–CO_2)$. . . .	1220	2.31	1.46	7.0
SiF_4 transport. . . .	500	2.23	1.45	6.7
Pyrolysis $C_2H_5Si(OC_2H_5)_3$	825	2.14	1.43	12
Pyrolysis $C_2H_5Si(OC_2H_5)_3$	90–150	1.63	1 43	550

10-7. SILICON NITRIDE FILMS

Silicon dioxide, while used almost exclusively throughout the semiconductor industry, has many disadvantages when used in device fabrication for surface passivation. Silica films do not mask against the diffusion of many elements such as gallium and zinc. Water is strongly adsorbed and has a high mobility through silicon films. Sodium ion has been shown to be mobile in silica films and, as will be discussed later, has been shown to be the source of early metal-oxide-semiconductor (MOS) device voltage-temperature instability. Bulk crystalline silicon nitride was known to have better chemical, electrical, and physical properties than quartz, but little was known about amorphous Si_3N_4.

Amorphous silicon nitride films were grown on silicon[24] by using a chemical technique where the reaction

$$3SiH_4(g) + 4NH_3(g) \rightarrow Si_3N_4(s) + 12H_2(g) \tag{10-6}$$

occurs between 700 and 1200°C. The effect of growth conditions on the properties of these Si_3N_4 films has been extensively studied.[25-29] The amorphous films were shown to have many of the desired properties that silica films do not have, including the much desired sodium-ion barrier for MOS devices. However, the electrical properties of Si_3N_4 appear to be far more sensitive than those of SiO_2. As a result, combinations of SiO_2–Si_3N_4 films on device structures look most attractive.

10-8. REACTIVELY SPUTTERED FILMS

In certain types of device fabrication it is desirable to deposit silica films at, or near, room temperature. Reactive sputtering answers many of these problems and is accomplished by using an apparatus like that shown in Fig. 10-6. An electric discharge is initiated between a pure silicon cathode and the anode which is the work to be coated. The apparatus shown is usually set up in a vacuum evaporator under a glass bell jar where the atmosphere can be controlled. Generally, mixtures of 50% argon and 50% oxygen are used at pressures of approximately 2.5×10^{-2} torr. The discharge is carried out at 3,000 to 4,000 volts and a current density of

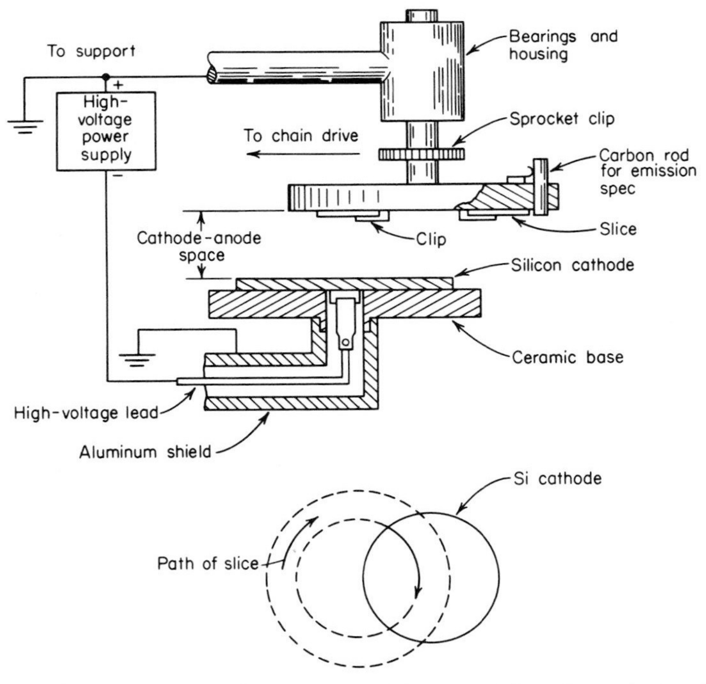

Fig. 10-6. Typical apparatus used for sputtering silicon dioxide onto a semiconductor surface.

around 1 ma/cm^2. Under these conditions the silicon cathode is bombarded with ionized gas molecules, causing free silicon atoms to be ejected. These silicon atoms react with the oxygen and move to the anode, where they deposit as a dense SiO_2 film. Clark[30] deposited SiO_2 films at 50 Å/min to make thin-film capacitors for integrated circuits.

Valletta et al.[31] characterized reactively sputtered films as a function of sputtering conditions. They observed that the properties of the oxide were strongly dependent on the deposition rate. With infrared techniques, it was shown that at rates above 250 Å/min the films contained appreciable amounts of water and were highly porous. High oxygen pressures also caused porous films.

10-9. EVAPORATED FILMS

Conventional vacuum-evaporation techniques have been used to deposit thin dielectric films of SiO–SiO_2.[32,33] Either silicon monoxide is evaporated directly, or a silicon electrode is heated in a low-pressure oxygen environment. The technique is difficult to control, offers no advantages over the previously mentioned methods, and has received little attention as a method for device production.

10-10. CHARACTERIZATION PROBLEMS

The analysis of dielectric films has several connotations for the analyst. Both the bulk of the film and the surface can be analyzed for chemical and physical imperfections, and these will be discussed later. However, there are other characteristics which are just as important to the final device characteristics. Often it is possible to determine several of these parameters in a single measurement, such as ellipsometry, where both thickness and refractive index are measured.

10-11. FILM THICKNESS

Gillespie[34] has carried out an excellent review of measurement methods for the determination of film thickness. Gillespie's review is a general survey of methods for all films, including epitaxial layers and thin metal films, as well as dielectric films. It briefly describes each method with a list of advantages and disadvantages and gives the limits of sensitivity. The survey is highly recommended to the reader to provide groundwork for thickness measurements in thin films.

10-12. COLOR CHARTS

One of the earliest methods of estimating the thickness of SiO_2 films was a visual color examination. This technique is still used and under controlled conditions can be reasonably accurate. This color phenomenon is caused when white light interacts with the SiO_2 film on the reflecting silicon substrate and a portion of the reflected white light is lost because of destructive interference. This destructive interference occurs when

$$d = \frac{(2k - 1)\lambda}{2n} \tag{10-7}$$

where d = film thickness
k = the order 1, 2, 3, . . .
n = refractive index of the film

Generally in a production area, a set of calibrated samples is prepared by the same oxidation process to ensure the same index of refraction and covering the thickness range of interest. These standards are physically mounted on a white card and labeled. The thickness of the sample is determined by visually observing the sample and the standards obliquely and adjacent to each other.

Pliskin and Conrad[35] have published a carefully devised table of colors of silicon dioxide films on silicon as a function of film thickness. These data are shown in Table 10-3. Since the determination of color is a highly subjective judgment, it is difficult to utilize these color-vs.-thickness charts without some knowledge of the approximate thickness (order) of the film, and even then considerable operator experience is required. Generally, the method is useful only in the 500 to 4000 Å range.

10-13. VAMFO

Pliskin and Conrad[35] developed a simple apparatus to determine silicon dioxide film thickness, which is based on the destructive interference of white light on reflection by the film. However, only one wavelength is monitored, as shown in the schematic in Fig. 10-7. The sample, mounted on a stage, is rotated, and the number of interference fringes are counted while the sample is being rotated through a measured angle from a known initial angle of incidence. The technique is described as Variable-Angle Monochromatic Fringe Observation and hence VAMFO. The film thickness is calculated by

$$d = \frac{\Delta N \lambda}{2n(\cos r_2 - \cos r_1)} \tag{10-8}$$

$$d = \frac{\lambda}{2n(\Delta \cos r)} \tag{10-9}$$

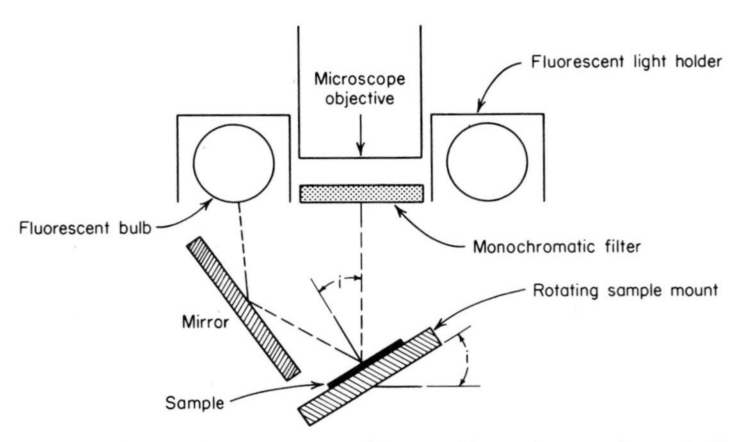

Fig. 10-7. Schematic showing the apparatus used for variable-angle monochromatic fringe observation (VAMFO) to determine the thickness of oxide films. *(From Pliskin and Conrad.[35])*

Table 10-3. Color Chart for Thermally Grown SiO_2 Films Observed Perpendicularly under Daylight Fluorescent Lighting[35]

Film thickness, μ	Order (5450 Å)	Color and comments	Film thickness, μ	Order (5450 Å)	Color and comments
0.05_0		Tan	0.63_0		Violet-red
0.07_5		Brown	0.68		"Bluish." (Not blue but borderline between violet and blue-green. It appears more like a mixture between violet-red and blue-green and looks grayish.
0.10_0		Dark violet to red-violet			
0.12_5		Royal blue			
0.15_0		Light blue to metallic blue			
0.17_5	I	Metallic to very light yellow-green			
			0.72	VI	Blue-green to green (quite broad)
0.20_0		Light gold or yellow—slightly metallic	0.77		"Yellowish"
			0.80		Orange (rather broad for orange)
0.22_5		Gold with slight yellow-orange	0.82		Salmon
0.25_0		Orange to melon	0.85		Dull, light red-violet
0.27_5		Red-violet	0.86		Violet
0.30_0		Blue to violet-blue	0.87		Blue-violet
0.31_0		Blue	0.89		Blue
0.32_5		Blue to blue-green	0.92	V	Blue-green
0.34_5		Light green	0.95		Dull yellow-green
0.35_0		Green to yellow-green	0.97		Yellow to "yellowish"
0.36_5	II	Yellow-green	0.99		Orange
0.37_5		Green-yellow	1.00		Carnation pink
0.39		Yellow	1.02		Violet-red
0.41_2		Light orange	1.05		Red-violet
0.42_5		Carnation pink	1.06		Violet
0.44_2		Violet-red	1.07		Blue-violet
0.46_5		Red-violet	1.10		Green
0.47_5		Violet	1.11		Yellow-green
0.48_0		Blue-violet	1.12	VI	Green
0.49_2		Blue	1.18		Violet
0.50_2		Blue-green	1.19		Red-violet
0.52_0		Green (broad)	1.21		Violet-red
0.54_0		Yellow-green	1.24		Carnation pink to salmon
0.56	III	Green-yellow	1.25		Orange
0.57_4		Yellow to "yellowish." (Not yellow but is in the position where yellow is to be expected. At times it appears to be light creamy gray or metallic.)	1.28		"Yellowish"
			1.32	VII	Sky blue to green-blue
			1.40		Orange
			1.45		Violet
			1.46		Blue-violet
0.58_5		Light-orange or yellow to pink borderline	1.50	VIII	Blue
			1.54		Dull yellow-green
0.60_0		Carnation pink			

where d = film thickness
λ = wavelength of monochromatic light
n = refractive index of film
r_j = angle of refraction at fringe for which angle of incidence is i_j and sin $r_j = \sin i_j/n$
ΔN = number of fringes between i_j and i_2
$\Delta \cos r = (\cos r_2 - \cos r_1)/\Delta N$ averaged for both maxima and minima

VAMFO is best suited to the measurement of films thicker than 10,000 Å, which is fairly thick for films normally used in device production. However, it can be used with films as thin as 500 to 800 Å with the aid of a small step etched through the film to the substrate and an interference-pattern chart. VAMFO assumes a known, constant index of refraction, which might not hold for oxides grown by several different methods.

10-14. ULTRAVIOLET-VISIBLE INTERFERENCE

Corl and Wimpfheimer[36] utilized the same principle of destructive interference of light passed through and reflected from the silicon substrate as that used in VAMFO. However, in this case the angle of incidence was held constant, and the wavelength of light interacting with the film was varied. With a standard specular reflectance attachment on an ultraviolet-visible spectrophotometer, the reflectance spectrum from an SiO_2 film on silicon is recorded in the same manner as described in Sec. 8-9 for the determination of the thickness of epitaxial films using infrared radiation. An interference pattern, as shown in Fig. 10-8, is obtained. The thickness of the film is then calculated:

$$t = \frac{N(\lambda, \lambda_m)}{2(\lambda - \lambda_m)(n^2 - \sin^2 \theta)^{1/2}} \tag{10-10}$$

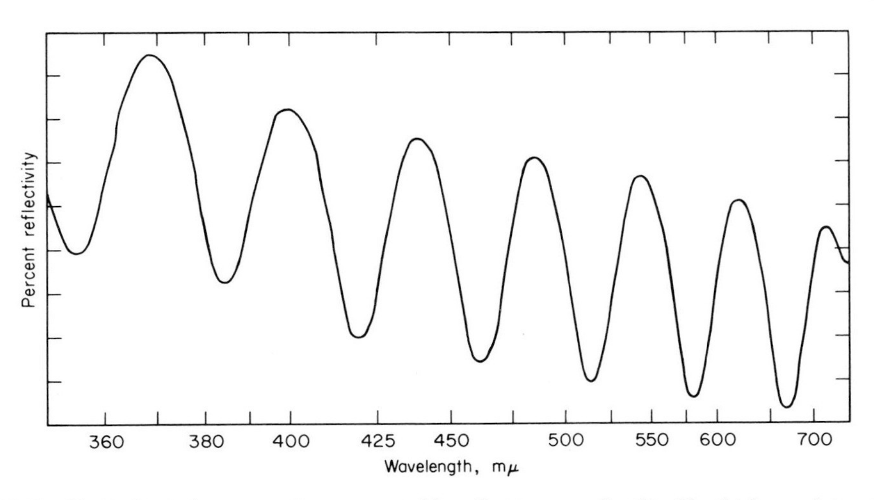

Fig. 10-8. Typical interference pattern on a visible reflection scan for thin-film thickness determination.

where t = thickness of oxide film
N = number of complete cycles between λ and λ_m
λ = convenient maximum wavelength
λ_m = convenient minimum wavelength
n = refractive index of film
θ = angle of incidence

A number of other workers[37-40] have modified and improved this technique to determine oxide film thickness where corrections for the variation of the index of refraction and certain phase-shift problems were performed. Reizman and Van Gelder[41] have applied this interference technique to silicon nitride as well as silicon dioxide films and can determine the thickness of double films of silicon nitride on silicon dioxide and vice versa.

Reizman and Van Gelder have suggested the name CARIS (Constant-Angle Reflection Interference Spectrum) for this general technique, which would be analogous to the VAMFO name for the variable-angle method discussed earlier.

The CARIS method of determining film thickness assumes a known index of refraction which does not hold for oxides formed by different methods. The technique works well down to film thicknesses of 1000 Å, is nondestructive, and is easily performed. However, in semiconductor research, a sensitivity limit of 1000 Å is not acceptable, and one cannot assume a known index of refraction. Lukes and Schmidt[42] have reported a further modification of this general technique down to 250 Å by measuring the intensity of reflected light at λ = 4000 Å both before and after oxidation of the sample. Measurements of this type in production are of course not possible. Good agreement between their method and ellipsometry was reported.

10-15. ELLIPSOMETRY

Ellipsometry provides the most accurate analytical technique for determining dielectric film thicknesses on semiconductor substrates. It also has the distinct advantage, over the earlier techniques, that it determines the refractive index and consequently does not have to rely on a constant value for the film-thickness determination.

Ellipsometry is in reality polarization spectrometry since one measures the change in polarization of light upon reflection from a film-covered surface. The method involves only measurements of polarization and not the intensity or changes in the intensity of the reflected light. The apparatus used to measure this change in polarization is shown in Fig. 10-9. The system is reasonably simple and involves a monochromatic light source (either a laser or a mercury lamp with a filter) and a polarizer section to produce a linearly polarized wave (polarized at 45° to the plane of incidence) which interacts with the oxide layer and is reflected off the semiconductor surface back through the oxide to the analyzer section. The instrument then indirectly measures two parameters Ψ and Δ, where

Ψ = azimuth angle
Δ = phase difference

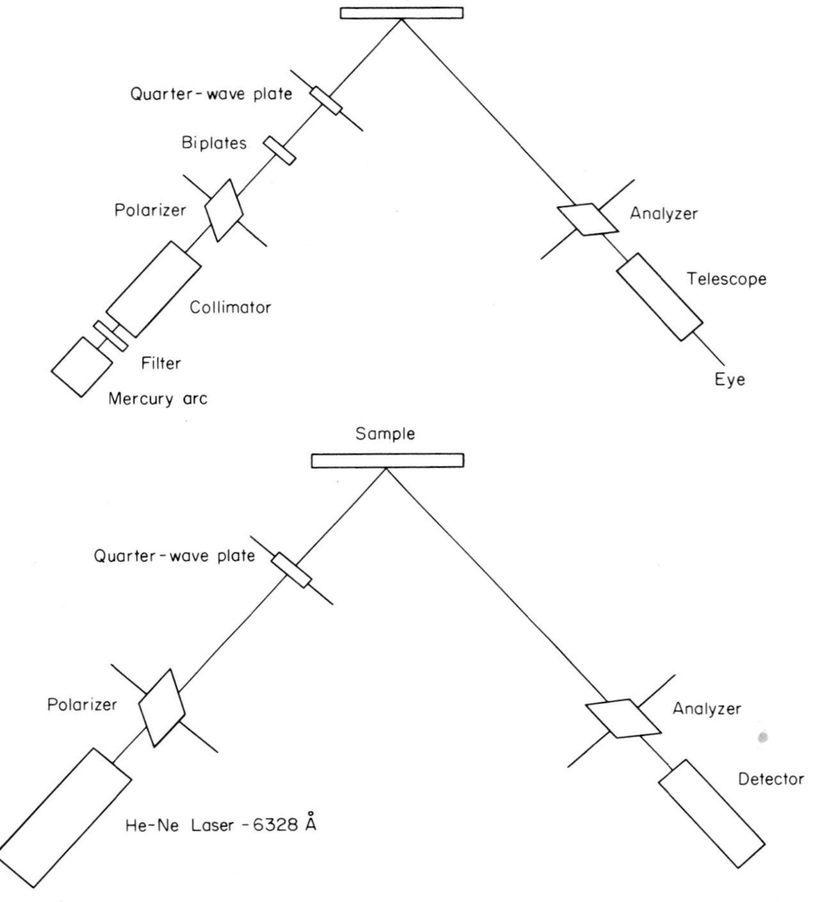

Fig. 10-9. Schematic showing the construction of an ellipsometer with a mercury arc (*top*) **and a helium neon laser** (*bottom*).

In thin-film measurements, Ψ and Δ are functions of

n_1, k_1 = optical constants of film
d = thickness of film
n_2, k_2 = optical constants of reflecting surface under film
λ = wavelength of light used to make the measurement
ϕ_0 = angle of incidence

Under the conditions of measurement, the SiO_2 films are transparent and $k_1 = 0$. All other parameters are known except d and n_1 for the film. These are then calculated from the basic ellipsometer equations.[43-45]

$$R^p = \frac{r_{12}^p + r_{23}^p \exp D}{1 + r_{12}^p r_{23}^p \exp D} \tag{10-11}$$

$$R^s = \frac{r_{12}^s + r_{23}^s \exp D}{1 + r_{12}^s r_{23}^s \exp D} \tag{10-12}$$

where $$D = -4\pi i n_2 \cos (\phi_2 d_2/\lambda) \qquad (10\text{-}13)$$

Subscripts 1, 2, and 3 refer to the medium (usually air), film, and substrate, respectively.

r_{12}^p, r_{12}^s = Fresnel coefficients for reflection between medium and film
r_{23}^p, r_{23}^s = Fresnel coefficients for reflection between film and substrate
p, s = superscripts referring to light polarized with its electric vector parallel (p) or perpendicular (s) to plane of incidence
n_2 = refractive index of film
ϕ_2 = angle of incidence
d = thickness of film
R^p, R^s = reflection coefficients

The ratio of the reflection coefficients, ρ, is the quantity measured by the ellipsometer, where

$$\rho = \frac{R^p}{R^s} = \tan \Psi e^{i\Delta} \qquad (10\text{-}14)$$

and Ψ and D are defined above. These equations are rather complex, difficult, and tedious to calculate by hand and are therefore usually programmed for computer calculations. Copies of these computer programs are available from many workers in the field, including the authors of this book. Sets of curves are usually calculated relating Ψ and Δ to the film thickness at various values for the film refractive index. A set of these curves is shown in Fig. 10-10 for SiO_2 films on silicon with refractive indices of 1.40, 1.50, and 1.60. In actual measurement and calculation, the computer program yields the best value for film thickness and the refractive index through a 20-step iterative process. Archer[44] has related film density to refractive index, as shown in Table 10-4.

Table 10-4. Relationship between Film Density and Refractive Index for Various Oxide Films[44]

Method of film formation	Index of refraction	Density, g/cm³
Oxygen	1.450	2.23 ± 0.02
Steam	1.475	2.32 ± 0.02
Anodization	1.362	1.80 ± 0.05
Decomposition of silane	1.430	2.14 ± 0.07
Oxygen + P_2O_5	1.495	

Traditionally, workers in the field have used the 5461-Å line from mercury light to make the measurements. Generally this single line is weak, and obtaining nulls on the analyzer and polarizer of the ellipsometer is a little difficult until the operator acquires considerable experience. Jones,[46] in the Texas Instrument laboratories, has applied a 0.3-mw He–Ne laser as a light source (6328 Å radiation) with an inexpensive cadmium sulfide photocell as a detector. Significantly sharper nulls were obtained with resulting better accuracy and reproducibility. It now appears that the new argon-krypton-xenon ion lasers can provide all the necessary monochromatic light lines to cover the entire visible spectrum.

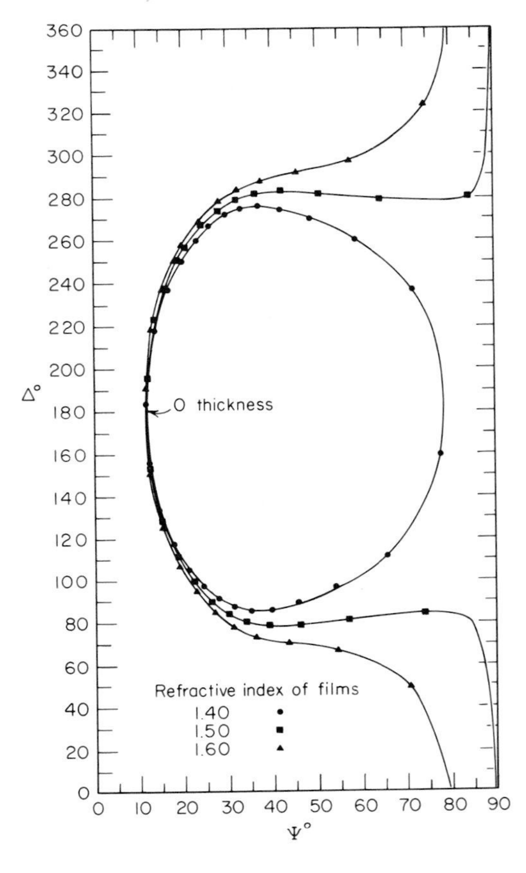

Fig. 10-10. Typical ellipsometer curves for SiO_3 films on silicon showing the effect of refractive index on the shape of the curves.

Ellipsometry is the most sensitive and accurate technique for determining dielectric film thicknesses on semiconductors. The method is nondestructive, and films as thin as 10 Å can be measured, with no upper limit on thickness. Precisions of ±10 Å are claimed for the method.

10-16. INTERFEROMETRY

The technique of optical interferometry, utilizing a monochromatic light source and a microscope-mounted interferometer, is described in Secs. 8-7 and 9-9. The application of this technique to the measurement of the thickness of oxide films was described in detail by Booker and Benjamin.[47] The technique is of necessity a destructive one since a step must be etched through the oxide film to the substrate. The height of this step is measured by using two-beam interference for thick films (>3000 Å) and multiple-beam interference for thin films (<3000 Å). The multiple-beam technique[48] requires that the sample with the etched step be silvered to be 100 percent reflecting, while one surface of a glass slide is silvered to be approximately 80 percent reflecting. The two silvered surfaces are then put close together with the glass slide on top and viewed with a metallurgical microscope using monochromatic light. The number of fringes is then determined in the conventional manner.

While these two interference methods have the disadvantage of being destructive methods, they have fairly good sensitivity, accuracy, and precision. Booker and Benjamin[47] estimate an accuracy of ±35 Å for the multiple-beam technique for films as thin as 320 Å and ±350 Å for the two-beam technique on the thick films.

10-17. STYLUS TECHNIQUES

Frequently during the fabrication of integrated-circuit devices, there will be multiple steps in the oxide as a result of the various processing steps. To measure the height of these steps it is convenient to use an electromechanical stylus instrument such as the Taylor-Hobson,® Model 4 Talysurf.® In this instrument, a diamond stylus with a tip radius of 1 to 10 mils and with a 50- to 100-mg pressure is drawn across the surface of the specimen. The mechanical motions of the stylus are converted to electrical signals which are recorded by a rectilinear recorder. A typical recording is shown in Fig. 10-11. Schwartz and Brown[49] have described the application of this technique to various oxide films and report an accuracy of ±50 to ±100 Å in the 100- to 5000-Å range. The technique is simple, rapid, and is nondestructive except for the step examined, which is scratched. However, a step in the oxide is required before this method is applicable.

10-18. FILM DENSITY

The measurement of the density of a dielectric film is often required as a measure of the compactness of the film. This is usually evaluated by comparing the measured density with the theoretical density, and a value of 2.20 g/cm³ is usually used for silica glass.[44] There really is no direct, nondestructive method of determining density. Deal[1] weighed the silicon wafer before and after oxidation and calculated the density, knowing the thickness and surface area of the slice. The reverse of this technique is frequently used, where a control slice is etched with hydrofluoric acid and the weight of oxide removed determined by weight difference. Deal's method is nondestructive but weighs only the oxygen added to the film, while the second method is destructive but provides a total weight of oxide film.

Pliskin and Lehman[50] devised an empirical relationship between density and measured refractive index at a wavelength of 5460 Å. The relationship was given by

$$\rho = -4.784 + 4.785n \tag{10-15}$$

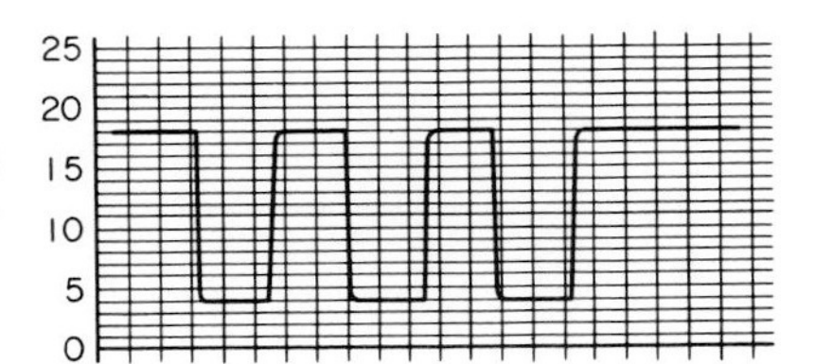

Fig. 10-11. A Talysurf® trace over a series of steps etched into an oxide film. Each vertical division represents 400 Å.

®The Rank Organization.

where ρ was density in grams per cubic centimeter and n was the refractive index at 5460 Å. It was sometimes necessary to correct for the water content of the oxides, although details were not given on how to apply these corrections. Some typical values of density for various types of oxide are given in Table 10-5. The

Table 10-5. Typical Densities for Oxides Grown on Silicon by Various Methods

Growth technique	*Density, g/cm³*
Steam, open-tube	2.00–2.20
Dry oxygen	2.24–2.27
Wet oxygen	2.18–2.21
Anodic oxides	1.80
Alkoxysilane decomposition	2.09–2.15

more dense values are after densification, which is a technique whereby an oxide is treated after growth to increase the density of the film. Typically, a 15-min heat treatment in steam at 800°C will achieve this densification.

10-19. REFRACTIVE INDEX

The refractive index is a parameter that is frequently used as an indicator of the overall quality of an oxide or dielectric film. One of the main reasons for this is that it is an easily measured parameter and one that can be determined by nondestructive techniques. There is of course sound scientific basis for using the refractive index as the indicator since it is a ratio of the velocity of light in a vacuum to the velocity of light in the dielectric film. The velocity of light in the film is controlled by its structure and composition. Empirically it has been observed[50] that the refractive index of oxide films on silicon is affected by the density and stoichiometry. Since the velocity of light in any medium is a function of the wavelength, refractive indices are always quoted at a particular wavelength. In the case of SiO_2 films on silicon the refractive index is usually compared with the value 1.460 at 5460 Å for fused quartz.[51]

10-20. REFRACTIVE INDEX BY INTERFEROMETRY

Booker and Benjamin[47] used an optical interferometer to determine the refractive index simultaneously with film thickness for SiO_2 films on silicon. By using a sample with a step in the oxide and half metallized, as shown in Fig. 10-12, the

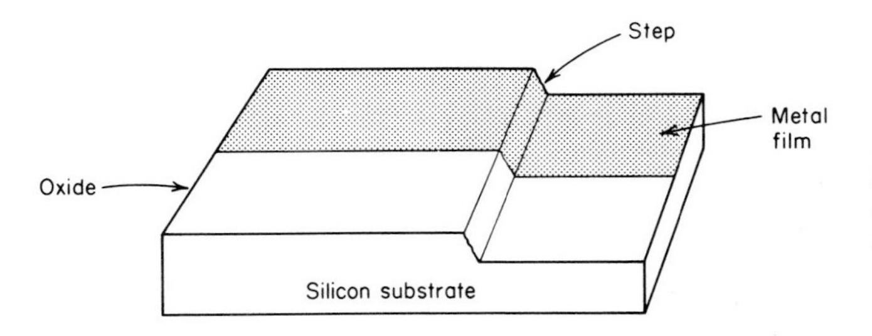

Fig. 10-12. An oxide sample prepared for the determination of refractive index and thickness by optical interferometry.

fringe displacement is measured on the metallized portion q and on the nonmetallized portion p. The thickness d on each section is given by

$$d = p\frac{\lambda}{2} \tag{10-16}$$

$$d = q\frac{\lambda}{2(n-1)} \tag{10-17}$$

where λ is the wavelength of light used for the measurement. Since these two equations have the same value, the refractive index n can be calculated from

$$n = 1 + \frac{q}{p} \tag{10-18}$$

Booker and Benjamin estimate an accuracy of 1 to 2 percent for a 15,000-Å film in the range 1.48 to 1.50.

Pliskin[35,50] used the VAMFO technique (Sec. 10-13) to determine the refractive index of oxide films. For any given fringe order N_1' observed on the VAMFO instrument at an angle i of incidence or reflection, it can be shown that

$$N_i' = \frac{\cos r_i}{\Delta \cos r} \tag{10-19}$$

where r_i is the angle of refraction at that fringe and $\sin r_i = \sin i/n$ and $\Delta \cos r = (\cos r_2 - \cos r_1)/\Delta N$. The value of N_i' must be an integer for maxima and a half integer for minima. When there is a step in the film, an approximate N_0, where $i = 0$, can be obtained by counting fringes at the step. Then a value for the refractive index is obtained by iterative interpolation of values until one is obtained that gives agreement between the N_i' set and N_0.

It is not always necessary to have a step in the film to obtain N_0. If the film is not too thick and an approximate value of the refractive index is known (and this would be the case for SiO_2 films on silicon), then an iterative interpolation of n would be used to obtain the proper integral value for a maximum and the half value for the closest minimum.

This technique is reported to be capable of determining refractive indices with an accuracy of 0.2 percent. However, this method is highly dependent on the thickness of the film under study, and a step is frequently required to obtain a value for the refractive index.

10-21. REFRACTIVE INDEX BY LIQUIDS

Lewis[52] applied a well-established technique of matching the refractive index of a set of oils to that of a transparent solid. The solid is successively immersed in known oils, and the relief of the interface observed until the interface disappears. The refractive index of the solid is then the same as that of the liquid.

For thin dielectric films on silicon, the same technique is applied, except that a step in the film to the substrate is required. A drop of oil is applied at this step and the interface observed. When the refractive index of the oil and the oxide film are the same, the interface disappears optically, and the film edge is no longer visible.

The factor that controls accuracy in this method is the availability of a useful set of calibrated oils. A set of oils in the range 1.300 to 1.700 with intervals of 0.002 and accurate to 0.0002 is commercially available. Lewis feels that an accuracy of 0.001 to 0.006 is attainable with this technique. This technique must have a step etched in the oxide and must be considered a destructive method.

10-22. REFRACTIVE INDEX BY ELLIPSOMETRY

As in the case of thickness measurements (Sec. 10-15), ellipsometry offers the best nondestructive method for determining the refractive index of dielectric films on semiconductors. However, the computations are difficult, and a computer is necessary to perform the iterative process.

Archer[44] described the technique of determining Δ and Ψ for oxide films on silicon. Since Δ and Ψ are unique values for each thickness and refractive index of the film, it is possible to have the computer iteratively calculate thickness over a given range of refractive indices to obtain the correct value of n for that Δ and Ψ. The technique is nondestructive and is estimated to be accurate to ± 0.004.

10-23. ELECTRICAL EVALUATION

The electrical properties of the dielectric films grown on semiconductors are vitally important in the finished device. Surprisingly little work has been published on the evaluation of the electrical characteristics of these films. Guidelines can be obtained from examining the electrical measurements made on silica glasses.[53] Generally it can be stated that most workers in semiconductor-device fabrication evaluate the electrical characteristics of the oxide film by looking at the final electrical properties of the device. It is then possible to relate empirically some property of the oxide to one of the electrical properties of the device. This device parameter is monitored while the properties of the oxide are changed.

10-24. RESISTIVITY

The dielectric films currently in use in device fabrication have resistivities as high as 10^{17} ohm-cm with a range down to 10^{12}, depending on the method of growth. Electrical measurement of the resistivity of these films is complicated by several factors, including making good electrical contact and preventing surface leakage. Wellard[54] described a guard-ring technique, shown schematically in Fig. 10-13, in which the guard circuit prevents surface leakage. The measurement circuit also provides a means for discharging the sample. This latter step is necessary because all these films are dielectrics and act as capacitors.

The electrical properties of evaporated SiO films have been more extensively studied[55-58] as capacitance structures by using evaporated aluminum electrodes. Siddall[55] observed that the conductivity ($\sigma = 1/\rho$) of oxide films was made up of ionic conductivity by impurities, small leakage paths caused by physical defects, and slow polarization effects which show up as a slow decay of leakage current with time. Because all these factors are dependent upon film-growth conditions, the

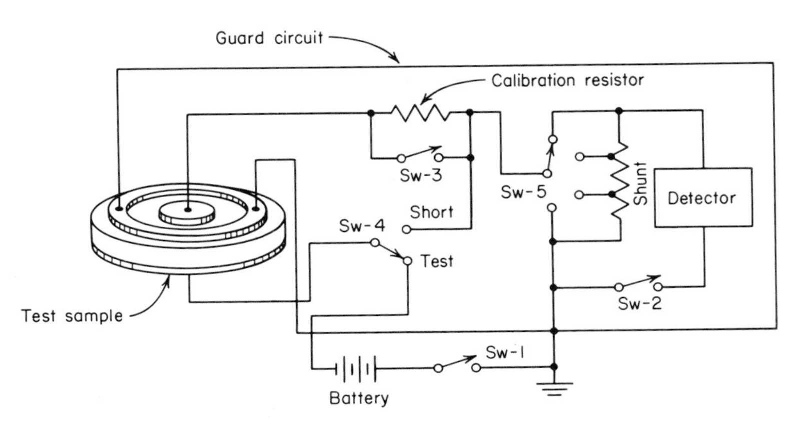

Fig. 10-13. Direct-deflection method used for making high-range resistance measurements. (*From Wellard.*[54])

electrical properties observed are not the intrinsic SiO properties, but depend on the conditions of deposition. The resistivity of the films is often studied by observing the current-voltage (I-V) curves, and Ohm's law is obeyed only at very low voltages (up to 0.1 volt). The nonlinear behavior continues up to the breakdown voltage and must be attributed to the three factors discussed by Siddall.[55]

Table 10-6 gives some typical resistivities of oxide films grown by different

Table 10-6. Resistivities of Oxide Films Prepared by Various Methods

Oxide film	*Resistivity, ohm-cm*
Thermal oxide (steam)	10^{15}–10^{17}
Anodic oxides	10^{12}–10^{16}
Evaporated SiO	10^{12}–10^{13}
Chemical deposited (CO_2)	10^{14}–10^{15}

techniques. Oxide films with resistivities in the 10^{16} to 10^{17} ohm-cm range are readily attainable. The electrical evaluation of the resistivity of these films is more difficult.

10-25. DIELECTRIC STRENGTH, DIELECTRIC CONSTANT, DISSIPATION FACTOR

The other electrical properties that are used to characterize oxide films all make use of the dielectric properties of the film.[59,60,30] In these tests ohmic contact is made to the silicon substrate and to an evaporated metal-dot electrode on the surface of the SiO_2 film. Dielectric-strength or breakdown-voltage studies are always performed with the silicon electrode positive. In this configuration, the breakdown voltage is not time-dependent and appears to be of an intrinsic nature. The time-dependent breakdown occurs with the silicon negative (metal dots on oxide positive) and obeys an empirical relationship known as Peek's law, where there is a linear relationship between time t to breakdown and applied voltage ($V \propto t^{-1/4}$).

The dielectric constant and dissipation factor are obtained from measurement of capacitance on a standard capacitance bridge. The dielectric constant is always quoted with the measuring frequency and compared with pure quartz (3.78 at 10 khz). Some typical values are shown in Table 10-7 along with some values of dielectric strength for the different types of oxides.

Table 10-7. Dielectric Strength and Dielectric Constant of Various Oxides

Oxide	Dielectric strength $\times 10^6$ volts/cm	Dielectric constant, 10 khz
Thermal oxide (steam)	6.8–9	3.2
Anodic oxide	5.2–20	3.8 (1 Mhz)
Evaporated SiO	1–5	4.1–80 (1 Mhz)
Chemically deposited (CO_2)	5–6	3.54

10-26. FILM ANALYSIS

In previous sections of this chapter, techniques were described for the measurement of the properties of dielectric films. These intrinsic properties of the films gave little insight into the chemical and physical imperfections that caused variation of these properties. The following sections will deal with characterization techniques for studying these imperfections.

10-27. CHEMICAL IMPERFECTIONS

The basic problem in the analysis of dielectric films used on semiconductor materials is the extremely small sample available for analysis. The problem here is more acute than with epitaxial layers because the dielectric films are at least an order of magnitude thinner. A typical range of film thickness is 1000 to 10,000 Å. If we assume a 5000-Å SiO_2 film on a 1-in.-diameter slice, there is 550 μg of sample per side available for chemical analysis. Frequently it is necessary to analyze incremental parts of the film. For example, it may be necessary to determine the impurity content of the outer few hundred angstroms, then the center bulk of the film, and finally the last few hundred angstroms at the oxide-silicon interface. Under these conditions, the sample size can be prohibitively small except for the most sensitive analytical techniques.

10-28. ACTIVATION ANALYSIS

Dielectric films such as SiO_2 and Si_3N_4 on undoped or boron-doped silicon substrates provide an ideal medium for neutron activation analysis. Moreover, activation analysis is the only technique with sufficient sensitivity to analyze these films and obtain information on the distribution of impurities through the film. Virtually all the published work on neutron activation analysis of this type of film

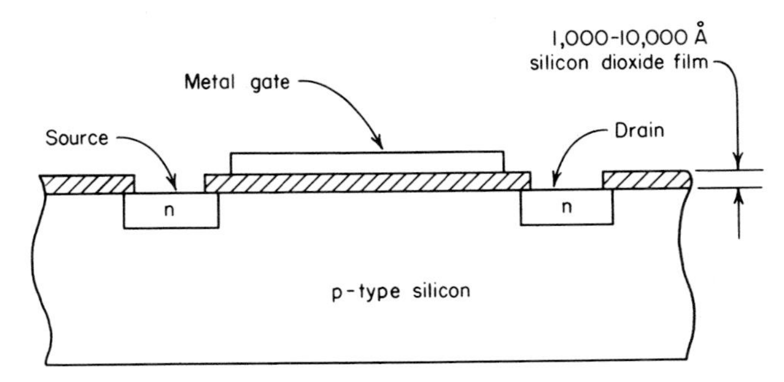

Fig. 10-14. Schematic of a metal-oxide-semiconductor (MOS) field-effect transistor. (*From Osborne, Larrabee, and Harrap.*[62])

is restricted to sodium in SiO_2 films on silicon.[62-64] However, everything that is described here for sodium is directly applicable to any other chemical impurity in the oxide film, provided it has a sufficiently long-lived radioactive species or a gamma spectrum with energies higher than the 2.6-hr silicon-31 1.27-Mev peak. Workers in the Texas Instruments laboratories[61] have successfully determined 2.6-hr manganese-56 and 12.6-hr copper-64 in SiO_2 films.

Analysis for sodium in SiO_2 films on silicon became important in the semiconductor industry with the introduction of metal-oxide-semiconductor (MOS) technology. Figure 10-14 shows a schematic of an MOS field-effect transistor. In this structure the oxide film becomes an active part of the device, which puts far more stringent requirements on the oxide than when it is used to passivate a device. In Fig. 10-14 the oxide film supports a thin metal film which in turn is used to supply charge through the oxide to the semiconductor substrate. In this manner the metal oxide can act as a gate in the same way as a grid in a vacuum tube. The silicon dioxide film must be virtually free of any mobile ions which can drift through the oxide when a bias is applied to the metal film, since this can cause device instability. Sodium ion has been identified as the principal source of mobile ions[65,66] in these films. The electrical effects and distribution of sodium in oxide films have been described by Carlson et al.,[67,68] Yon et al.,[64] and Buck et al.[63]

Basically the analytical procedure[62] consisted of a carefully controlled sample packaging for irradiation, involving wrapping in reactor-grade zirconium foil or placing in special quartz irradiation containers. The irradiation was carried out for 24 hr at a flux of 1×10^{13} neutrons/(sec)(cm^2). The silicon slices were returned to the laboratory within 6 to 7 hr, unwrapped, and rinsed in 6 *N* hydrochloric acid. The oxide film was then incrementally etched in 50 to 500 Å steps using 5% HF, and each etch lap was analyzed radiochemically. While sodium-24 was specifically sought, any impurity with a radioactive species could be studied. The oxide thickness before and after each etch was measured with a Gaertner Model L-119 ellipsometer. A typical sodium impurity profile through an oxide film is shown in Fig. 10-15. The high sensitivity of this analytical approach can be seen in Fig. 10-15, where as little as 1 ppm sodium was determined in 3.5 to 70 μg of SiO_2 removed in an etch lap. The detection limit was shown to be 7.6×10^{-6} μg at the 95 percent confidence level. The detection limit for any impurity in the oxide film is controlled

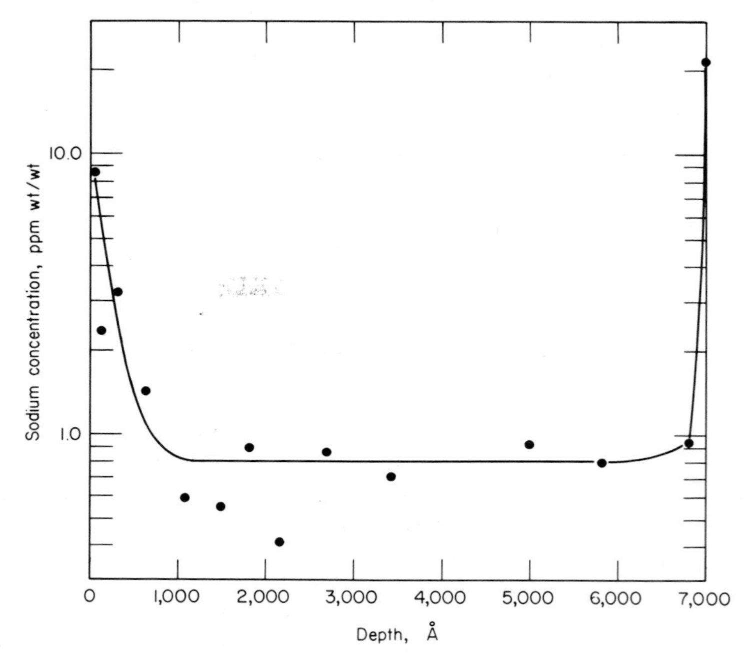

Fig. 10-15. Typical sodium concentration profile through a silicon dioxide film (7000 Å). (*From Osborne, Larrabee, and Harrap.*[62])

by the flux, cross section, and irradiation as discussed under activation analysis of bulk silicon in Sec. 5-5.

One of the solutions to the problem of sodium migration through oxide films has been the introduction of large concentrations of phosphorus into the outer surface of the film to form a glaze which getters the sodium. Neutron activation has been applied to the analysis of oxide films to determine the phosphorus distribution[62,70-72] and when combined with autoradiography yields valuable information on phase segregation. Neutron activation analysis of oxide films for phosphorus is ideal for the radiochemist because the 14-day half-life of phosphorus-32 is sufficiently long to allow the decay of shorter-lived interfering activities such as ^{31}Si, ^{24}Na, ^{82}Br, ^{198}Au, and ^{64}Cu.

10-29. RADIOACTIVE TRACERS IN FILM FORMATION

In the characterization of oxide films for chemical imperfections, it is sometimes more convenient to use radioactive tracers than activation analysis. There are generally two reasons for choosing this approach. Firstly, the chemical impurity may not yield a suitable radioactive species (half-life too short, poor yield due to small capture cross section or very long half-life, or the radioactive species does not lend itself readily to counting). Secondly, there may be a predominance of other impurities, yielding radioactive species that obscure the activity of the impurity of interest.

A good example of such a short-lived activity is the study of sodium in SiO_2 films. In the neutron activation analysis of sodium utilizing the $^{23}Na\ (n, \gamma)^{24}Na$ reaction, the sodium-24 half-life (15 hr) precludes further experimental work since all available time is used in the analysis. However, sodium does have a long-lived radioactive species that can be used as a tracer, sodium-22, with a half-life of 2.6 years. The isotope is fairly expensive but does allow the experimenter considerable latitude in studying the behavior of sodium in dielectric films. Buck et al.[63] used this approach in certain parts of their studies of sodium in SiO_2 films.

Water in oxide films is of considerable interest from the device standpoint. Since most oxide films are grown in steam or in wet oxygen in production processes, there is need to analyze these films for water and to determine the behavior of this water under various thermal and electrical treatments of the oxide. Kuper and Nicollian,[69] in a study of the electrical behavior of steam-grown oxides, observed the presence of an n-type inversion layer in the p-type silicon substrate. This n-type channel was significantly reduced when the oxide was baked at elevated temperatures in a dry ambient. The process was reversible, and hydration with resultant n-type channeling was easily achieved. Kuper and Nicollian postulated a water species acting as a donor located on the oxide side of the Si–SiO_2 interface.

Burkhardt[73] utilized tritium (^3H, half-life 12.3 years) labeled water to steam-grow oxides. The experimental difficulties associated with the handling, counting, and interpretation of the data are large. Tritium emits only a very weak beta (0.018 Mev) whose range is very small in any medium. Exchange of the labeled water with atmospheric water in subsequent handling is always a possibility. There are associated health hazards that are difficult to assess because of the very weak emanating radiation. However, Burkhardt's execution of the experimental study and subsequent interpretation were excellent.

Burkhardt showed that the tritium concentration profile through the oxide approached a complementary error function with a surface concentration (C_0) of 4×10^{19} molecules H_2O per cubic centimeter. Even more significant was the presence of a minimum in the tritium concentration profile at around 600 Å from the Si–SiO_2 interface. This minimum is shown in Fig. 10-16 for three different thicknesses of oxide; it always occurred at around 600 Å. The water concentration then rose sharply at the Si–SiO_2 interface. This region of higher water concentration probably strongly influences the silicon surface and could affect the segregation of other species in this region. Sodium ion has been observed to exhibit this same behavior[54,67,68] with a sharp increase in concentration from a minimum to the Si–SiO_2 interface.

Burkhardt's data also showed the removal of water from the oxide with baking. An activation energy for the outdiffusion of the water from the oxide was found to be 15.7 kcal/mole.

10-30. MASS SPECTROSCOPY

The spark-source solids mass spectrometer could be used to analyze oxide films if the energy of the spark were controlled so that the film was not penetrated. The techniques described in Sec. 7-21 for the mass spectrometric analysis of thin epitaxial

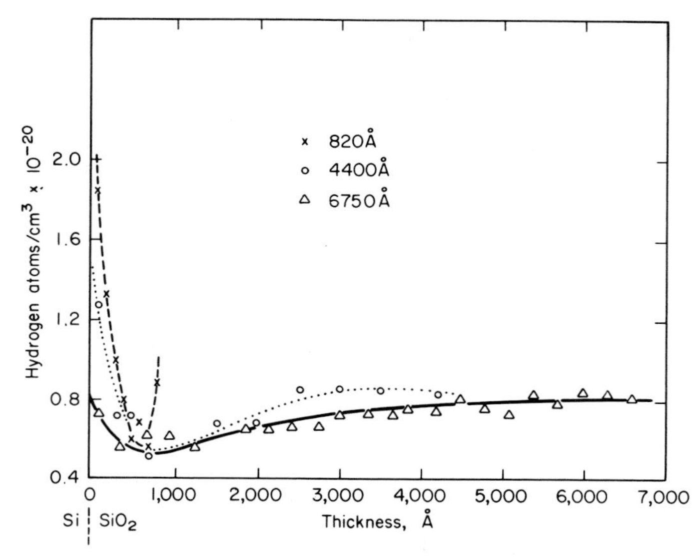

Fig. 10-16. Water concentration profiles through silicon dioxide films as measured by tritium (^{3}H) tracer. (*From Burkhardt.*[73])

films would be directly applicable. There appears to be no published work on this particular application of the spark-source spectrometer. However, when the new pulsed-laser sources and ion-bombardment sources are developed further, this method of analyzing dielectric films may become more attractive.

10-31. INFRARED SPECTROSCOPY

Infrared spectroscopy has been widely used in the study of dielectric films on semiconductors. Utilization of reflection attachments eliminates any effect of the semiconductor substrate and allows examination of the film alone.

In studies of silicon dioxide films, the Si–O bands at 1,090 and 805 cm^{-1} provide information on the oxide film itself. Pliskin and Lehman[50] have studied the Si–O stretching bands in the densification of oxide films. Both the position and half-width of the bands were shown to be related to the density, porosity, and bonding character in the films. The O–H stretching bands at 3,650 and 3,410 cm^{-1} provided valuable information on the water content of these films as well as bonded hydrogen and hydroxyl groups in the films. Pliskin and Castrucci[74] used infrared spectroscopy to study the reactivity and bond strain of films formed by electron-gun evaporation of silicon dioxide.

Valletta et al.[31] used the techniques described by Pliskin and Lehman to study reactively sputtered silicon dioxide films.

Silicon nitride films on silicon can be analyzed for Si–O content and water content by using these same stretching frequencies. Bean et al.[28] have recorded the infrared transmission of Si_3N_4 on silicon from 0.2 to 24 μ, and Levitt and Zwicker[75] studied the infrared spectrum on gallium arsenide.

Certain impurities or dopants have characteristic spectra when incorporated in oxide films. Pliskin[76,50] studied the silica and phosphosilicate structures in SiO_2 films. Corl et al.[77] used infrared spectroscopy to study the phosphosilicate layer in phosphorus-doped SiO_2 films and were able to measure the amount and uniformity of P–O bonding in the oxide structure. Using this analytical technique to monitor the process, they were able to reduce failure incidence due to inversion in diodes and n-p-n transistors.

10-32. PHYSICAL IMPERFECTIONS

The physical structure and perfection of dielectric films on semiconductors have a profound effect both on the production yield and on the ultimate electrical characteristics of the finished device. Knoop and Stickler[78] and Stickler and Faust[79] have extensively studied the nature of these imperfections and their effect on devices and device processes. It is obvious that pinholes or cracks in the amorphous films would be highly detrimental to yields in processes where the film must act as a diffusion mask. Similarly, device characteristics would be adversely affected where the oxide is used to passivate the surface or as an insulator to support evaporated metal leads in integrated circuits. However, there are other types of oxide defect that are not as obvious. The film may not be uniform in thickness; and when a phosphorus or boron glaze is applied to the outer layer of oxide, the subsequent diffusion processes will be nonuniform and the devices will have inversion or depletion layers caused by the proximity of the impurity in the thin areas. Similarly, crystalline areas in the amorphous film can have deleterious effects on the devices. There can also be mechanical strain both in the oxide and in the silicon at the silicon–SiO_2 interface. The device implications of this strain are discussed by Jaccodine and Schlegel,[80] who point out that this stress can affect the breakdown voltage of p-n junctions and adversely influence the density and distribution of surface states, as well as act as a source of dislocation generation.

10-33. ELECTROCHEMICAL TESTS

McCloskey[81] applied an electrograph technique, described in detail by Hermance and Wadlow,[82] to the detection of pinholes in silicon dioxide films. The apparatus used is simple and easily constructed. The electrolyte used in the "Millipore" type membrane paper is a benzidine–hydrochloric acid solution. Where there is a hole in the oxide, the benzidine will be electrochemically oxidized to a blue product and the paper will become an electrograph revealing pinholes in the oxide, as shown in Fig. 10-17.

The procedure works best on wafers that have etched patterns in the oxide film which serve as reference guidelines to aid in locating the pinhole. The procedure will only detect pinholes that go completely through the oxide to the silicon substrate. To repeat the test on the same wafer it is necessary to etch the wafer for 1 min in 0.5% HF to remove the 25 Å of SiO_2 formed anodically during the electrograph formation.

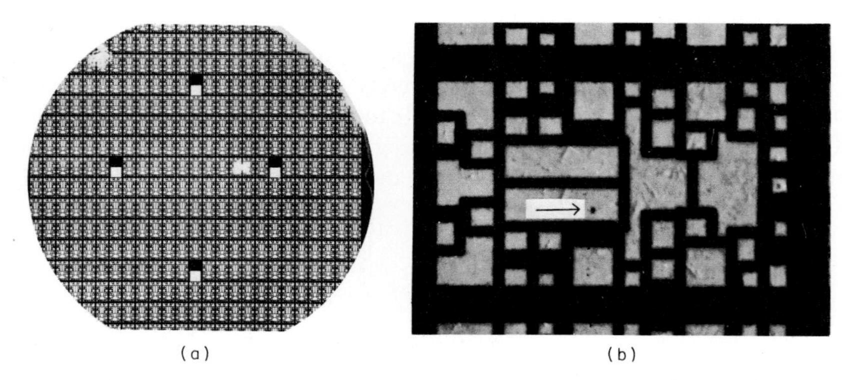

Fig. 10-17. Electrographs used to detect the presence of pinholes in oxide films. (*a*) **Electrograph of a silicon wafer containing etched microcircuit patterns;** (*b*) **one of the microcircuits on the electrograph enlarged, showing a pinhole location.** (*From McCloskey.*[81])

Accordingly, the electrograph procedure will record pinholes as small as 1000 Å in diameter but only on holes that penetrate through the oxide to the substrate. Since as little as 25 Å of oxide will prevent the electrochemical oxidation of the benzidine, it is likely that the test will be valid only on reasonably freshly etched slices. Silicon grows a native oxide of 15 to 25 Å in thickness on exposure to the atmosphere, which would inhibit the test.

10-34. ELECTRON MICROSCOPY

Electron microscopy is the most widely used analytical technique for studying physical defects in dielectric films. Balk et al.[83] used collodion surface-replication electron microscopy to investigate the microstructural properties of thermally grown SiO_2 films. By mechanically scribing a small v mark in the original silicon substrate surface, it was possible to grow oxide on damaged and damage-free areas near this reference point. Excellent electron micrographs were obtained which showed the adverse effect of substrate mechanical damage on the subsequent quality of the oxide films.

Stickler and Faust[79,84] have had good success using transmission electron microscopy to look at oxide films on silicon. One distinct advantage of this approach is the ability to examine the bulk of the film, while replica techniques only yield information on the surface morphology. The SiO_2 films were separated from the silicon substrates by a hot chlorine technique similar to that described by Tannenbaum.[85] The oxidized silicon slices were placed in quartz boats in a tube furnace, and chlorine was passed over the samples at 800°C for about 20 min. The silicon was converted to $SiCl_4$ and distilled off, while the SiO_2 was left behind since it does not react with chlorine. The quartz boat was removed, and the film was floated off and mounted on a standard electron-microscope grid. The thin films were then examined by transmission at 100 kv in an electron microscope. Stickler and Faust[84] were able to show the presence of embedded abrasive particles as well as other "dirt" of an unidentified nature and source.

In subsequent work,[79] these same workers combined electron diffraction and optical microscopy with their transmission electron microscopy studies to define the effect of device processes on the SiO_2 films. It was shown that thermal cycling caused a sharpening of the electron diffraction rings, indicative of some initial ordering in the oxide. This start toward crystallinity occurred after 90 hr at 450°C in a vacuum. Since many processing steps utilize more severe time-temperature cycles than this, film crystallinity can be a serious problem. The diffusion process was also shown to adversely affect the amorphous oxide by the formation of "tubes" in the film. The reason for the formation of these "tubes" or the origin of particles associated with the bottom of these "tubes" was not reported. However, these imperfections could be detrimental in subsequent process steps and affect both device yield and reliability.

Knoop and Stickler[78] used electron microscopy and electron diffraction to examine the structure of thermally grown oxide films as a function of various heat treatments. These workers showed the importance of preoxidation handling and surface preparation. Uniform amorphous oxide films of high perfection were grown in open-tube systems between 900 and 1200°C. Further heat treatments at 1235°C in a quartz tube in either vacuum or argon-oxygen mixtures caused localized cracking of the oxide films. Similarly, heat treatment at 1235°C in sealed quartz ampules containing oxygen resulted in uniform crystalline oxide films (α-crystobalite).

10-35. ETCH RATE

The rate of dissolution of a dielectric film is strongly dependent upon the film perfection, density, bond strain, and stoichiometry. Pliskin and Lehman,[50] in their structural analysis of silicon oxide films, describe the use of "P" etch (15 parts HF, 10 parts HNO_3, 300 parts H_2O, by volume). Table 10-8 shows the gross differences in etch rate observed by these workers on oxide films grown by different methods. As can be seen, there is a wide spread in etch rate between different types of oxide films, but more significant is the spread in etch rates for the same oxide. This latter difference in etch rates can be used to evaluate the oxide growth process, where the effect of any change in the process can be monitored by observing any difference in etch rate. The lower etch rates indicate higher film perfection.

Table 10-8. Etch Rate of "P" Etch of Various Types of Oxide Film

Film	*Etch rate, Å/sec*
Pyrolytic	6–20
Pyrolytic densified	2.0
CO_2, chemical	2.4–4.2
Reactive sputtering	3.8–5.2
Anodic	18–228
PbO, catalyzed	600
Evaporated	20–70

Bean et al.[28] used etch-rate studies of Si_3N_4 films to determine the optimum growth conditions. The etch used in this work was Bell No. 2 (200 ml H_2O, 200 g

NH_4F, 45 ml HF). When the SiH_4 and NH_3 flow rates were not optimized, a silicon-rich Si_3N_4 film was formed which gave a slower etch rate than the pure film. The etch rate of the optimized films as a function of temperature yielded an apparent activation energy of 15 kcal/mole.

Lopez[86] combined etching studies with electrical measurements to examine fast etching imperfections in SiO_2 films. These imperfections produced pinholes in the films when the oxide was partially thinned with a buffered hydrofluoric acid etch. A metal contact was evaporated onto the thinned oxide, and the dielectric breakdown was measured between the silicon substrate and the oxide film. Good oxide films have typical breakdown fields of 80 to 90 volts per 1000 Å. However, those oxides with fast etching imperfections, when thinned, yielded breakdown fields near zero, which were attributed to pinholes in the film. Lopez showed that the fast etching imperfections were probably due to surface contamination prior to oxidation.

10-36. STRAIN MEASUREMENTS

The direct measurement of strain at the silicon–silicon dioxide interface is very difficult. Jaccodine and Schlegel[80] measured the stress in this system by two techniques. In the first, the silicon sample (less than 2 mils in thickness) with oxide on one side was used as a beam, and the amount of bowing, caused by the SiO_2 strain, was measured. In the second technique, the SiO_2 film was separated from the silicon by hot chlorine etching and the SiO_2 film used as a balloon. The strain was measured as a function of the air pressure inflating the balloon. Both techniques yielded comparable results for a 1200°C oxide, with a measured stress of 3.1×10^9 dynes/cm^2.

Howard et al.[87] reported on the use of transmission x-ray topography to examine the strain at the edges of windows etched in SiO_2 films. X-ray topography, of course, cannot be used to look at the amorphous oxide but can detect strain in the silicon by looking in the window which is free of oxide and strain and comparing this with the edge where strain will be maximized.

10-37. ANALYSIS OF THE FILM SURFACE

The analysis of the surface of a 1000 to 10,000 Å film presents a real problem to the analyst, both with regard to available sample size and, more particularly, with regard to available analytical techniques. Since virtually all dielectric films used in the semiconductor industry are silicon dioxide, it is possible to restrict our consideration to this, realizing that all analytical techniques can be used on the other dielectric films.

Amorphous silicon dioxide films are best thought of as silica or hydrated silica. The surface chemistry of silica has been widely studied and is well understood.[88] Silica surfaces will both physically and chemically adsorb inorganic and organic impurities, and these surface residues will have adverse effects on device characteristics and long-term device stability.

10-38. INORGANIC SURFACE RESIDUES

The most obvious analytical tools for inorganic surface impurities are emission spectroscopy and mass spectroscopy. Unfortunately, the technology for excitation of only surface contaminants has not been sufficiently developed to allow application of these techniques to insulating films. There is some doubt that either of these approaches has the required sensitivity because of the small amount of available sample on the surface of the film. In some preliminary studies, Werner[89] used a solids mass spectrometer and "sputtered" the surface of an SiO_2 film on silicon, using an 11-kev beam of positive argon ions. It was possible to monitor the sodium concentration from the surface into the oxide film. The interpretation of the results was open to question because the $(SiOH)^+$ peak decreases with depth into the film. However, the technique offers promise of a high-sensitivity analytical technique for the analysis of surfaces, because no electrical continuity is required, as is usually the case in spark-source mass spectroscopy.

10-39. RADIOACTIVE TRACERS IN ADSORPTION STUDIES

Radioactive tracers appear to offer the most powerful tool to the analyst for this type of surface study. It is possible to introduce a radioactive tracer into the process of interest, count the silica sample, and then determine the amount of impurity adsorbed. By determining the activation energy associated with either the adsorption or the desorption, it is possible to distinguish between physical and chemical adsorption. This radiotracer approach was applied to the problem of etching SiO_2 films with fluoride-containing etches.[90] It was shown, in agreement with similar studies on silica gel,[88,91] that the fluoride ion was chemically adsorbed at discrete sites on the SiO_2 surface. Other radiotracers could be used in the same way to study surface contamination resulting from any particular treatment.

10-40. WATER CONTACT ANGLE

When these dielectric films are used as diffusion barriers, it is necessary to etch windows through the film to allow the diffusant to enter the substrate and form active components. These windows are formed by coating the slice with a photosensitive plastic, as described in Sec. 10-1. This photoresist when properly applied must adhere firmly to the oxide film or the etch will react under the photoresist as well as in the window, and serious undercutting will occur. Bergh[92] reported a correlation between the contact angle of water on silicon oxide and the adherence of Kodak Photoresist (KPR). Bergh utilized the fact that silicon oxide films can be hydrophobic or hydrophilic, depending on previous surface treatment. By using a reflection goniometer,[93] the contact angle of a 4-μl drop of water on the oxide surface was measured. KPR was then applied, exposed, and developed, and the uncoated oxide areas were etched away. It was observed that low water contact angle consistently showed poor KPR adherence. Surprisingly, the contact angle for KPR was found to be identical ($\sim 20°$) on hydrophobic and hydrophilic oxides, which implies that a hydrophobic surface is not necessarily an organophilic surface.

Bergh interpreted his results to indicate that the high-contact-angle oxides did not indicate better coupling between KPR and oxide, but rather a decrease in the tendency of water to penetrate the KPR–oxide interface. Rand and Ashworth[18] used Bergh's technique to evaluate H_2–CO_2–$SiBr_4$ chemically deposited silicon oxide films on germanium for potential KPR adherence.

Lussow et al.[94] developed a technique to measure the wettability of SiO_2 films thermally grown on epitaxial silicon. In this method, the sample was placed in an environmental chamber with air saturated with water and allowed to equilibrate. The chamber was opened, and small droplets of water were placed on the surface. The chamber was closed, and equilibrium within the system was reestablished. The wettability of the sample was determined by measuring the average contact diameter of four droplets. A reproducibility of ± 2.5 percent at the 95% confidence level was reported. Subsequently Lussow[95] studied the adherence of photoresists on thermally grown SiO_2 surfaces which were modified by various treatments. Lussow observed that each type of photoresist adhered differently to the different SiO_2 surfaces, and water wettability alone was not sufficient to predict adhesion performance.

10-41. ORGANIC SURFACE RESIDUES

The techniques described above provide a means of ensuring that the polymerized photoresist film adheres strongly to the oxide film. As described in Sec. 10-1,

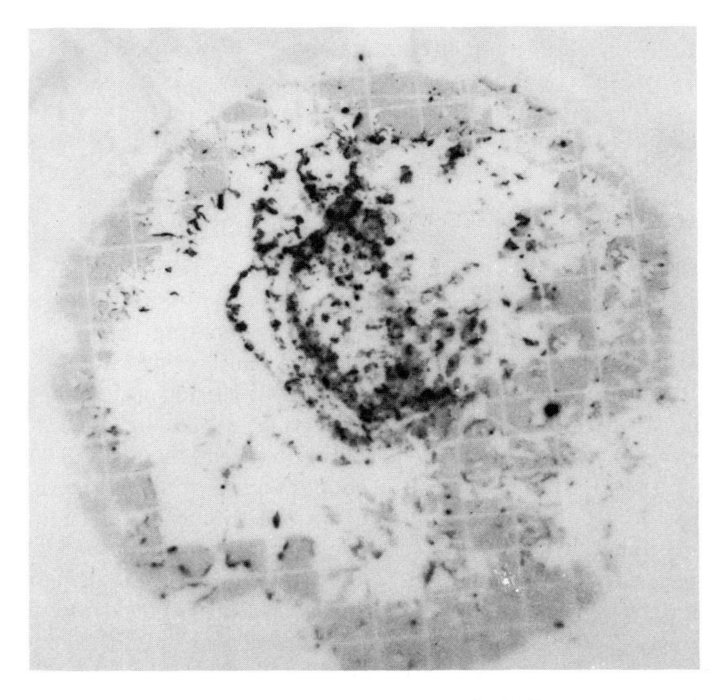

Fig. 10-18. Autoradiogram showing the distribution of residual photoresist on a silicon slice using iodine-131. *(From Heinen and Larrabee.[96])*

device fabrication involves multiple sequential steps. The photoresist must be completely removed between these steps to allow subsequent uniform regrowth of oxide, application of evaporated metal contacts, or similar operations. A radiotracer technique for residual resist was developed in the Texas Instruments laboratories[96] since existing methods[94,97] lacked sensitivity and did not yield quantitative results.

Basically, the procedure involves the addition of radioactive iodine-131 to inactive KI and oxidation of all iodide ion to iodine. The tagged iodine is allowed to react with any residual photoresist on a cleaned slice. The slice is beta-counted and an autoradiogram obtained of the surface distribution of organic contaminant (Fig. 10-18). The procedure can determine 20 ng of residual photoresist on a slice with a 90% confidence error of 3.4%.

REFERENCES

1. Deal, B. E.: *J. Electrochem. Soc.*, **110**:527 (1963).
2. Evitts, H. C., H. W. Cooper, and S. S. Flaschen: *J. Electrochem. Soc.*, **111**:688 (1964).
3. Deal, B. E., and M. Sklar: *J. Electrochem. Soc.*, **112**:430 (1965).
4. Deal, B. E., and A. S. Grove: *J. Appl. Phys.*, **36**:3770 (1965).
5. Nakayama, T., and F. C. Collins: *J. Electrochem. Soc.*, **113**:706 (1966).
6. Deal, B. E., E. H. Snow, and A. S. Grove: *Semicond. Prod. Solid State Tech.*, **9**:25 (November, 1966).
7. Collins, F. C., and T. Nakayama: *J. Electrochem. Soc.*, **114**:167 (1967).
8. Schmidt, P. F., and W. Michel: *J. Electrochem. Soc.*, **104**:230 (1957).
9. Young, L.: "Anodic Oxide Films," Academic Press, Inc., New York, 1961.
10. Duffek, E. F., E. A. Benjamini, and C. Mylroie: *Electrochem. Tech.*, **4**:75 (1965).
11. Duffek, E. F., C. Mylroie, and E. A. Benjamini: *J. Electrochem Soc.*, **111**:1042 (1964).
12. Schmidt, P. F., and A. E. Owen: *J. Electrochem. Soc.*, **111**:682 (1964).
13. Kraitchman, J., and J. Oroshnick: *J. Electrochem. Soc.*, **114**:405 (1967).
14. Schmidt, P. F., and D. R. Wonsidler: *J. Electrochem. Soc.*, **114**:603 (1967).
15. Chu, T. L.: *Semicond. Prod. Solid State Tech.*, **10**:36 (May, 1967).
16. Steinmaier, W., and J. Bloem: *J. Electrochem. Soc.*, **111**:206 (1964).
17. Tung, S. K., and R. E. Caffrey: *Trans. Met. Soc, AIME*, **233**:572 (1965).
18. Rand, M. J., and J. L. Ashworth: *J. Electrochem. Soc.*, **113**:48 (1966).
19. Rand, M. J.: *J. Electrochem. Soc.*, **114**:274 (1967).
20. Chu, T. L., J. R. Gavaler, G. A. Gruber, and Y. C. Kao: *J. Electrochem. Soc.*, **111**:1433 (1964).
21. Klerer, J.: *J. Electrochem. Soc.*, **112**:503 (1965).
22. Secrist, D. R., and J. D. MacKenzie: *J. Electrochem. Soc.*, **113**:914 (1966).
23. Sterling, H. F., and R. C. G. Swann: *Solid-State Electron.*, **8**:653 (1965).
24. Doo, V. Y.: *IEEE. Trans. Electron Devices*, **ED-13**:561 (1966).
25. Hu, S. M.: *J. Electrochem. Soc.*, **113**:693 (1966).
26. Doo, V. Y., D. R. Nichols, and G. A. Silvey: *J. Electrochem. Soc.*, **113**:1279 (1966).
27. Chu, T. L., C. H. Lee, and G. A. Gruber: *J. Electrochem. Soc.*, **114**:717 (1967).
28. Bean, K. E., P. S. Gleim, R. L. Yeakley, and W. R. Runyan: *J. Electrochem. Soc.*, **114**:733 (1967).
29. Doo, V. Y., D. R. Kerr, and D. R. Nichols: *J. Electrochem. Soc.*, **115**:61 (1968).
30. Clark, R. S.: *Trans. Met. Soc., AIME*, **233**:592 (1965).
31. Valletta, R. M., J. A. Perri, and J. Riseman: *Electrochem. Tech.*, **4**:402 (1966).
32. York, D. B.: *J. Electrochem. Soc.*, **110**:271 (1963).

33. Nesh, F.: *Rev. Sci. Instrum.*, **34**:1437 (1963).
34. Gillespie, D. J.: in B. Schwartz and N. Schwartz (eds.), "Measurement Techniques for Thin Films," Electrochemical Society, New York, 1967.
35. Pliskin, W. A., and E. E. Conrad: *IBM J. Res. Dev.*, **8**:43 (1964).
36. Corl, E. A., and H. Wimpfheimer: *Solid-State Electron.*, **7**:755 (1964).
37. McCallum, J. D.: *Semicond. Prod. Solid State Tech.*, **10**:67 (January, 1967).
38. Reizman, F.: *J. Appl. Phys.*, **36**:3804 (1965).
39. Goldsmith, N., and L. A. Murray: *Solid-State Electron.*, **9**:331 (1966).
40. Wesson, R. A., R. P. Phillips, and W. A. Pliskin: *J. Appl. Phys.*, **38**:2455 (1967).
41. Reizman, F., and W. Van Gelder: *Solid-State Electron.*, **10**:625 (1967).
42. Lukes, F., and E. Schmidt: *Solid-State Electron.*, **10**:264 (1967).
43. Passaglia, E., R. R. Stromberg, and J. Kruger (eds.): Ellipsometry in the Measurement of Surfaces and Thin Films, Symposium Proceedings, Washington, 1963, *Nat. Bur. Stand. Misc. Publ.* 256, 1964.
44. Archer, R. J.: *J. Opt. Soc. Amer.*, **52**:970 (1962).
45. Heavens, O. S.: "Optical Properties of Thin Solid Films," Butterworth & Co. (Publishers), Ltd., London, 1955.
46. Jones, C. E.: Private communication, Texas Instruments Incorporated, Dallas, Texas, 1967.
47. Booker, G. R., and C. E. Benjamin: *J. Electrochem. Soc.*, **109**:1206 (1962).
48. Tolansky, S.: "Multiple Beam Interferometry of Surfaces and Films," Oxford University Press, Fair Lawn, N.J., 1948.
49. Schwartz, N., and R. Brown: in "Transactions of the 8th National Vacuum Symposium, 1961," p. 836, Pergamon Press, London, 1962.
50. Pliskin, W. A., and H. S. Lehman: *J. Electrochem. Soc.*, **112**:1013 (1965).
51. Morey, G. W.: "The Properties of Glass," Reinhold Publishing Corporation, New York, 1954.
52. Lewis, A. E.: *J. Electrochem. Soc.*, **111**:1007 (1964).
53. Guyer, E. M.: *Proc. IRE*, **32**:743 (1944).
54. Wellard, C. L.: "Resistance and Resistors," McGraw-Hill Book Company, New York, 1960.
55. Siddall, G.: *Vacuum*, **9**:274 (1960).
56. York, D. B.: *J. Electrochem. Soc.*, **110**:271 (1963).
57. Hirose, H., and Y. Wada: *Japan J. Appl. Phys.*, **3**:179 (1964).
58. Hirose, H., and Y. Wada: *Japan J. Appl. Phys.*, **4**:639 (1965).
59. Worthing, F. L.: *J. Electrochem. Soc.*, **115**:88 (1968).
60. Chaikin, S. W., and G. A. St. John: *Electrochem. Tech.*, **1**:291 (1963).
61. Osborne, J. F., and G. B. Larrabee: Unpublished work, 1967.
62. Osborne, J. F., G. B. Larrabee, and V. Harrap: *Anal. Chem.*, **39**:1144 (1967).
63. Buck, T. M., F. G. Allen, J. V. Dalton, and J. D. Struthers: *J. Electrochem. Soc.*, **114**:862 (1967).
64. Yon, E., W. H. Ko, and A. B. Kuper: *IEEE Trans. Electron Devices*, **ED-13**:276 (1966).
65. Snow, E. H., A. S. Grove, B. E. Deal, and C. T. Sah: *J. Appl. Phys.*, **36**:1664 (1965).
66. Reverz, A. G.: *IEEE Trans. Electron Devices*, **ED-12**:97 (1965).
67. Carlson, H. G., G. A. Brown, C. R. Fuller, and J. F. Osborne: in M. E. Goldberg and J. Vaccaro (eds.), "Physics of Failure in Electronics," vol. 4, p. 390, Rome Air Development Center, 1966.
68. Carlson, H. G., C. R. Fuller, D. E. Meyer, J. F. Osborne, V. Harrap, and G. A. Brown: in T. S. Shilliday and J. Vaccaro (eds.), "Physics of Failure in Electronics," vol. 5, p. 265, Rome Air Development Center, 1967.

69. Kuper, A. B., and E. H. Nicollian: *J. Electrochem. Soc.*, **112**:528 (1965).
70. Kooi, E.: *J. Electrochem. Soc.*, **111**:1383 (1964).
71. Schmidt, P. F., W. van Gelder, and J. Drobek: *J. Electrochem. Soc.*, **115**:79 (1968).
72. Duffy, M. C., F. Barson, J. M. Fairfield, and G. H. Schwuttke: *J. Electrochem. Soc.*, **115**:84 (1968).
73. Burkhardt, P. J.: *J. Electrochem. Soc.*, **114**:196 (1967).
74. Pliskin, W. A., and P. P. Castrucci: *Electrochem. Tech.*, **6**:85 (1968).
75. Levitt, R. S., and W. K. Zwicker: *J. Electrochem. Soc.*, **114**:1192 (1967).
76. Pliskin, W. A.: *Appl. Phys. Letters*, **7**:158 (1965).
77. Corl, E. A., S. L. Silverman, and Y. S. Kim: *Solid-State Electron.*, **9**:1009 (1966).
78. Knoop, A. N., and R. Stickler: *Electrochem. Tech.*, **5**:37 (1967).
79. Stickler, R., and J. W. Faust, Jr.: *Electrochem. Tech.*, **4**:277 (1966).
80. Jaccodine, R. J., and W. A. Schlegel: *J. Appl. Phys.*, **37**:2429 (1966).
81. McCloskey, J. P.: *J. Electrochem. Soc.*, **114**:643 (1967).
82. Herrmance, H. W., and H. V. Wadlow: in F. J. Welcher (ed.), "Standard Methods of Chemical Analysis," vol. 3, pt. A, p. 500, D. Van Nostrand Company, Inc., New York, 1966.
83. Balk, P., C. F. Aliotta, and L. V. Gregor: *Trans. AIME*, **233**:563 (1965).
84. Stickler, R., and J. W. Faust, Jr.: *Electrochem. Tech.*, **2**:298 (1964).
85. Tannenbaum, E.: *J. Appl. Phys.*, **31**:940 (1960).
86. Lopez, A. D.: *J. Electrochem. Soc.*, **113**:89 (1966).
87. Howard, J. K., and G. H. Schwuttke: Presented at the 1966 Pittsburgh Diffraction Conference, Pittsburgh, Pa., 1966.
88. Iler, R. K.: "The Colloid Chemistry of Silica and Silicates," Cornell University Press, Ithaca, N.Y., 1955.
89. Werner, H. W.: *Philips Tech. Rev.*, **27**:344 (1966).
90. Larrabee, G. B., K. G. Heinen, and S. A. Harrell: *J. Electrochem. Soc.*, **114**:867 (1967).
91. Kimberlen, C. N.: U.S. Patent 2,477,695, Standard Oil Development Co., 1949.
92. Bergh, A. A.: *J. Electrochem. Soc.*, **112**:457 (1965).
93. Fort, T., Jr., and H. T. Patterson: *J. Colloid Sci.*, **18**:217 (1963).
94. Lussow, R. O., L. H. Wirtz, and H. A. Levine: *J. Electrochem. Soc.*, **114**:877 (1967).
95. Lussow, R. O.: *J. Electrochem. Soc.*, **115**:660 (1968).
96. Heinen, K. G., and G. B. Larrabee: *Solid State Tech.*, **12**:44 (April, 1969).
97. Anderson, J. L.: *J. Amer. Ass. Contam. Contr.*, **2**(6):9 (1963).

Author Index

In this index, nonitalic figures indicate text pages on which the author is cited by name. Italic figures indicate the pages on which complete publication data are given in the References at the end of each chapter.

Subject Index